Molecular Therapeutics for Diabetes and Related Complications

Molecular Therapeutics for Diabetes and Related Complications

Guest Editor

Kota V. Ramana

Basel • Beijing • Wuhan • Barcelona • Belgrade • Novi Sad • Cluj • Manchester

Guest Editor
Kota V. Ramana
Department of Biomedical Sciences
Noorda College of
Osteopathic Medicine
Provo, UT
USA

Editorial Office
MDPI AG
Grosspeteranlage 5
4052 Basel, Switzerland

This is a reprint of the Special Issue, published open access by the journal *International Journal of Molecular Sciences* (ISSN 1422-0067), freely accessible at: https://www.mdpi.com/journal/ijms/special_issues/diabetes_related_complications.

For citation purposes, cite each article independently as indicated on the article page online and as indicated below:

Lastname, A.A.; Lastname, B.B. Article Title. *Journal Name* **Year**, *Volume Number*, Page Range.

ISBN 978-3-7258-5879-8 (Hbk)
ISBN 978-3-7258-5880-4 (PDF)
https://doi.org/10.3390/books978-3-7258-5880-4

Contents

About the Editor

Kota V. Ramana

Kota V. Ramana is a Professor at the Department of Biomedical Sciences, Noorda College of Osteopathic Medicine, Provo, UT-USA. He is also an adjunct professor at Roseman University of Health Sciences. The chief objectives of his lab investigations are to examine the involvement of cellular metabolism and oxidative stress signals in inflammation. He uses various genetic, biochemical, and cell biological approaches to analyze inflammatory responses regulated by cellular lipid metabolites that lead to secondary diabetic complications and cancer. He has published more than 140 peer-reviewed articles. He is also serving as an Editorial Board Member of several journals

International Journal of
Molecular Sciences

Editorial

Special Issue "Molecular Therapeutics for Diabetes and Related Complications"

Kota V. Ramana

Department of Biomedical Sciences, Noorda College of Osteopathic Medicine, Provo, UT-84606, USA; karamana@noordacom.org

Diabetes mellitus is a significant public health challenge worldwide, and in many developing countries, diabetes and its complications are the number one cause of morbidity and mortality. Besides genetic predisposition, unhealthy lifestyle habits, unhealthy diet, stress, and lack of appropriate physical activity could lead to increased and uncontrolled blood glucose levels, insulin resistance, and obesity, which eventually lead to diabetic mellitus. There are two major types of diabetes. Type 1 diabetes results from genetic predisposition and autoimmune destruction of pancreatic beta cells. Type 2 diabetes is due to increased stress, insulin resistance, obesity, and a sedentary lifestyle. Although hyperglycemia can be controlled with pharmaceutical drugs, lifestyle adjustments, increased physical activity, and reduced stress, the underlying molecular causes of diabetes and its complications still need to be explored. Recent advances in molecular therapeutics target molecular mechanisms involved in glucose metabolism, insulin signaling, and pancreatic β-cell function.

Recent studies suggest that glucagon-like peptide-1 (GLP-1) receptor agonists and dipeptidyl peptidase-4 (DPP-4) inhibitors enhance insulin secretion, offering cardiovascular benefits [1,2]. Similarly, sodium–glucose co-transporter 2 (SGLT 2) inhibitors have been shown to lower blood sugar by promoting urinary glucose excretion and protecting renal and heart health [3]. Gene and stem cell therapies are also emerging, with gene editing tools like clustered regularly interspaced short palindromic repeats (CRISPR) aiming to correct insulin-related defects, and stem cells being used to generate insulin-producing cells [4]. However, conventional treatments such as insulin and glucose-lowering drugs may also come with potential side effects. For example, GLP-1 receptor agonists may cause gastrointestinal issues and, in rare cases, pancreatitis or thyroid concerns [5]. DPP-4 inhibitors could also lead to respiratory symptoms and joint pain [6]. Similarly, SGLT2 inhibitors have been shown to increase the risk of urinary infections and diabetic ketoacidosis [7]. Therefore, careful monitoring and effective dosage adjustment are crucial when using these drugs. Further, recent advances in precision medicine and pharmacogenomics studies could also help to personalize treatment strategies and individual drug responses. Reducing oxidative stress, increasing antioxidant defenses, and modulating the pathways that regulate key molecular pathways in innate immune and inflammatory responses are also emerging approaches to control diabetes and its associated complications such as retinopathy, neuropathy, nephropathy, and cardiovascular diseases. Thus, developing novel therapeutics holds significant promise for controlling diabetes and improving long-term patient outcomes.

In this Special Issue, we compile recent findings on the role of novel molecular therapeutics in ameliorating diabetes and related complications. The compiled review and

 https://doi.org/10.3390/ijms26125585

research articles discuss how current molecular therapeutics represent a paradigm shift in treating diabetes and its complications.

An excellent article by Muntean et al. (contribution 1) investigated whether specific gene polymorphisms associated with type 2 diabetes (ADIPOQ rs266729, CDKN2A/2B rs10811661, and SSR1 rs9505118) are linked to gestational diabetes mellitus (GDM) and perinatal outcomes in Romanian pregnant women. This study reported that among 213 participants, those with GDM had significantly higher pre-pregnancy body mass index (BMI), insulin resistance, and lower adiponectin levels, as well as more frequent gestational hypertension and perineal lacerations. However, no significant associations were found between the studied gene polymorphisms and GDM. Only weak correlations were observed between specific genetic models and certain birth outcomes, suggesting that these polymorphisms are not major contributors to GDM in this population. Similarly, Guardiola et al. (contribution 2) showed the relationship between specific growth differentiation factor 15 (GDF15) gene variants and metabolic diseases, focusing on three SNPs in a cohort of 153 individuals. Among these, only the rs1054564 variant was significantly associated with elevated GDF15 serum levels and a higher frequency of diabetes and atherosclerotic carotid plaque. These studies thus suggest that the rs1054564 variant independently increased the risk of diabetes and subclinical atherosclerosis, suggesting its potential as a genetic marker for these conditions. Further, a 12-year follow-up METSIM cohort study by SFernandes Silva et al. (contribution 3) identified the metabolic markers associated with impaired kidney function by analyzing 1,009 metabolites in 10,159 Finnish men. They found 108 metabolites significantly associated with decreased eGFR, including 28 novel ones across various categories such as amino acids, lipids, and xenobiotics. These findings offer new insights into metabolic pathways linked to kidney function decline and may help in the early detection and prevention of kidney-related complications.

Esteves-Monteiro et al. (contribution 4) examined the gastrointestinal complications of diabetes in type 2 diabetic Goto-Kakizaki (GK) rats. They showed a significant thickening of the intestinal wall and muscular layers and a reduction in myenteric neuronal density, suggesting structural and neurological alterations in GK rats. Diabetic rats showed decreased glutathione (GSH) levels and a lower GSH/GSSG ratio in most gut segments. This indicates that diabetes induces substantial intestinal remodeling and oxidative damage, which may underlie the gastrointestinal dysfunction frequently observed in diabetic patients. Another animal study by Drygalski et al. (contribution 5) showed the potential role of phloroglucinol (PHG) in preventing non-alcoholic fatty liver disease (NAFLD) and insulin resistance caused by a high-fat diet in Wistar rats. PHG treatment improved fasting glucose levels, insulin sensitivity, and liver health. PHG also influenced gastric motility through potassium channel activation and nitric oxide signaling, suggesting PHG as a promising candidate for further clinical investigation.

Interestingly, a case study by Mikami-Saito et al. (contribution 6) reported an association between oral Semaglutide and carnitine depletion. A 34-year-old male with multiple acyl-CoA dehydrogenase deficiency developed hypoglycemia and significantly reduced blood-free carnitine levels after switching from injectable to oral Semaglutide. This finding highlights the need for clinicians to monitor carnitine levels in at-risk patients taking salcaprozate sodium-containing medications.

This Special Issue also published four informative review articles on recent molecular therapeutics in diabetes. Puscasu et al. (contribution 7) discussed the potential of N-methyl-D-aspartate receptor (NMDAR) antagonists such as ketamine, memantine, and methadone as therapeutic options for neuropathic pain. They indicated these agents' efficacy in managing various forms of neuropathic pain, which is a common diabetes complication. Another article by Cheng and Zhang discussed how mesenchymal stem cell (MSC) therapy is a promising approach for reversing renal damage and restoring kidney function in diabetic kidney disease (DKD). Specifically, they reported current research on MSC-based therapies, their mechanisms of action, and their integration with conventional treatments. Similarly, Guo et al. (contribution 8) discussed recent advancements in molecular therapeutics such as miRNA therapy, stem cell therapy, gene therapy, gut microbiota-targeted treatment, and lifestyle interventions as promising new avenues for DKD management. Finally, a systematic review by Keskesiadou et al. (contribution 9) explored the potential link between endocrine-disrupting chemicals (EDCs) and the development of type 1 diabetes mellitus. This study identified a correlation between type 1 diabetes and specific EDCs, including bisphenol A, bisphenol S, phthalates, dioxins, and persistent organic pollutants (POPs). Thus, this report supports a correlation between EDCs and type 1 diabetes.

In conclusion, the articles published in this Special Issue highlight molecular therapeutics as a transformative approach to treat diabetes and its complications. By targeting the genetic and biochemical pathways involved in the pathophysiology of diabetes, these advanced therapies could significantly improve patient outcomes. Further, continued research is essential to fully understand how these novel treatments will facilitate a future where diabetes is manageable and potentially curable.

Acknowledgments: I would like to thank all the authors and reviewers who contributed to the success of this Special Issue. I would also like to acknowledge the great support from the journal's editorial staff.

Conflicts of Interest: The editor declares no conflicts of interest.

List of Contributions:

1. Muntean, M.; Mărginean, C.; Bernad, E.S.; Bănescu, C.; Nyulas, V.; Muntean, I.E.; Săsăran, V. Adiponectin C1Q and Collagen Domain Containing rs266729, Cy-clin-Dependent Kinase Inhibitor 2A and 2B rs10811661, and Signal Sequence Re-ceptor Subunit 1 rs9505118 Polymorphisms and Their Association with Gesta-tional Diabetes Mellitus: A Case-Control Study in a Romanian Population. *Int. J. Mol. Sci.* **2025**, *26*, 1654. https://doi.org/10.3390/ijms26041654.
2. Guardiola, M.; Girona, J.; Barroso, E.; García-Altares, M.; Ibarretxe, D.; Plana, N.; Ribalta, J.; Correig, X.; Vázquez-Carrera, M.; Masana, L.; et al. The GDF15 3′ UTR Polymorphism rs1054564 Is Associated with Diabetes and Subclinical Athero-sclerosis. *Int. J. Mol. Sci.* **2024**, *25*, 11985. https://doi.org/10.3390/ijms252211985.
3. Fernandes Silva, L.; Vangipurapu, J.; Oravilahti, A.; Laakso, M. Novel Metabolites Associated with Decreased GFR in Finnish Men: A 12-Year Follow-Up of the METSIM Cohort. *Int. J. Mol. Sci.* **2024**, *25*, 10044. https://doi.org/10.3390/ijms251810044.
4. Esteves-Monteiro, M.; Ferreira-Duarte, M.; Vitorino-Oliveira, C.; Costa-Pires, J.; Oliveira, S.; Matafome, P.; Morato, M.; Dias-Pereira, P.; Costa, V.M.; Duarte-Araújo, M. Oxidative Stress and Histomorphometric Remodeling: Two Key Intestinal Features of Type 2 Diabetes in Goto–Kakizaki Rats. *Int. J. Mol. Sci.* **2024**, *25*, 12115. https://doi.org/10.3390/ijms252212115.
5. Drygalski, K.; Maciejczyk, M.; Miksza, U.; Ustymowicz, A.; Godzień, J.; Buczyńska, A.; Chomentowski, A.; Walczak, I.; Pietrowska, K.; Siemińska, J.; et al. New Ap-plication of an Old Drug: Anti-Diabetic Properties of Phloroglucinol. *Int. J. Mol. Sci.* **2024**, *25*, 10291. https://doi.org/10.3390/ijms251910291.
6. Mikami-Saito, Y.; Maekawa, M.; Watanabe, M.; Hosaka, S.; Takahashi, K.; Totsune, E.; Arai-Ichinoi, N.; Kikuchi, A.; Kure, S.; Katagiri, H.; et al. Carnitine Deficiency Caused by Salcaprozic Acid Sodium Contained in Oral Semaglutide in a Patient with Multiple Acyl-CoA Dehydrogenase Deficiency. *Int. J. Mol. Sci.* **2025**, *26*, 2962. https://doi.org/10.3390/ijms26072962.
7. Pușcașu, C.; Chiriță, C.; Negreș, S.; Blebea, N.M. Exploring the Therapeutic Poten-tial of N-Methyl-D-Aspartate Receptor Antagonists in Neuropathic Pain Man-agement. *Int. J. Mol. Sci.* **2024**, *25*, 11111. https://doi.org/10.3390/ijms252011111.
8. Guo, M.; He, F.; Zhang, C. Molecular Therapeutics for Diabetic Kidney Disease: An Update. *Int. J. Mol. Sci.* **2024**, *25*, 10051. https://doi.org/10.3390/ijms251810051.

9. Keskesiadou, G.-N.; Tsokkou, S.; Konstantinidis, I.; Georgaki, M.-N.; Sioga, A.; Pa-pamitsou, T.; Karachrysafi, S. Endocrine-Disrupting Chemicals and the Develop-ment of Diabetes Mellitus Type 1: A 5-Year Systematic Review. *Int. J. Mol. Sci.* **2024**, *25*, 10111. https://doi.org/10.3390/ijms251810111.

References

1. Gieroba, B.; Kryska, A.; Sroka-Bartnicka, A. Type 2 diabetes mellitus—Conventional therapies and future perspectives in innovative treatment. *Biochem. Biophys. Rep.* **2025**, *42*, 102037. [CrossRef] [PubMed]
2. Bae, J.H. SGLT2 Inhibitors and GLP-1 Receptor Agonists in Diabetic Kidney Disease: Evolving Evidence and Clinical Application. *Diabetes Metab. J.* **2025**, *49*, 386–402. [CrossRef] [PubMed]
3. Moustafa, B.; Trifan, G. The Role of Diabetes and SGLT2 Inhibitors in Cerebrovascular Diseases. *Curr. Neurol. Neurosci. Rep.* **2025**, *25*, 37. [CrossRef] [PubMed]
4. Arivarasan, V.K.; Diwakar, D.; Kamarudheen, N.; Loganathan, K. Current approaches in CRISPR-Cas systems for diabetes. *Prog. Mol. Biol. Transl. Sci.* **2025**, *210*, 95–125. [CrossRef] [PubMed]
5. Xie, Y.; Choi, T.; Al-Aly, Z. Mapping the effectiveness and risks of GLP-1 receptor agonists. *Nat. Med.* **2025**, *31*, 951–962. [CrossRef] [PubMed]
6. Rai, P.; Dwibedi, N.; Rowneki, M.; Helmer, D.A.; Sambamoorthi, U. Dipeptidyl Peptidase-4 Inhibitors and Joint Pain: A Retrospective Cohort Study of Older Veterans with Type 2 Diabetes Mellitus. *Am. Health Drug Benefits* **2019**, *12*, 223–231. [PubMed]
7. Hamblin, P.S.; Wong, R.; Ekinci, E.I.; Fourlanos, S.; Shah, S.; Jones, A.R.; Hare, M.J.L.; Calder, G.L.; Epa, D.S.; George, E.M.; et al. SGLT2 Inhibitors Increase the Risk of Diabetic Ketoacidosis Developing in the Community and During Hospital Admission. *J. Clin. Endocrinol. Metab.* **2019**, *104*, 3077–3087. [CrossRef] [PubMed]

International Journal of
Molecular Sciences

MDPI

Article

Adiponectin C1Q and Collagen Domain Containing rs266729, Cyclin-Dependent Kinase Inhibitor 2A and 2B rs10811661, and Signal Sequence Receptor Subunit 1 rs9505118 Polymorphisms and Their Association with Gestational Diabetes Mellitus: A Case-Control Study in a Romanian Population

Mihai Muntean [1], Claudiu Mărginean [1,*], Elena Silvia Bernad [2,3,4], Claudia Bănescu [5], Victoria Nyulas [6], Irina Elena Muntean [7] and Vladut Săsăran [1]

1 Department of Obstetrics and Gynecology 2, George Emil Palade University of Medicine, Pharmacy, Science, and Technology of Târgu Mureș, 540142 Târgu Mureș, Romania; mihai.muntean@umfst.ro (M.M.); vladut.sasaran@umfst.ro (V.S.)
2 Department of Obstetrics and Gynecology, Faculty of Medicine, "Victor Babes" University of Medicine and Pharmacy, 300041 Timisoara, Romania; bernad.elena@umft.ro
3 Clinic of Obstetrics and Gynecology, "Pius Brinzeu" County Clinical Emergency Hospital, 300723 Timisoara, Romania
4 Center for Laparoscopy, Laparoscopic Surgery and In Vitro Fertilization, "Victor Babes" University of Medicine and Pharmacy, 300041 Timisoara, Romania
5 Genetics Laboratory, Center for Advanced Medical and Pharmaceutical Research, George Emil Palade University of Medicine, Pharmacy, Science, and Technology of Târgu Mureș, 540142 Târgu Mureș, Romania; claudia.banescu@umfst.ro
6 Departament of Informatics and Medical Biostatistics, George Emil Palade University of Medicine, Pharmacy, Science, and Technology of Târgu Mureș, 540142 Târgu Mureș, Romania; victoria.rus@umfst.ro
7 Algcocalm SRL, 540360 Târgu Mureș, Romania; irinagauca@yahoo.com
* Correspondence: marginean.claudiu@gmail.com; Tel.: +40-722505311

Abstract: Gestational diabetes mellitus (GDM) and type 2 diabetes mellitus (T2DM) are public health concerns worldwide. These two diseases share the same pathophysiological and genetic similarities. This study aimed to investigate the T2DM known single nucleotide polymorphisms (SNPs) of the adiponectin C1Q and collagen domain containing (*ADIPOQ*), cyclin-dependent kinase inhibitor 2A and 2B (*CDKN2A/2B*), and signal sequence receptor subunit 1 (*SSR1*) genes in a cohort of Romanian GDM pregnant women and perinatal outcomes. DNA was isolated from the peripheral blood of 213 pregnant women with (n = 71) or without (n = 142) GDM. Afterward, *ADIPOQ* (rs266729), *CDKN2A/2B* (rs10811661), and *SSR1* (rs9505118) gene polymorphisms were genotyped using TaqMan Real-Time PCR analysis. Women with GDM had a higher pre-pregnancy body mass index (BMI) ($p < 0.0001$), higher BMI ($p < 0.0001$), higher insulin resistance homeostatic model assessment (IR-HOMA) ($p = 0.0002$), higher insulin levels ($p = 0.003$), and lower adiponectin levels ($p = 0.004$) at birth compared to pregnant women with normoglycemia. GDM pregnant women had gestational hypertension (GH) more frequently during pregnancy ($p < 0.0001$), perineal lacerations more frequently during vaginal birth ($p = 0.03$), and more macrosomic newborns ($p < 0.0001$) than pregnant women from the control group. We did not find an association under any model (allelic, genotypic, dominant, or recessive) of *ADIPOQ* rs266729, *CDKN2A/2B* rs10811661, and *SSR1* rs9505118 polymorphisms and GDM. In correlation analysis, we found a weak positive correlation (r = 0.24) between the dominant model GG + CG vs. CC of rs266729 and labor induction failure. In the dominant model TT vs. CC + CT of rs10811661, we found a weak negative correlation between this model and perineal lacerations. Our results suggest that the *ADIPOQ* rs266729, the *CDKN2A/2B* rs10811661, and the *SSR1* rs9505118 gene polymorphisms are not associated with GDM in a cohort of Romanian pregnant women.

Keywords: SNPs; gestational diabetes mellitus; ADIPOQ; CDKN2A/2B; SSR1; gene

1. Introduction

The World Health Organization (WHO) report from 2022 showed that the incidence of overweight adults and obesity in adults over 20 years has increased in Europe in recent years [1]. Unfortunately, childhood and adolescent overweight and obesity incidences have also increased, occurring in up to 21% of American adolescents [2,3]. Obesity in adolescents is associated with an increased risk of insulin resistance, type 2 diabetes mellitus (T2DM), hypertension, cardiovascular diseases, musculoskeletal problems, sleep disorders, depression, and cancer later in life [2,4].

T2DM and GDM are two forms of diabetes mellitus (DM).

T2DM is defined as a value of glycosylated hemoglobin A1c (HbA1c) $\geq$ 6.5%, fasting plasma glucose $\geq$ 126 mg/dL, or 2 h plasma glucose $\geq$ 126 mg/dL during an oral glucose tolerance test, or a patient with symptoms of hyperglycemia or hyperglycemic crisis and a random glucose $\geq$ 200 mg/dL [5].

GDM is defined as diabetes diagnosed in the second or third trimester of pregnancy that was not diagnosed before gestation as T2DM or another type of diabetes (type 1 diabetes mellitus) throughout the first trimester of pregnancy [5]. These two forms of DM have almost the same risk factors, such as a high glycemic index diet, overweight, obesity, impaired glucose tolerance, physical inactivity [6–8], family history of DM, and specific ethnicity (non-Hispanic Black women, Pima Indians) [6–8]. GDM and T2DM also share the same metabolic abnormalities, such as increased insulin resistance, decreased insulin secretion, and altered levels of adipokines like adiponectin [6,8–10], and consequently, 40% of women with a history of GDM will develop T2DM in the next 15 years after a pregnancy affected by GDM [7].

In addition to environmental and lifestyle factors, the genetic component of T2DM occurrence is highlighted by the 72% heritability in monozygotic twin pairs [11]. There is increasing evidence that GDM has both epigenetic and genetic components, with certain genetic variants linked to GDM also being prevalent in T2DM [12–14]. DNA methylation of genes like insulin-like growth factor-2 (*IGF-2*) related to insulin and glucose metabolism is one of the epigenetic mechanisms involved in the development of GDM [14,15]. The polymorphisms of the insulin receptor gene (*INSR*) [16], and of transcription factor 7-like 2 (*TCF7L2*) [17] are examples of gene polymorphisms associated with GDM.

Some genome-wide association studies (GWAS) showed that rs266729 single nucleotide polymorphisms (SNP) in the promoter of the *ADIPOQ* gene [18,19] are associated with GDM and adiponectin serum levels in GDM patients, suggesting that genetics play a role in circulating adiponectin in pregnant women with GDM [19]. The *ADIPOQ* gene is located on chromosome 3q27 and consists of three exons and two introns. SNPs in the *ADIPOQ* gene are also associated with T2DM and type 1 DM (T1DM) [20].

Rs10811661 polymorphism of the gene *CDKN2A/2B*, localized on chromosome 9p21.3, is associated with altered beta cell function, impaired insulin release, and impaired glucose tolerance (IGT) [21]. *CDKN2A/2B* encodes two kinase inhibitors essential for beta cell function. GWASs show that rs10811661 polymorphism was associated with GDM and T2DM risk in Chinese populations [22,23]. Conversely, some studies show that a higher number of C alleles of rs10811661 were protective against GDM [24].

Another SNP associated with GDM [25] and T2DM [26] is rs9505118 of the *SSR1* gene, on chromosome 6p24.3. The *SSR1* gene is involved in regulating fasting insulin and fasting glucose levels [27].

Based on the above findings, we hypothesized that SNP rs266729 in the *ADIPOQ* gene, rs10811661 in the *CDKN2A/2B* gene, and rs9505118 in the *SSR1* gene are associated with GDM and perinatal outcomes.

This study focused on comparing the single nucleotide polymorphisms (SNPs) rs266729 in the *ADIPOQ* gene, rs10811661 in the *CDKN2A/2B* gene, and rs9505118 in the *SSR1* gene among a cohort of Romanian GDM patients versus healthy pregnant women. Additionally, it explored the perinatal outcomes and the relationship between these polymorphisms and perinatal outcomes.

2. Results

2.1. Maternal Demographic, Anthropometric, and Biochemical Parameters at Birth

Table 1 presents the demographic, anthropometric, and biochemical parameters of the pregnant women included in the study.

Table 1. Demographic, anthropometric, and biochemical parameters of GDM and control cases at birth.

Parameters	GDM Group (n = 71)	Control Group (n = 142)	p-Value
Maternal age at delivery, Median (IQR)	33.0 (31.0–34.0)	31.0 (30.0–32.0)	0.051
Heredo-colateral history of T2DM, %	25 (35.2%)	15 (10.6%)	<0.0001
Gestation, Median (IQR)	2.0 (1.0–3.0)	2.0 (1.0–3.0)	0.2
Parity, Median (IQR)	2.0 (1.0–4.0)	1.0 (1.0–4.0)	0.2
Gestational age at delivery, weeks, Median (IQR)	38.6 (38.2–39.3)	39.2 (38.5–39.5)	0.001
Pre-pregnancy BMI, Kg/m^2, Median (IQR)	27.58 (25.7–29.0)	22.1 (21.7–22.76)	<0.0001
GWG, Mean (SD)	12.7 ± 7.13	15.3 ± 5.49	0.004
BMI at birth, Kg/m^2, Mean (SD)	33.1 ± 5.52	28.5 ± 3.8	<0.0001
MUAC, cm, Median (IQR)	31.5 (28.4–33.9)	28.0 (26.2–41.2)	<0.0001
TST, mm, Median (IQR)	22.2 (18.8–25.8)	19.6 (14.8–24.2)	0.001
CRP, mg/dL, Median (IQR)	0.76 (0.4–1.18)	0.62 (0.3–1.0)	0.68
HgbA1c, %, Median (IQR)	5.6 (5.4–6.1)	5.4 (5.2–5.5)	<0.0001
IR-HOMA, Median (IQR)	3.23 (2.0–4.2)	2.3 (1.6–3.4)	0.0002
Insulin, mUI/L, Median (IQR)	14.8 (10.1–20.8)	11.7 (8.4–16.3)	0.003
C-peptide, Median (IQR)	3.2 (2.4–3.9)	2.7 (2.0–3.4)	0.008
Adiponectin, ng/mL, Median (IQR)	6020 (4346–7306)	7131 (5272–8724)	0.004

Note: Data are presented as medians (standard deviation or interquartile range), counts, and percentages. T2DM = type 2 diabetes mellitus; BMI = body mass index; GWG = gestational weight gain; IQR = interquartile range, MUAC = mid-upper arm circumference; TST = tricipital skinfold thickness; CRP = C reactive protein; HbA1c = glycosylated hemoglobin; IR HOMA = insulin resistance homeostatic model assessment; SD = standard deviation.

We found that GDM patients had a significantly higher incidence of a heredo-collateral history of T2DM ($p < 0.0001$) and a lower gestational age at delivery than the control group ($p = 0.001$).

In the case of anthropometric parameters, GDM patients were overweight before pregnancy and became obese grade 1 at birth. They had a greater pre-pregnancy BMI ($p < 0.0001$), a greater BMI at birth ($p < 0.0001$), and a lower GWG ($p = 0.004$) than the control group. The values of MUAC and TST were also greater than those of the controls ($p < 0.0001$ and $p = 0.001$, respectively). CRP values were not different between the groups ($p = 0.68$). Regarding glucose homeostasis parameters and insulin resistance parameters at birth, the GDM group had significantly higher values than the control group (HgbA1c ($p < 0.0001$), insulin ($p = 0.003$), and C-peptide ($p = 0.008$)). Serum adiponectin values at birth in the GDM group were significantly lower than in the control group ($p = 0.004$). However, in the GDM group, there were no differences in adiponectin values at birth between different rs266729 genotypes (CC vs. CG $p = 0.9$, CC vs. GG $p = 0.8$).

Table 2 shows newborn anthropometric characteristics. Newborns from GDM were heavier than those from the control group ($p = 0.01$). The MUAC values of newborns from GDM mothers were greater than those of newborns from the control group mothers ($p = 0.03$). There was no difference between groups regarding the 5 min APGAR score or newborn gender.

Table 2. Newborn anthropometric characteristics.

Newborn Characteristics	GDM Group ($n = 71$)	Control Group ($n = 142$)	p-Value
Weight, g, Median	3470 (3170–3850)	3350 (3108–3603)	0.01
APGAR 5 min			
≥7	71 (100%)	142 (100%)	N/A
<7	0	0	N/A
Newborn gender, n, % Female, Male	 38 (53.5%) 33 (46.5%)	 68 (47.9%) 74 (52.1%)	0.47
MUAC, mm, Mean, (SD)	11.26 ± 1.08	10.9 ± 0.89	0.03
TST, mm, Mean (SD)	6.12 ± 1.45	5.67 ± 1.29	0.051

Note: Data are presented as medians (standard deviation or interquartile range), counts, and percentages. IQR = interquartile range; MUAC = mid-upper arm circumference; TST = tricipital skinfold thickness; SD = standard deviation.

In a multivariate logistic regression analysis, we found that a higher pre-pregnancy BMI independently predicted an increased risk of GDM (Table 3), with an OR of 1.247 for pre-pregnancy BMI.

Table 3. Multivariate logistic regression analysis assessing predictors of higher hazard of GDM.

Parameters	OR (95% CI)	p-Value
Maternal age	1.043 (0.973–1.118)	0.2
Pre-pregnancy BMI	1.247 (1.151–1.352)	0.0001
Parity (≥2)	1.057 (0.646–1.730)	0.8

Note: BMI = body mass index.

2.2. Maternal-Fetal Outcomes

Maternal-fetal outcomes are shown in Table 4. Pregnant women with GDM from our cohort had GH more frequently during pregnancy ($p < 0.0001$), experienced perineal lacerations during vaginal birth ($p = 0.03$), and delivered macrosomic newborns ($p < 0.0001$)

more often than healthy pregnant women from the control group. There were no differences in the incidence of cesarean section rates between groups ($p = 0.76$).

Table 4. Maternal-fetal outcomes.

Parameters	GDM Group ($n = 71$)	Control Group ($n = 142$)	p-Value
Gestational hypertension, n, %	12 (16.9%)	3 (2.1%)	<0.0001
Preterm birth (<37 weeks), n, %	5 (7.0%)	8 (5.6%)	0.4
Failure of labor induction, n, %	3 (4.2%)	2 (1.4%)	0.2
Cesarean section, n, %	45 (63.38%)	93 (67.39%)	0.76
Perineal lacerations, n, %	8 (11.3%)	5 (3.5%)	0.03
Macrosomia (≥4000 g), n, %	15 (21.1%)	5 (3.5%)	<0.0001

Note: data are presented as counts and percentages.

2.3. *Association Between Maternal Studied Gene Polymorphisms and the Risk of GDM*

Table 5 shows the frequencies and distribution of alleles and genotypes in the rs266729, rs10811661, and rs9505118 polymorphisms in GDM and control women. Under any model (the allele model, the genotypic model, the codominant model, the dominant model, and the recessive model), we did not find significant differences in allele and genotype frequencies of all studied SNPs between the GDM and control groups ($p > 0.05$).

Table 5. Comparison of maternal genotype frequencies between GDM cases and control group.

Parameters	GDM Group ($n = 71$) %		Control Group ($n = 142$) %		p-Value
rs 266729					
Alele					
C	109	76.7%	215	75.7%	0.9
G	33	23.2%	69	24.2%	
Genotype					
CC	41	57.7%	85	59.9%	0.4
CG	27	38.0%	45	31.7%	
GG	3	4.2%	12	8.5%	
rs 9505118					
Alele					
A	89	62.6%	164	57.7%	0.3
G	53	37.3%	120	42.2%	
Genotype					
AA	23	32.4%	44	31.0%	0.3
AG	42	59.2%	76	53.5%	
GG	6	8.5%	22	15.5%	

Table 5. *Cont.*

Parameters	GDM Group (*n* = 71) %		Control Group (*n* = 142) %		*p*-Value
rs10811661					
Alele					
C	29	20.4%	53	18.6%	0.6
T	113	79.5%	231	81.3%	
Genotype					
CC	4	5.6%	8	5.6%	0.6
CT	23	32.4%	38	26.8%	
TT	44	62.0%	96	67.6%	

The rs 266729 polymorphism was in Hardy–Weinberg equilibrium (HWE) in GDM cases ($p = 0.57$), and control cases ($p = 0.09$). The rs 9505118 polymorphism was out of HWE in GDM cases ($p = 0.03$), but in control cases, it was in HWE ($p = 0.24$). The rs10811661 polymorphism was in HWE in GDM cases ($p = 0.66$), and control cases ($p = 0.11$).

2.4. Correlations Between rs266729, rs10811661, and rs9505118 Polymorphisms and Maternal-Fetal Outcomes

In correlation analysis, we found a weak positive correlation (r = 0.24) between the dominant model GG+CG vs. CC of rs266729 and labor induction failure. In the dominant model TT vs. CC+CT of rs10811661, we found a weak negative correlation between the rs10811661 SNP and perineal lacerations. Table 6 shows correlations between the studied SNPs and maternal-fetal outcomes.

Table 6. Correlations between rs266729, rs10811661, and rs9505118 polymorphisms and maternal-fetal outcomes in pregnant women with GDM (*n* = 71).

Variables/Genotypes ↓	→	rs266729-CG+GG vs. CC	rs266729-CG+CC vs. GG	rs9505118-AA+AG vs. GG	rs9505118-GG+AG vs. AA	rs10811661-CT+TT vs. CC	rs10811661-TT vs. CC+CT
Preterm birth	r	−0.124	0.058	0.084	0.073	−0.171	0.011
	p-value	0.303	0.632	0.488	0.546	0.153	0.926
Labor induction failure	r	0.246 *	0.044	−0.188	−0.154	0.051	−0.165
	p-value	0.039	0.715	0.117	0.200	0.671	0.170
Perineal lacerations	r	−0.034	−0.147	−0.052	0.151	0.087	−0.279 *
	p-value	0.776	0.223	0.667	0.207	0.470	0.018
Macrosomia (≥4000 g)	r	−0.093	−0.063	−0.215	0.063	0.126	0.092
	p-value	0.438	0.603	0.072	0.600	0.293	0.445
Gestational hypertension	r	0.071	0.095	−0.133	0.152	−0.053	0.189
	p-value	0.558	0.432	0.268	0.207	0.662	0.115

*. Correlation is significant at the 0.05 level (2-tailed). Spearman correlation.

3. Discussion

In our study, we found that women from the GDM group had a higher incidence of a heredo-collateral history of T2DM and delivered at a lower gestational age compared to the control group. They exhibited greater adiposity, insulin resistance, and lower adiponectin levels at birth than the control group. Pregnant women with GDM in our cohort experienced GH more frequently during pregnancy, suffered perineal lacerations during vaginal births, and delivered macrosomic newborns more often than those in the control group. We did not observe significant differences in allele and genotype frequencies of all examined SNPs between the GDM and control groups. In correlation analysis, we identified a weak positive correlation between the dominant model GG+CG versus CC of rs266729 and labor induction failure. In the dominant model TT versus CC+CT of rs10811661, we found a weak negative correlation between the rs10811661 SNP and perineal lacerations.

Concerning the higher incidence of heredo-collateral history of T2DM from our cohort, this aligns with the findings of McIntyre et al.'s [8] findings. Additionally, Monod et al. [28] found that pregnant women with both parents diagnosed with T2DM exhibited worse glucometabolic profiles at the end of the first trimester and were more frequently diagnosed with GDM during pregnancy, indicating the influence of genetic and environmental factors on GDM occurrence. We collected data regarding the heredo-collateral history of T2DM without distinguishing between first- or second-degree relatives affected by T2DM or between pregnant women with affected mothers, fathers, or both.

In the case of greater adiposity, Zhu et al. [29], Dias et al. [30], and Tangjittipokin et al. [19] also found that GDM pregnant women have higher pre-pregnancy BMI and greater BMI at birth compared to the control pregnant group. In obese pregnant women, weight gain during gestation leads to an increase in visceral and subcutaneous adipose tissue. This results in visceral adipose tissue dysfunction and the release of inflammatory cytokines (tumor necrosis factor—TNF-α, interleukin-6 (IL-6), and CRP), which can alter hepatic glucose production. Consequently, this leads to increased glucose output from the liver, reduced glucose uptake in muscle tissue, and heightened insulin resistance [31–33].

During a healthy pregnancy, increased insulin resistance is counterbalanced by an upregulation of beta cell function, which maintains normoglycemia. We found that our GDM pregnant women had higher levels of insulin and IR-HOMA at birth compared to the control group. Ellerbrock et al. [34] and Pan et al. [35] also discovered that in obese pregnant women with GDM, insulin resistance plays a more significant role in the occurrence of GDM than beta cell dysfunction.

Adipose tissue has endocrine functions. Adiponectin is a peptide hormone secreted by adipose tissue. It is found in multiple multimeric complexes and has insulin-sensitizing, anti-atherogenic, and anti-inflammatory properties [36]. Low levels of adiponectin throughout all stages of gestation are linked to a higher risk of metabolic dysfunction during pregnancy and an increased incidence of GDM [37]. Our findings indicate that serum adiponectin levels at birth were significantly lower in the GDM group compared to the control group, which is consistent with our results from a previous study [9]. One explanation for lower adiponectin levels in obese patients comes from Kim et al. [38]'s work. They demonstrated that epigenetic changes, such as DNA methylation of the adiponectin R2 promoter, inhibit adiponectin expression and worsen metabolic disturbances in obese individuals. Moyce et al. [39] showed in a study on mice that adiponectin deficiency led to increased hepatic lipid accumulation during pregnancy; consequently, this deficiency contributed to glucose intolerance, dysregulated gluconeogenesis, and hyperglycemia.

Regarding the higher incidence of GH in our GDM group compared to the control group, Parrettini et al. [40] could provide one explanation. They emphasized in their review the crucial role of maternal obesity and excessive weight gain in fostering a pro-inflammatory state and endothelial dysfunction within the fetoplacental unit. This condition can lead to insulin resistance and an exaggerated vascular response to vasoconstrictors, which are common pathogenic factors for GDM and GH. Carpenter et al. [41] suggest that GDM is associated with an overexpressed innate immune response related to vascular dysfunction and disease.

Concerning the higher macrosomic newborns in our GDM group, Parettini et al. [40] revealed in their work that secondary to systemic low-grade inflammation, insulin resistance, increased maternal insulinemia and hyperglycemia, increased placental volume, and impaired placental genes that regulate cell cycle parameters, lipid metabolism, and mitochondrial activity, fetuses of GDM mothers had increased growth and adiposity, leading to macrosomia. Li et al. [42] also concluded in their study that increased insulin resistance during pregnancy was associated with macrosomia in Chinese women with GDM rather than beta cell dysfunction.

Our findings regarding higher perineal tears in GDM patients during vaginal births align with those of Fabricius et al. [43]. They conducted a systematic review and meta-analysis, revealing that primiparous women with GDM face a higher risk of obstetric anal sphincter injury compared to those without the condition. Von Theobald et al. [44] also a strong association between GDM and grades 3–4 of deep perineal trauma in cases of operative delivery. Conversely, Strand-Holm et al. [45] did not find a higher risk of lower genital tract tears among women diagnosed with diabetes (type 1 Diabetes Mellitus, T2DM, and GDM) compared to those without diabetes. It is important to mention that in our unit, there is routine use of episiotomy during the vaginal births of primiparous women and selective use of episiotomy for multiparous women. Our higher rate of perineal tears in the GDM group may be related to a higher incidence of newborn macrosomia, and possibly to a deficiency of perineal protection techniques.

Previous studies have shown that GDM and T2DM share similar metabolic abnormalities, including insulin resistance and β-cell dysfunction [6–8]. Additionally, some genetic variants associated with GDM are also common to T2DM [12,13], highlighting the genetic component in the emergence of these diseases.

The *ADIPOQ* gene encodes adiponectin, and SNPs of the *ADIPOQ* gene, such as rs266729, have been associated with the occurrence of GDM [20]. We found no differences between the groups regarding the allele and genotype distribution of rs266729 in the *ADIPOQ* gene. Pawlik et al. [18] found that the G allele of rs 266729 was an independent predictor of an increased risk of GDM. Beltcheva et al. [46] reported that the C allele of rs 266729 is associated with GDM, likely influencing the transcription process of adiponectin. Consistent with our results, Zhu et al. [29], and Dias et al. [30] found no associations between the rs266729 polymorphism and GDM. The differences in these results may arise from the varying G allele frequency across different populations and the fact that GDM is a multifactorial disease in which diet and physical activity play significant roles [47].

Hribal et al. [21] found that the rs10811661 polymorphism in the *CDKN2A/2B* gene is associated with altered beta cell function, impaired insulin release, and IGT. In their meta-analysis of multiethnic studies, Guo et al. [22] demonstrated that carriers of the T allele of the *CDKN2A/2B* rs10811661 have a moderate risk of developing GDM. Additionally, Li et al. [23] found that the rs10811661 polymorphism was significantly associated with the risk of T2DM. On the contrary, Tarnowski et al. [24] demonstrated that a higher number of C alleles of the rs10811661 SNP offer protection against GDM in a cohort of 411 pregnant women, both with and without GDM. Conversely, we did not find an association between the *CDKN2A/2B* rs10811661 polymorphism and GDM. Noury et al. [48] reported similar findings in a cohort of 98 Egyptian pregnant women, regardless of their GDM status.

Scott et al. [27] found that the *SSR1* gene regulates fasting insulin and fasting glucose by influencing preproinsulin translocation across the endoplasmic reticulum membrane for proinsulin biosynthesis. Data available for the rs9505118 polymorphism's role in T2DM and GDM are limited and conflicting. Kasuga et al. [25], in a cohort of 299 Japanese pregnant women with and without GDM, found that *SSR1* rs9505118, *ADIPOQ* rs266729, and *CDKN2A/2B* rs10811661 are associated with the development of GDM. In a GWAS, Mahajan et al. [26] found that the *SSR1* rs9505118 polymorphism, among other polymorphisms, had a role in T2DM susceptibility. Contrary to this, Matsuba et al. [49] in 7620 Japanese patients with and without T2DM found that the *SSR1* rs9505118 polymorphism is not associated with T2DM. Also, we did not find an association between the *SSR1* rs9505118 polymorphism and GDM. The discrepancy in results among these studies may be attributed to variations in the frequencies of this polymorphism across different ethnicities, the number of participants included in each study, and the influence of environmental factors on the polymorphism expression.

In correlation analysis, we found a weak positive correlation between the dominant model GG + CG versus CC of rs266729 and labor induction failure. Several factors increase the likelihood of failed labor induction in obese patients, including lower cervical dilation at admission, nulliparity, fetal weight above 4000 g, and pregnancies complicated by GH and GDM, which raise the necessity for labor induction before term. The proposed pathophysiologic mechanism for failed labor induction in obese patients includes decreased myometrial contractile function. This is caused by the inhibition of intramyometrial calcium influx by higher levels of leptin and cholesterol in obese pregnant women, which antagonizes the effect of oxytocin [50]. Since we did not assess the leptin and cholesterol levels of the pregnant women included in the study, nor their relation to the rs266729 polymorphism, and given the low r-value ($r = 0.24$), we cannot conclude that the statistical significance of the association between the dominant model GG + CG versus CC of rs266729 and failed labor induction is clinically relevant for these patients. This warrants further studies in the future.

Regarding protective factors of perineal tears, obesity [44,51], modified Ritgen maneuver [52], perineal massage, the application of warm compresses during the second stage of labor [53], and side-lying position for birth [54] are protective against perineal tears during vaginal birth. In the dominant model TT versus CC + CT of rs10811661, we found a weak negative correlation between the rs10811661 SNP and perineal lacerations. More extensive prospective studies are needed to verify whether these genotypes influence the incidence of perineal tears during vaginal birth.

To the best of our knowledge, this is the first study to investigate the association among the *ADIPOQ* gene rs266729 polymorphism, the *CDKN2A/2B* gene rs10811661 polymorphism, and the *SSR1* gene rs9505118 polymorphism, and the development of GDM in a Romanian population.

We acknowledge that our study has several limitations. First, the number of our participants was a relatively small sample size for testing the associations of these SNPs with the disease phenotype. Second, we did not assess the patients' diets and physical activity, which can significantly influence GDM occurrence, and we also did not evaluate the patients' lipid profiles, which could affect GDM risk. Third, all participants in this study were recruited from a secondary maternity hospital in Târgu Mureș and thus may not be representative of pregnant Romanian women in Romania. Fourth, we collected data regarding the heredo-collateral history of T2DM without differentiating between first- or second-degree relatives affected by T2DM or between pregnant women with affected mothers, fathers, or both.

What are the implications of these findings for clinical practice and further research? Our results align with existing literature regarding the higher risk of gestational hypertension, macrosomia, and perineal tears during vaginal births in women with GDM. The SNPs we studied should not be used as markers for an increased risk of developing GDM.

Further studies are necessary to investigate the connection between diet, physical activity, adipokines, these SNPs, and GDM. Larger prospective studies are essential to clarify whether the *ADIPOQ* gene rs266729 and *CDKN2A/2B* gene rs10811661 genotypes are associated with perinatal adverse outcomes in GDM patients.

4. Materials and Methods

4.1. Study Design

The University of Medicine, Pharmacy, Science, and Technology "G. E. Palade" of Târgu-Mures Ethics Committee has authorized this study (decision number 1557/2022) following the principles of the Declaration of Helsinki (1964).

4.2. Description of Study Area and Duration of Study

This prospective case–control study was conducted between 1 February 2022, and 31 August 2024, in the Obstetrics–Gynecology Clinic 2 unit of County Hospital Mureș in Târgu Mureș, Romania.

4.3. Inclusion and Exclusion Criteria

The inclusion criteria were singleton pregnancy, diagnosis of GDM at 24–28 weeks of pregnancy, Romanian ethnicity, age above 18 years, and delivery at the Obstetrics and Gynecology Clinic 2 Târgu Mureș. The exclusion criteria were patients with T1DM or T2DM diagnosed before pregnancy, GDM diagnosis before 24 weeks of pregnancy, pregnancies with chromosomal anomalies or fetal malformations, cases of intrauterine fetal death, chronic infections, autoimmune and inflammatory diseases, neoplastic diseases, and those who lacked informed consent.

Before enrollment in the study, written informed consent was obtained from all pregnant women.

After applying the inclusion and exclusion criteria during the time mentioned above, we consecutively included 213 pregnant women in the study, divided into two groups based on the oral glucose tolerance test (OGTT) results: 71 with GDM and 142 healthy pregnant women as a control group. Structured questionnaires were used to obtain demographics (maternal age, gestation, parity, and first-degree family history of T2DM) and medical and reproductive history.

In all cases, the gestational age was determined using the date of the last menstrual period and a first-trimester ultrasound.

4.4. Diagnosis of GDM

For the diagnosis of GDM, we used The International Association of Diabetes and Pregnancy Study Groups (IADPSG) criteria [55]. One or more abnormal glucose values above $\geq$92 mg/dL ($\geq$5.2 mmol/L) fasting, 1 h $\geq$ 180 mg/dL ($\geq$10 mmol/L), or 2 h $\geq$ 153 mg/dL ($\geq$ 8.5 mmol/L) after 75 g glucose ingestion were used for diagnosis.

Patients with GDM were instructed to participate in moderate exercise for 30 min per day, adhere to nutritional therapy (1600–1800 kcal per day with 35–40% carbohydrates), and track their glycemic levels by conducting three daily checks of fasting and postprandial glucose for two weeks as a part of our local protocol. The target glucose levels were set at fasting below 95 mg/dL and postprandial levels under 120 mg/dL, measured two hours after eating [56]. A diabetologist prescribed insulin therapy at 0.7–1.0 units/kg of body weight per day for women who did not achieve glycemic control through exercise and diet. Pregnant women continued to monitor their glucose levels until birth under the supervision of the diabetologist. All pregnant women had appointments every two weeks, or more frequently if necessary, until delivery as part of their routine prenatal care.

4.5. Anthropometric Measurements

We performed a series of maternal anthropometric measurements at admission to the hospital before birth on all pregnant women included in the study, including weight, height, mid-upper arm circumference (MUAC), tricipital skinfold thickness (TST), body mass index

(BMI), and total weight gain during pregnancy. We used the patient's pre-pregnancy weight, which was reported at the first prenatal visit, to calculate her pre-pregnancy BMI.

The patient's height was measured in centimeters without shoes using a wall-mounted tape measure. The obtained value was estimated to the nearest 1 mm.

We used a Beurer PS digital scale (Beurer GmbH, Ulm, Germany) to assess the patient's weight (kg), deducting 0.5 kg for the clothing.

The BMI was calculated by dividing the patient's weight by the square of the height (kg/m^2).

All newborns were measured for weight, length, MUAC, and TST during the first hour after birth.

The newborns' weight was measured using the U-Grow electronic baby scale, U001-BS (Guangzhou Berrcom Medical Device Co., Ltd., Guangzhou, China), and their length was measured using an inextensible tape measure.

Using an inextensible tape measure, the MUAC was measured halfway between the acromion and olecranon of the posterior left upper arm. We used a Harpenden Skinfold Caliper (Baty International, West Sussex, UK) for TST measurement, calibrated to the nearest 0.2 mm. The measurements were conducted at the same location where the MUAC measurement was performed. Two measurements were obtained, and the average was computed and noted [57].

Information about newborn gender and 5 min APGAR scores was obtained from medical records.

4.6. Biochemical Analyses

We collected maternal blood samples upon hospital admission during the prepartum period. Shortly after collection, we assessed them for C-reactive protein (CRP), glycosylated hemoglobin A1c (HbA1c), insulinemia, C peptide, insulin resistance homeostatic model assessment (IR HOMA), and adiponectin levels.

HbA1c and CRP values were determined via turbidimetry, while insulinemia and C-peptide were measured by chemiluminescence with the Atellica Solution CH 930 device (Siemens Healthcare GmbH, Forchheim, Germany).

The formula used to estimate IR HOMA was [(fasting insulin (mU/L) × 209 fasting glucose (mmol/L)]/22.5 [58].

For adiponectin assessment, the blood samples were left in a serum separator tube at room temperature for 30 min to allow the serum to clot. The blood samples were centrifuged at 6000 rev/min for 4 min at room temperature. The serum was then separated and stored at −20 °C until assayed.

An automated enzyme immunoassay analyzer (DYNEX DSX Automated ELISA System, DYNEX Technologies Inc., Chantilly, VA, USA) was used to assess adiponectin levels using the Human Total Adiponectin/ACRP30 ELISA kits (PDRP 300, R&D Systems, Biotechne, Minneapolis, MN, USA), adhering to the manufacturer's protocol. The intra-assay coefficient of variation for adiponectin was <4.8% and the inter-assay coefficient of variation was <7.0%. The manufacturer states that the sensitivity of the assays for adiponectin is 0.246 ng/mL.

4.7. Genotyping Analysis

We collected maternal blood for genotyping and biochemical analysis simultaneously. Blood samples were collected in tubes containing ethylenediaminetetraacetic acid (EDTA) and stored at −20 °C until assayed. DNA was extracted using a PureLink™ kit (Invitrogen, Life Technologies Corp, Carlsbad, CA, USA).

We genotyped all samples by using the TaqMan genotyping methodology, TaqMan™ Fast Advanced Master Mix (ThermoScientific LSG, Waltham, MA, USA), and specific TaqMan® pre-designed TaqMan® SNP Genotyping Assays (Applied Biosystems, Foster City, CA, USA) to discriminate the *ADIPOQ* rs266729 (C_2412786_10), *CDKN2A/2B* rs10811661 (C_31288917_10), and *SSR 1* rs9505118 (C_1945197_10).

Genotyping was performed on the 7500 Fast DX Real-Time polymerase chain reaction (PCR) system (Applied Biosystems, Foster City, CA, USA).

4.8. Maternal and Neonatal Complications

We recorded the gestational age at birth, the mode of delivery, the anthropometric measurements of the mother and newborn described above, and adverse events during pregnancy (premature birth and gestational hypertension) and at birth (abdominal delivery, failed induction of labor (IOL), perineal lacerations, and macrosomia).

We defined maternal and neonatal complications as follows:

Spontaneous or medically indicated birth at less than 37 completed weeks of gestation, according to ICD-10 definitions (O60), was defined as premature birth.

A systolic blood pressure reading of 140 mm Hg or higher, or a diastolic blood pressure reading of 90 mm Hg or higher, or both, on two separate occasions at least 4 h apart after 20 weeks of pregnancy in a woman who previously had normal blood pressure was defined as gestational hypertension (GH) [59]. We initiated antihypertensive therapy with nifedipine at a persistent systolic blood pressure of 160 mmHg or more, diastolic blood pressure of 110 mmHg or more, or both [59].

Failed IOL is defined as not entering the active phase of labor after 24 h of prostaglandin administration or after 12 h of oxytocin infusion [60].

Perineal lacerations are classified according to the anatomical structures involved, from first-degree lacerations where only perineal skin is involved to fourth-degree lacerations where are involved anal sphincter complex and anal epithelium [61].

A newborn weight of $\geq$4000 g at birth was used to diagnose macrosomia [62].

4.9. Statistical Analysis

Statistical analyses were performed by using GraphPad Prism version 9.0 (GraphPad Software, Boston, MA, USA). Continuous variables were expressed as mean $\pm$ standard deviation and median (IQR). Categorical variables were represented as percentages. Student's *t*-test was utilized for normally distributed continuous variables, while the Mann–Whitney test was used for non-normally distributed data. The chi-square test for categorical variables assessed clinical characteristics between subjects with GDM and the control group. We performed a Spearman correlation analysis to identify significant correlations between polymorphisms and perinatal outcomes. Furthermore, we applied logistic regression analysis to identify potential predictors for GDM. A two-sided $p < 0.05$ was considered statistically significant.

We conducted a priori power analysis with the program G Power Version 3.1.9.6 from Faul et al. [63] using data from Beltcheva et al. [46]. Based on these data, we estimated a medium effect size of 0.4, assuming a two-tailed *t*-test with at least 80% power and alpha = 0.05. The total number of 217 patients will be the minimum required to sample for sufficient power, n = 145 in the control group and n = 72 in the GDM group. We assumed that 213 patients divided into 142 control patients and 71 GDM patients (2:1 ratio) would be enough for our study to have sufficient power.

5. Conclusions

Based on our results, we suggest that the ADIPOQ gene rs266729 polymorphism, the CDKN2A/2B gene rs10811661 polymorphism, and the SSR1 gene rs9505118 polymorphism are not associated with GDM in a cohort of Romanian pregnant women. Women in the GDM group had lower adiponectin levels at birth and experienced higher rates of gestational hypertension, perineal lacerations during vaginal delivery, and macrosomic newborns compared to those in the control group. Pre-pregnancy BMI acts as an independent predictor of an increased risk of GDM.

Author Contributions: Conceptualization, M.M., C.M. and V.S.; Methodology, M.M. and C.M.; Software, V.N.; Validation, M.M., C.B. and V.N.; Formal analysis, M.M., C.M., E.S.B., C.B., V.N. and V.S.; Investigation, M.M., I.E.M. and V.S.; Resources, M.M., C.B. and I.E.M.; Data curation, V.N. and V.S.; Writing—original draft, M.M., E.S.B. and I.E.M.; Writing—review and editing, M.M., C.M., E.S.B., C.B., V.N. and V.S.; Visualization, Muntean Mihai; Supervision, M.M., C.M. and V.S.; Project administration, M.M. and C.M.; Funding acquisition, M.M. and I.E.M. All authors have read and agreed to the published version of the manuscript.

Funding: This research received no external funding.

Institutional Review Board Statement: The University of Medicine, Pharmacy, Science and Technology "G. E. Palade" of Târgu–Mures Ethics Committee authorized this study (decision number 1557/2022) in accordance with the principles of the Declaration of Helsinki (1964).

Informed Consent Statement: Informed consent was obtained from all subjects involved in the study.

Data Availability Statement: The data that support the findings of this study are available from the corresponding author C.M. upon reasonable request.

Conflicts of Interest: The authors declare no conflicts of interest.

References

1. Boutari, C.; Mantzoros, C.S. A 2022 Update on the Epidemiology of Obesity and a Call to Action: As Its Twin COVID-19 Pandemic Appears to Be Receding, the Obesity and Dysmetabolism Pandemic Continues to Rage On. *Metabolism* **2022**, *133*, 155217. [CrossRef] [PubMed]
2. Kelly, A.S.; Armstrong, S.C.; Michalsky, M.P.; Fox, C.K. Obesity in Adolescents: A Review. *JAMA* **2024**, *332*, 738–748. [CrossRef]
3. Cunningham, S.A.; Hardy, S.T.; Jones, R.; Ng, C.; Kramer, M.R.; Narayan, K.V. Changes in the Incidence of Childhood Obesity. *Pediatrics* **2022**, *150*, e2021053708. [CrossRef] [PubMed]
4. Bendor, C.D.; Bardugo, A.; Pinhas-Hamiel, O.; Afek, A.; Twig, G. Cardiovascular Morbidity, Diabetes and Cancer Risk among Children and Adolescents with Severe Obesity. *Cardiovasc. Diabetol.* **2020**, *19*, 79. [CrossRef]
5. American Diabetes Association Professional Practice Committee. 2. Diagnosis and Classification of Diabetes: Standards of Care in Diabetes—2024. *Diabetes Care* **2023**, *47* (Suppl. S1), S20–S42. [CrossRef]
6. Wu, Y.; Ding, Y.; Tanaka, Y.; Zhang, W. Risk Factors Contributing to Type 2 Diabetes and Recent Advances in the Treatment and Prevention. *Int. J. Med. Sci.* **2014**, *11*, 1185. [CrossRef]
7. Fletcher, B.; Gulanick, M.; Lamendola, C. Risk Factors for Type 2 Diabetes Mellitus. *J. Cardiovasc. Nurs.* **2002**, *16*, 17–23. [CrossRef]
8. McIntyre, H.D.; Catalano, P.; Zhang, C.; Desoye, G.; Mathiesen, E.R.; Damm, P. Gestational Diabetes Mellitus. *Nat. Rev. Dis. Primer* **2019**, *5*, 47. [CrossRef]
9. Muntean, M.; Săsăran, V.; Luca, S.-T.; Suciu, L.M.; Nyulas, V.; Mărginean, C. Serum Levels of Adipolin and Adiponectin and Their Correlation with Perinatal Outcomes in Gestational Diabetes Mellitus. *J. Clin. Med.* **2024**, *13*, 4082. [CrossRef]
10. Mir, M.M.; Mir, R.; Alghamdi, M.A.A.; Wani, J.I.; Sabah, Z.U.; Jeelani, M.; Marakala, V.; Sohail, S.K.; O'haj, M.; Alharthi, M.H.; et al. Differential Association of Selected Adipocytokines, Adiponectin, Leptin, Resistin, Visfatin and Chemerin, with the Pathogenesis and Progression of Type 2 Diabetes Mellitus (T2DM) in the Asir Region of Saudi Arabia: A Case Control Study. *J. Pers. Med.* **2022**, *12*, 735. [CrossRef]
11. Willemsen, G.; Ward, K.J.; Bell, C.G.; Christensen, K.; Bowden, J.; Dalgård, C.; Harris, J.R.; Kaprio, J.; Lyle, R.; Magnusson, P.K.; et al. The Concordance and Heritability of Type 2 Diabetes in 34,166 Twin Pairs from International Twin Registers: The Discordant Twin (DISCOTWIN) Consortium. *Twin Res. Hum. Genet.* **2015**, *18*, 762–771. [CrossRef] [PubMed]

12. Robitaille, J.; Grant, A.M. The Genetics of Gestational Diabetes Mellitus: Evidence for Relationship with Type 2 Diabetes Mellitus. *Genet. Med.* **2008**, *10*, 240–250. [CrossRef] [PubMed]
13. Pervjakova, N.; Moen, G.-H.; Borges, M.-C.; Ferreira, T.; Cook, J.P.; Allard, C.; Beaumont, R.N.; Canouil, M.; Hatem, G.; Heiskala, A.; et al. Multi-Ancestry Genome-Wide Association Study of Gestational Diabetes Mellitus Highlights Genetic Links with Type 2 Diabetes. *Hum. Mol. Genet.* **2022**, *31*, 3377–3391. [CrossRef]
14. Linares-Pineda, T.; Peña-Montero, N.; Fragoso-Bargas, N.; Gutiérrez-Repiso, C.; Lima-Rubio, F.; Suarez-Arana, M.; Sánchez-Pozo, A.; Tinahones, F.J.; Molina-Vega, M.; Picón-César, M.J.; et al. Epigenetic Marks Associated with Gestational Diabetes Mellitus across Two Time Points during Pregnancy. *Clin. Epigenetics* **2023**, *15*, 110. [CrossRef]
15. Zhang, Q.; Su, R.; Qin, S.; Wei, Y. High Glucose Increases IGF-2/H19 Expression by Changing DNA Methylation in HTR8/SVneo Trophoblast Cells. *Placenta* **2022**, *118*, 32–37. [CrossRef]
16. Ober, C.; Xiang, K.; Thisted, R.A.; Indovina, K.A.; Wason, C.J.; Dooley, S.; Rao, D. Increased Risk for Gestational Diabetes Mellitus Associated with Insulin Receptor and Insulin-like Growth Factor II Restriction Fragment Length Polymorphisms. *Genet. Epidemiol.* **1989**, *6*, 559–569. [CrossRef]
17. Wu, L.; Cui, L.; Tam, W.H.; Ma, R.C.; Wang, C.C. Genetic Variants Associated with Gestational Diabetes Mellitus: A Meta-Analysis and Subgroup Analysis. *Sci. Rep.* **2016**, *6*, 30539. [CrossRef]
18. Pawlik, A.; Teler, J.; Maciejewska, A.; Sawczuk, M.; Safranow, K.; Dziedziejko, V. Adiponectin and Leptin Gene Polymorphisms in Women with Gestational Diabetes Mellitus. *J. Assist. Reprod. Genet.* **2017**, *34*, 511–516. [CrossRef]
19. Tangjittipokin, W.; Narkdontri, T.; Teerawattanapong, N.; Thanatummatis, B.; Wardati, F.; Sunsaneevithayakul, P.; Boriboonhirunsarn, D. The Variants in ADIPOQ Are Associated with Maternal Circulating Adipokine Profile in Gestational Diabetes Mellitus. *J. Multidiscip. Healthc.* **2023**, *16*, 309–319. [CrossRef]
20. Howlader, M.; Sultana, M.I.; Akter, F.; Hossain, M.M. Adiponectin Gene Polymorphisms Associated with Diabetes Mellitus: A Descriptive Review. *Heliyon* **2021**, *7*, e07851. [CrossRef]
21. Hribal, M.; Presta, I.; Procopio, T.; Marini, M.; Stančáková, A.; Kuusisto, J.; Andreozzi, F.; Hammarstedt, A.; Jansson, P.-A.; Grarup, N.; et al. Glucose Tolerance, Insulin Sensitivity and Insulin Release in European Non-Diabetic Carriers of a Polymorphism Upstream of CDKN2A and CDKN2B. *Diabetologia* **2011**, *54*, 795–802. [CrossRef] [PubMed]
22. Guo, F.; Long, W.; Zhou, W.; Zhang, B.; Liu, J.; Yu, B. FTO, GCKR, CDKAL1 and CDKN2A/B Gene Polymorphisms and the Risk of Gestational Diabetes Mellitus: A Meta-Analysis. *Arch. Gynecol. Obstet.* **2018**, *298*, 705–715. [CrossRef] [PubMed]
23. Li, H.; Tang, X.; Liu, Q.; Wang, Y. Association Between Type 2 Diabetes and Rs10811661 Polymorphism Upstream of CDKN2A/B: A Meta-Analysis. *Acta Diabetol.* **2013**, *50*, 657–662. [CrossRef]
24. Tarnowski, M.; Malinowski, D.; Safranow, K.; Dziedziejko, V.; Pawlik, A. CDC123/CAMK1D Gene Rs12779790 Polymorphism and Rs10811661 Polymorphism Upstream of the CDKN2A/2B Gene in Women with Gestational Diabetes. *J. Perinatol.* **2017**, *37*, 345–348. [CrossRef]
25. Kasuga, Y.; Hata, K.; Tajima, A.; Ochiai, D.; Saisho, Y.; Matsumoto, T.; Arata, N.; Miyakoshi, K.; Tanaka, M. Association of Common Polymorphisms with Gestational Diabetes Mellitus in Japanese Women: A Case-Control Study. *Endocr. J.* **2017**, *64*, 463–475. [CrossRef]
26. Mahajan, A.; Go, M.; Zhang, W.; Below, J.; Gaulton, K.; Ferreira, T.; Horikoshi, M.; Johnson, A.; Ng, M.; Prokopenko, I. DIAbetes Genetics Replication And Meta-Analysis (DIAGRAM) Consortium; Asian Genetic Epidemiology Network Type 2 Diabetes (AGEN-T2D) Consortium; South Asian Type 2 Diabetes (SAT2D) Consortium; Mexican American Type 2 Diabetes (MAT2D) Consortium; Type 2 Diabetes Genetic Exploration by Nex-Generation Sequencing in Muylti-Ethnic Samples (T2D-GENES) Consortium. Genome-Wide Trans-Ancestry Meta-Analysis Provides Insight into the Genetic Architecture of Type 2 Diabetes Susceptibility. *Nat. Genet.* **2014**, *46*, 234–244.
27. Scott, R.A.; Lagou, V.; Welch, R.P.; Wheeler, E.; Montasser, M.E.; Luan, J.; Mägi, R.; Strawbridge, R.J.; Rehnberg, E.; Gustafsson, S.; et al. Large-Scale Association Analyses Identify New Loci Influencing Glycemic Traits and Provide Insight into the Underlying Biological Pathways. *Nat. Genet.* **2012**, *44*, 991–1005. [CrossRef]
28. Monod, C.; Kotzaeridi, G.; Linder, T.; Eppel, D.; Rosicky, I.; Filippi, V.; Tura, A.; Hösli, I.; Göbl, C.S. Prevalence of Gestational Diabetes Mellitus in Women with a Family History of Type 2 Diabetes in First-and Second-Degree Relatives. *Acta Diabetol.* **2023**, *60*, 345–351. [CrossRef]
29. Zhu, M.; Lv, Y.; Peng, Y.; Wu, Y.; Feng, Y.; Jia, T.; Xu, S.; Li, S.; Wang, W.; Tian, J.; et al. GCKR and ADIPOQ Gene Polymorphisms in Women with Gestational Diabetes Mellitus. *Acta Diabetol.* **2023**, *60*, 1709–1718. [CrossRef]
30. Dias, S.; Adam, S.; Rheeder, P.; Pheiffer, C. No Association Between ADIPOQ or MTHFR Polymorphisms and Gestational Diabetes Mellitus in South African Women. *Diabetes Metab. Syndr. Obes.* **2021**, *14*, 791–800. [CrossRef]
31. Alejandro, E.U.; Mamerto, T.P.; Chung, G.; Villavieja, A.; Gaus, N.L.; Morgan, E.; Pineda-Cortel, M.R.B. Gestational Diabetes Mellitus: A Harbinger of the Vicious Cycle of Diabetes. *Int. J. Mol. Sci.* **2020**, *21*, 5003. [CrossRef]
32. Pantham, P.; Aye, I.L.H.; Powell, T.L. Inflammation in Maternal Obesity and Gestational Diabetes Mellitus. *Placenta* **2015**, *36*, 709–715. [CrossRef] [PubMed]

33. Šimják, P.; Cinkajzlová, A.; Anderlová, K.; Pařízek, A.; Mráz, M.; Kršek, M.; Haluzík, M. The Role of Obesity and Adipose Tissue Dysfunction in Gestational Diabetes Mellitus. *J. Endocrinol.* **2018**, *238*, R63–R77. [CrossRef] [PubMed]
34. Ellerbrock, J.; Spaanderman, B.; van Drongelen, J.; Mulder, E.; Lopes van Balen, V.; Schiffer, V.; Jorissen, L.; Alers, R.-J.; Leenen, J.; Ghossein-Doha, C.; et al. Role of Beta Cell Function and Insulin Resistance in the Development of Gestational Diabetes Mellitus. *Nutrients* **2022**, *14*, 2444. [CrossRef] [PubMed]
35. Pan, Q.; Yang, Y.; Cao, H.; Xu, Z.; Tian, Z.; Zhan, Y.; Li, Z.; Lu, M.; Gu, F.; Lu, Q.; et al. Contribution of Insulin Resistance and β Cell Dysfunction to Gestational Diabetes Stratified for Pre-Pregnant Body Mass Index. *Reprod. Sci.* **2024**, *31*, 1151–1158. [CrossRef]
36. da Silva Rosa, S.C.; Liu, M.; Sweeney, G. Adiponectin Synthesis, Secretion and Extravasation from Circulation to Interstitial Space. *Physiology* **2021**, *36*, 134–149. [CrossRef]
37. Moyce Gruber, B.L.; Dolinsky, V.W. The Role of Adiponectin during Pregnancy and Gestational Diabetes. *Life* **2023**, *13*, 301. [CrossRef]
38. Kim, A.Y.; Park, Y.J.; Pan, X.; Shin, K.C.; Kwak, S.-H.; Bassas, A.F.; Sallam, R.M.; Park, K.S.; Alfadda, A.A.; Xu, A.; et al. Obesity-Induced DNA Hypermethylation of the Adiponectin Gene Mediates Insulin Resistance. *Nat. Commun.* **2015**, *6*, 7585. [CrossRef]
39. Moyce Gruber, B.L.; Cole, L.K.; Xiang, B.; Fonseca, M.A.; Klein, J.; Hatch, G.M.; Doucette, C.A.; Dolinsky, V.W. Adiponectin Deficiency Induces Hepatic Steatosis during Pregnancy and Gestational Diabetes in Mice. *Diabetologia* **2022**, *65*, 733–747. [CrossRef]
40. Parrettini, S.; Caroli, A.; Torlone, E. Nutrition and Metabolic Adaptations in Physiological and Complicated Pregnancy: Focus on Obesity and Gestational Diabetes. *Front. Endocrinol.* **2020**, *11*, 611929. [CrossRef]
41. Carpenter, M.W. Gestational Diabetes, Pregnancy Hypertension, and Late Vascular Disease. *Diabetes Care* **2007**, *30*, S246–S250. [CrossRef] [PubMed]
42. Li, J.; Leng, J.; Li, W.; Zhang, C.; Feng, L.; Wang, P.; Chan, J.C.; Hu, G.; Yu, Z.; Yang, X. Roles of Insulin Resistance and Beta Cell Dysfunction in Macrosomia among Chinese Women with Gestational Diabetes Mellitus. *Prim. Care Diabetes* **2018**, *12*, 565–573. [CrossRef] [PubMed]
43. Fabricius, E.E.; Bergholt, T.; Kelstrup, L.; Jangö, H. Gestational Diabetes Mellitus Affects the Risk of Obstetric Anal Sphincter Injury: A Systematic Review and Meta-Analysis of Cohort Studies. *Int. Urogynecol. J.* **2024**, *36*, 25–34. [CrossRef]
44. Von Theobald, P.; Bohrer, M.; Lorrain, S.; Iacobelli, S. Risk Factors Associated with Severe Perineal Tears: A Five-Year Study. *J. Gynecol. Obstet. Hum. Reprod.* **2020**, *49*, 101820. [CrossRef]
45. Strand-Holm, K.M.; Fuglsang, J.; Ovesen, P.G.; Maimburg, R.D. Diabetes Mellitus and Lower Genital Tract Tears after Vaginal Birth: A Cohort Study. *Midwifery* **2019**, *69*, 121–127. [CrossRef]
46. Beltcheva, O.; Boyadzhieva, M.; Angelova, O.; Mitev, V.; Kaneva, R.; Atanasova, I. The Rs266729 Single-Nucleotide Polymorphism in the Adiponectin Gene Shows Association with Gestational Diabetes. *Arch. Gynecol. Obstet.* **2014**, *289*, 743–748. [CrossRef] [PubMed]
47. Mizgier, M.; Jarzabek-Bielecka, G.; Mruczyk, K. Maternal Diet and Gestational Diabetes Mellitus Development. *J. Matern. Fetal Neonatal Med.* **2021**, *34*, 77–86. [CrossRef]
48. Noury, A.E.; Azmy, O.; Alsharnoubi, J.; Salama, S.; Okasha, A.; Gouda, W. Variants of CDKAL1 Rs7754840 (G/C) and CDKN2A/2B Rs10811661 (C/T) with Gestational Diabetes: Insignificant Association. *BMC Res. Notes* **2018**, *11*, 181. [CrossRef]
49. Matsuba, R.; Imamura, M.; Tanaka, Y.; Iwata, M.; Hirose, H.; Kaku, K.; Maegawa, H.; Watada, H.; Tobe, K.; Kashiwagi, A.; et al. Replication Study in a Japanese Population of Six Susceptibility Loci for Type 2 Diabetes Originally Identified by a Transethnic Meta-Analysis of Genome-Wide Association Studies. *PLoS ONE* **2016**, *11*, e0154093. [CrossRef]
50. Ruhstaller, K. Induction of Labor in the Obese Patient. In *Seminars in Perinatology*; Elsevier: Amsterdam, The Netherlands, 2015; Volume 39, pp. 437–440.
51. Garretto, D.; Lin, B.B.; Syn, H.L.; Judge, N.; Beckerman, K.; Atallah, F.; Friedman, A.; Brodman, M.; Bernstein, P.S. Obesity May Be Protective against Severe Perineal Lacerations. *J. Obes.* **2016**, *2016*, 9376592. [CrossRef]
52. Habek, D.; Tikvica Luetić, A.; Marton, I.; Prka, M.; Pavlović, G.; Kuljak, Ž.; Švanjug, D.; Mužina, Z. Modified Ritgen Maneuver in Perineal Protection–Sixty-Year Experience. *Acta Clin. Croat.* **2018**, *57*, 116–121. [CrossRef] [PubMed]
53. Le Ray, C.; Pizzagalli, F. Which Interventions during Labour to Decrease the Risk of Perineal Tears? CNGOF Perineal Prevention and Protection in Obstetrics Guidelines. *Gynecol. Obstet. Fertil. Senol.* **2018**, *46*, 928–936. [PubMed]
54. Hastings-Tolsma, M.; Vincent, D.; Emeis, C.; Francisco, T. Getting through Birth in One Piece: Protecting the Perineum. *MCN Am. J. Matern. Nurs.* **2007**, *32*, 158–164. [CrossRef]
55. Metzger, B.E.; Gabbe, S.G.; Persson, B.; Lowe, L.P.; Dyer, A.R.; Oats, J.J.; Buchanan, T.A. International Association of Diabetes and Pregnancy Study Groups Recommendations on the Diagnosis and Classification of Hyperglycemia in Pregnancy: Response to Weinert. *Diabetes Care* **2010**, *33*, e98. [CrossRef]
56. Mazaki-Tovi, S.; Romero, R.; Vaisbuch, E.; Erez, O.; Mittal, P.; Chaiworapongsa, T.; Kim, S.K.; Pacora, P.; Yeo, L.; Gotsch, F.; et al. Dysregulation of Maternal Serum Adiponectin in Preterm Labor. *J. Matern. Fetal Neonatal Med.* **2009**, *22*, 887–904. [CrossRef]

57. Kannieappan, L.M.; Deussen, A.R.; Grivell, R.M.; Yelland, L.; Dodd, J.M. Developing a Tool for Obtaining Maternal Skinfold Thickness Measurements and Assessing Inter-Observer Variability Among Pregnant Women Who Are Overweight and Obese. *BMC Pregnancy Childbirth* **2013**, *13*, 42. [CrossRef]
58. Wallace, T.M.; Levy, J.C.; Matthews, D.R. Use and Abuse of HOMA Modeling. *Diabetes Care* **2004**, *27*, 1487–1495. [CrossRef]
59. Espinoza, J.; Vidaeff, A.; Pettker, C.M.; Simhan, H. Gestational Hypertension and Preeclampsia. *Obstet. Gynecol.* **2019**, *133*, E1–E25.
60. Baños, N.; Migliorelli, F.; Posadas, E.; Ferreri, J.; Palacio, M. Definition of Failed Induction of Labor and Its Predictive Factors: Two Unsolved Issues of an Everyday Clinical Situation. *Fetal Diagn. Ther.* **2015**, *38*, 161–169. [CrossRef]
61. Cichowski, S.; Rogers, R. Prevention and Management of Obstetric Lacerations at Vaginal Delivery. *Obstet. Gynecol.* **2018**, *132*, E87–E102.
62. American College of Obstetricians and Gynecologists. Practice Bulletin No. 173: Fetal Macrosomia. *Obstet. Gynecol.* **2016**, *128*, e195–e209.
63. Faul, F.; Erdfelder, E.; Lang, A.-G.; Buchner, A. G* Power 3: A Flexible Statistical Power Analysis Program for the Social, Behavioral, and Biomedical Sciences. *Behav. Res. Methods* **2007**, *39*, 175–191. [CrossRef] [PubMed]

International Journal of
Molecular Sciences

MDPI

Article

The *GDF15* 3′ UTR Polymorphism rs1054564 Is Associated with Diabetes and Subclinical Atherosclerosis

Montse Guardiola [1,2,3,4], Josefa Girona [1,2,3,4], Emma Barroso [4,5,6,7], María García-Altares [4,8], Daiana Ibarretxe [1,2,3,4], Núria Plana [1,2,3,4], Josep Ribalta [1,2,3,4], Xavier Correig [4,8], Manuel Vázquez-Carrera [4,5,6,7], Lluís Masana [1,2,3,4] and Ricardo Rodríguez-Calvo [1,2,3,4,*]

1 Research Unit on Lipids and Atherosclerosis, Universitat Rovira i Virgili, 43201 Reus, Spain
2 Vascular Medicine and Metabolism Unit, "Sant Joan de Reus" University Hospital, 43204 Reus, Spain
3 Pere Virgili Health Research Institute (IISPV), 43007 Tarragona, Spain
4 Spanish Biomedical Research Centre in Diabetes and Associated Metabolic Disorders (CIBERDEM), Institute of Health Carlos III, 28029 Madrid, Spain
5 Pharmacology Unit, Department of Pharmacology, Toxicology and Therapeutic Chemistry, Faculty of Pharmacy and Food Sciences, University of Barcelona, 08028 Barcelona, Spain
6 Institut de Biomedicina de la Universidad de Barcelona (IBUB), University of Barcelona, 08028 Barcelona, Spain
7 Institut de Recerca Sant Joan de Déu (IR-SJD), 08950 Barcelona, Spain
8 Metabolomics Platform, Department of Electronic Engineering (DEEEA), Universitat Rovira i Virgili, 43007 Tarragona, Spain
* Correspondence: ricardo.rodriguez@ciberdem.org

Abstract: Growth differentiation factor 15 (GDF15) is a stress-response cytokine related to a wide variety of metabolic diseases. However, the impact of GDF15-specific genetic variants on the above-mentioned conditions is poorly known. The aim of this study was to assess the impact of selected GDF15 single-nucleotide polymorphisms (SNPs) on metabolic disturbances and subclinical atherosclerosis. A cross-sectional study involving 153 participants of a metabolic patient-based cohort was performed. Three selected SNPs (rs888663, rs1054564 and rs1059369) in a locus on chromosome 19 including the *GDF15* gene were genotyped by Polymerase Chain Reaction (PCR), and its relationship with the serum GDF15 levels, health status and clinical variables were analyzed. Of the three SNPs analyzed, only rs1054564 showed different distributions between the healthy volunteers and patients suffering lipid alterations and associated disorders. Accordingly, just the rs1054564 variant carriers showed a significant increase in GDF15 serum levels compared to the wild-type carriers. The group of variant carriers showed a higher frequency of individuals with diabetes, compared to the wild-type carrier group, without showing differences in other metabolic conditions. Additionally, the frequency of individuals with atherosclerotic carotid plaque was higher in the rs1054564 variant carriers than in the wild-type carriers. Logistic regression models identified that the presence of the rs1054564 variant carriers increase the likelihood for both diabetes and carotid plaque independently of confounding factors. Overall, the findings of this study identify the rs1054564 variant as a potential indicator for the likelihood of diabetes and subclinical atherosclerosis.

Keywords: GDF15; rs1054564; diabetes; atherosclerosis

1. Introduction

Growth differentiation factor 15 (GDF15), a divergent member of the transforming growth factor-β (TGF-β) superfamily [1–4], is a stress-response cytokine related to a wide variety of metabolic diseases, including type 2 diabetes mellitus (T2D) [5] and cardiovascular (CV) disease [6], among other conditions. Since both T2D and CV diseases share several common pathophysiological features [7,8], the common soil hypothesis postulated that both conditions share common genetic and environmental factors, which may explain the shared biomarkers for both conditions [9,10]. Among them, a proteomic study identified GDF15 as the highest-ranking protein associated with both T2D and coronary artery disease [11]. Additionally, GDF15 circulating levels have been proposed as a prognostic biomarker for insulin resistance and abnormal glucose control [12], and they have been found associated with T2D mortality [13]. Furthermore, GDF15 levels have been related to surrogate indicators of atherosclerosis and have been proposed as predictors of both all-cause and CV mortality [14–17]. Therefore, increasing evidence highlights the role of the GDF15 circulating levels as a relevant biomarker for both T2D and CV disease, beyond the well-established clinical risk factors for these conditions. Interestingly, genetic factors contribute to determine the GDF15 circulating concentrations similarly to cardiometabolic risk factors [18]. Nevertheless, the impact of *GDF15*-specific genetic variants on the above-mentioned conditions has only been partially explored. Since *GDF15* expression has been associated with certain genetic single-nucleotide polymorphisms (SNPs) [18,19] and given that genetic factors may explain between 21.5 and 38% of the phenotypic variation in the GDF15 blood concentration [18,20], genetic *GDF15* variants may underlie the development of these pathologies.

The human *GDF15* gene is located on chromosome 19p13.11, containing two exons interspaced by one intron, rendering an mRNA product of 1200 base pairs coding for a 308-amino acid pro-protein [4]. Interestingly, two meta-analyses of genome-wide association studies (GWASs) have identified a genome-wide significant locus on chromosome 19, including the *GDF15* gene, which influences blood concentrations of GDF15 [18,20]. Of note, eight SNPs on this locus were independently identified in both studies [18,20]. Nevertheless, the complete picture of how these *GDF15* genetic variants impact on T2D and CVD has not been fully described.

Given the role of *GDF15* genetic variants in influencing its circulating levels, we hypothesize that *GDF15* genetic variants are associated with metabolic alterations. In this study, we analyzed the impact of selected *GDF15* SNPs on metabolic disturbances and subclinical atherosclerosis.

2. Results

2.1. Genotype and Allele Frequencies of GDF15 SNPs in Both Healthy Volunteers and Patients with Lipid Alterations and Related Disorders

The genotype of the three studied GDF15 SNPs (rs888663, rs1054564 and rs1059369) is shown in both healthy volunteers and patients (Table 1). Of the three SNPs studied, only rs1054564 showed significantly different distributions between the patients and healthy volunteers ($p = 0.033$). While among the healthy volunteers we found 78.6% wild carriers (GG) and 21.4% carriers of the variant (19.6% GC and 1.8% CC), this latter percentage rose to 38.1% among the patient group (37.1% GC and 1.0% CC). The allelic frequencies were similar to those already described (source https://www.ebi.ac.uk/gwas/ (accessed on 2 February 2024)).

Table 1. Genotype frequencies of GDF15 SNPs in the study population.

SNPs	Healthy Volunteers	Patients	*p*-Value
rs888663			
TT	41 (73.2%)	69 (71.9%)	0.859
TG	12 (21.4%)	19 (18.8%)	
GG	3 (5.4%)	8 (8.3%)	
rs1054564			
GG	44 (78.6%)	60 (61.9%)	0.033
GC	11 (19.6%)	36 (37.1%)	
CC	1 (1.8%)	1 (1.0%)	
rs1059369			
TT	29 (51.8%)	59 (64.1%)	0.138
TA	24 (42.9%)	31 (33.7%)	
AA	3 (5.4%)	2 (2.2%)	

Data are shown as N (%). Differences in genotype distributions between groups were established by Chi-square tests.

2.2. Serum GDF15 Levels Are Increased in rs1054564 Variant Carriers

To explore whether the selected SNPs' variant carriers influence the GDF15 circulating levels, the serum levels of this cytokine were determined. No statistically significant differences were found in the serum GDF15 levels between the wild-type and variant carriers for both the rs888663 and rs1059369 SNPs. However, the GDF15 levels were found increased in the serum from the rs1054564 variant carriers compared to the wild-type carriers (rs1054564 wild-type carriers (GG): 752.6 [463.6–1213.1] pg/mL; rs1054564 variant carriers (GC, CC): 1227.0 [704.6–1833.0] pg/mL, $p = 0.002$) (Figure 1).

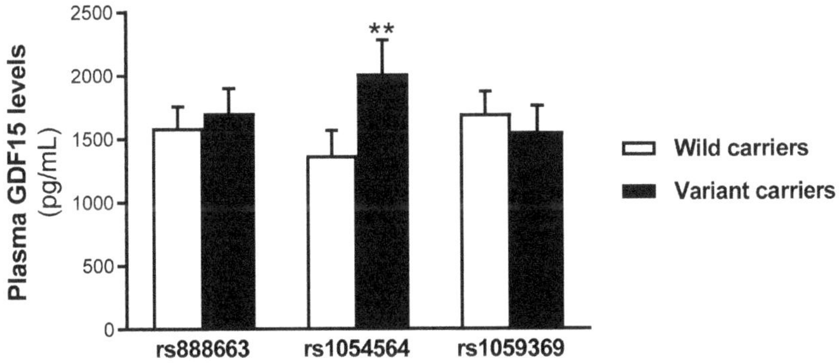

Figure 1. Serum GDF15 levels are increased in rs1054564 variant carriers compared to wild-type carriers. Data are expressed as the means ± SEM. ** $p < 0.01$ vs. rs1054564 wild-type carriers. p values are adjusted by age, glucose, insulin therapy, oral antidiabetic therapy and hypotensive therapy through the analysis of covariance (ANCOVA).

2.3. Characteristics of Study Populations According to rs1054564 Genotype

The characteristics of the study populations according to the rs1054564 variants are shown in Table 2. Of the 153 individuals, 104 were rs1054564 wild-type carriers (GG), whereas 49 were variant carriers (47 GC and 2 CC). Both groups show similar age and gender distributions. The group of variant carriers showed a higher frequency of individuals with diabetes, compared to the wild-type carrier group (wild-type carriers: 38.5%; variant carriers: 63.3%, $p < 0.004$). No significant differences were found between the groups in other pathologic conditions, including hypertension, obesity, metabolic syndrome or liver steatosis. Accordingly, the glucose concentration was 16% higher in the group of rs1054564 variant carriers [119.0 (91.0–153.0) mg/dL] than in the rs1054564 wild-type carriers [102.2 (85.8–133.0) mg/dL], not showing significant differences in any of the other analyzed parameters.

Table 2. Baseline characteristics of the study population according to rs1054564.

	Wild-Type Carriers (N = 104)	Variant Carriers (N = 49)	p-Value
	Clinical data		
Age (years)	55.2 ± 1.2	57.0 ± 1.5	0.374
Gender (F)	52.9%	51%	0.829
Hypertension	32.7%	42.9%	0.221
Diabetes	38.5%	63.3%	0.004
Obesity	32.7%	42.9%	0.221
Metabolic syndrome	53.8%	67.3%	0.114
Liver steatosis	43.7%	54.2%	0.230

Table 2. *Cont.*

	Wild-Type Carriers (N = 104)	Variant Carriers (N = 49)	*p*-Value
Anthropometric and analytical data			
Systolic BP (mmHg)	130.3 ± 2.0	130.8 ± 2.8	0.887
Diastolic BP (mmHg)	78.0 (70.0–83.0)	78.3 (72.5–80.8)	0.863
Weight (Kg)	78.0 ± 1.5	80.1 ± 2.3	0.429
Waist circumference (cm)	96.8 ± 1.4	99.1 ± 2.2	0.354
BMI (Kg/m^2)	28.6 ± 0.5	29.8 ± 0.8	0.175
Glucose (mg/dL)	102.2 (85.8–133.0)	119.0 (91.0–153.0)	0.040
Triglycerides (mmol/L)	1.4 (0.8–2.5)	1.6 (0.9–3.1)	0.390
Total cholesterol (mmol/L)	5.8 ± 0.1	6.1 ± 0.2	0.237
LDL (mmol/L)	3.5 ± 0.1	3.7 ± 0.2	0.358
HDL (mmol/L)	1.3 ± 0.0	1.3 ± 0.0	0.880
AST (U/L)	22.0 (20.0–26.0)	22.0 (20.0–30.0)	0.518
ALT (U/L)	16.0 (12.0–23.0)	19.0 (13.0–24.5)	0.310
GGT (U/L)	19.5 (14.0–30.8)	21.0 (14.0–38.5)	0.442
HsCRP (mg/L)	1.9 ± 0.1	2.1 ± 0.2	0.397
Glyc-A (μmol/L)	835.8 ± 28.9	908.2 ± 48.5	0.179
Glyc-B (μmol/L)	348.3 ± 7.4	361.8 ± 13.7	0.346
FIB-4	1.6 ± 0.0	1.6 ± 0.1	0.728
FLI (%)	43.1 (17.3–83.7)	68.9 (24.7–92.7)	0.136

Data are shown as percentage for categorical variables, mean ± SEM for continuous variables with normal distributions or median (interquartile range) for continuous variables with non-normal distributions. Normal distributions were analyzed by Student *t* test, non-normal distribution by U-Mann-Whitney and data gathered as categorical variables by Chi-square tests. Systolic BP: systolic blood pressure; Diastolic BP: diastolic blood pressure; BMI: body mass index; LDL: low-density lipoprotein; HDL: high-density lipoprotein; AST: aspartate aminotransferase; ALT: alanine aminotransferase; GGT: gamma-glutamyl transferase; HsCRP: high-sensitivity C-reactive protein; Glyc-A: glycoproteins A; Glyc-B: glycoproteins B; FLI: fatty liver index; FIB-4: fibrosis-4 score.

In order to explore the prevalence of rs1054564 variants on diabetes, univariate and multivariate logistic binary regression models were performed. The presence of rs1054564 variant carriers increased the likelihood for diabetes both in the crude (OR 2.75 and 95% CI 1.37–5.56; $p = 0.005$) and adjusted models (Model 1: OR 2.98 and 95% CI 1.35–6.59; $p = 0.007$. Model 2: OR 3.13 and 95% CI 1.36–7.19; $p = 0.007$) (Table 3).

Table 3. Crude and adjusted models used to assess the association between the rs1054564 and presence of diabetes.

	OR (95% CI)	*p*
Crude	2.75 (1.37–5.56)	0.005
Model 1	3.13 (1.36–7.19)	0.007
Model 2	4.22 (1.57–11.34)	0.007

Logistic regression models (odds ratio, OR; 95% confidence interval, CI). Model 1 was adjusted by age and gender, and Model 2 was adjusted by age, gender and glucose (METHOD = Enter).

2.4. Association of the rs1054564 Variants with the Carotid Plaque Burden

Finally, we explored the frequency of individuals with atherosclerotic carotid plaque according to the rs1054564 variants. Whereas 45.8% of the rs1054564 variant carriers had developed atherosclerotic plaque, this was only observed in 25.7% of the wild-type carriers ($p = 0.014$) (Figure 2). The logistic regression models identified that the likelihood of atherosclerotic plaque was ~2.4-fold higher in the rs1054564 variant carriers than in the wild-type carriers (crude model: OR 2.44 and 95% CI 1.19–5.03; $p = 0.015$. Model 1: OR 2.44 and 95% CI 1.11–5.37; $p = 0.026$. Model 2: OR 2.41 and 95% CI 1.08–5.37; $p = 0.032$) (Table 4).

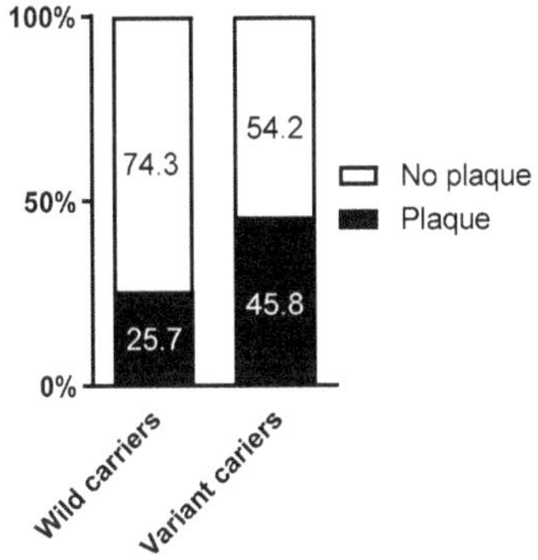

Figure 2. Percentage of plaque according to rs1054564. $p = 0.014$ between rs1054564 wild-type and variant carriers. Differences between groups were established by Chi-square tests.

Table 4. Crude and adjusted models used to assess the association between the rs1054564 and presence of atherosclerotic plaque.

	OR (95% CI)	*p*
Crude	2.44 (1.19–5.03)	0.015
Model 1	2.44 (1.11–5.37)	0.026
Model 2	2.41 (1.08–5.37)	0.032

Logistic regression models (odds ratio, OR; 95% confidence interval, CI). Model 1 was adjusted by age, and Model 2 was adjusted by age and gender (METHOD = Enter).

3. Discussion

Despite the well-known role of GDF15 in several (patho)physiological conditions [4], the influence of its genetic variants on these are still poorly studied. Given that certain genetic SNPs have been associated with *GDF15* expression [18,19] and its circulating levels may be influenced by genetic factors [18], we hypothesized that *GDF15* genetic variants may be associated with metabolic alterations. In this study, we explored the impact of selected SNPs from a chromosome 19 locus containing the *GDF15* gene [18,20] on metabolic disturbances and subclinical atherosclerosis.

Of the eight SNPs independently identified in two large GWASs [18,20], three selected SNPs (rs888663, rs1054564 and rs1059369) were chosen for genotyping. Of the three selected SNPs, only the rs1054564 variant carriers showed increased serum GDF15 levels compared to the wild-type carriers, without significant changes found in the GDF15 levels relative to the other two studied SNPs. These results conflict with previous studies suggesting that all these SNPs may significantly influence GDF15 blood concentrations [18,20]. However, our data cannot rule out the impact of these genetic variants (rs888663; rs1059369) on the blood concentrations of GDF15, probably due to the relatively small number of participants in our study cohort. Nevertheless, despite this limitation, the statistically significant increase in the serum GDF15 levels found in the rs1054564 variant carriers highlights the prominent role of this SNP in the regulation of GDF15 concentrations. Of note, previous studies have identified rs1054564 as the most significantly associated with circulating GDF15 levels among the SNPs near the 3′ untranslated region (UTR) of the *GDF15* locus [21,22]. The binding of miRNAs to 3′-UTR accelerates mRNA turnover. Therefore, genetic variants at the miRNA binding sites have a direct impact on miRNA-mRNA interactions, increasing mRNA stability and, therefore, protein translation. The rs1054564 contains binding sites for several miRNAs. Specifically, it has been previously reported that hsa-miR-1233-3p directly binds to rs1054564, and that the rs1054564-C allele partially abolishes the hsa-miR-1233-3p-mediated mRNA turnover and translational suppression of GDF15 [19]. Therefore, the increase in serum GDF15 levels found in the rs1054564 variant carriers in our population may be a consequence of the loss of translational repression of GDF15 by hsa-miR-1233-3p.

Because of the relevance of rs1054564 in the control of the GDF15 circulating levels, we classified our population into rs1054564 wild-type or variant carriers and analyzed its association with metabolic status in our population. The rs1054564 variant did not associate with hypertension, obesity, metabolic syndrome or liver steatosis. However, the percentage of diabetic patients was higher among carriers of the rs1054564 variant than in the wild-type group. Accordingly, the rs1054564 variant carriers showed higher glucose levels, without showing differences in any of the other analyzed variables. In line with these observations, the logistic regression models showed that the rs1054564 variant increased the likelihood of diabetes, even after adjusting for confounding factors. Our data are in line with a recent study reporting that the frequency of the rs1054564 variant carriers was significantly associated with an increased risk of T2DM compared to the wild-type genotype [23]. Therefore, the presence of the rs1054564 variant carriers may identify individuals at increased risk for diabetes. Additionally, given the role of GDF15 in atherosclerosis and CV disease [14–16], we analyzed the impact of the rs1054564 genetic variants on subclinical atherosclerosis. Our data revealed that in the group of variant carriers, there is a higher percentage of individuals with atherosclerotic plaque than in the wild-type group. Actually, the presence of the rs1054564 variant increases the likelihood of atherosclerotic plaque ~2.4-fold. To the best of our knowledge, this is the first study reporting a direct relationship between the rs1054564 variant carriers and atherosclerotic burden. Nevertheless, further studies are warranted in independent larger cohorts in order to confirm our data.

Despite no differences being found between the rs1054564 variant and wild-type carriers in any of the other studied variables, it is possible that the GDF15 influence on these is determined not just for rs1054564 but also for other parameters, both genetic and biochemical. However, the increase in GDF15 serum levels found in the rs1054564 variant carriers suggests that the impact of this SNP on diabetes and atherosclerosis may be due, at least partially, to the increase in GDF15 blood concentrations.

Our study has some limitations; some of them have already been mentioned. First, the relatively small sample size cannot rule out the impact of the rs888663 and rs1059369 variants on GDF15 blood concentrations. Additionally, it may attenuate the impact of the results found on the rs1054564 variant, and further studies in larger independent populations need to be carried out. Nevertheless, the relationships found among the rs1054564 variants and diabetes and carotid plaque are robust, despite the low number of participants in this study. Unfortunately, hemoglobin A1c (HbA1c) data were not available for all the participants in our cohort, so we do not have strict control of glycemic status. On the other hand, the cross-sectional nature of this study shows a fixed picture of the relationship between rs1054564 and the studied parameters at a given point and do not allow for exploring its impact over time. Additionally, it precludes establishing causal relationships between the rs1054564 and the studied outcomes. However, the increase in GDF15 serum level concentrations found in the rs1054564 variant carriers suggests that part of the observed effects may be mediated by the increase in the GDF15 blood concentrations.

4. Materials and Methods

4.1. Study Subjects

This cross-sectional study was performed in a subset of a well-characterized patient-based cohort attending the vascular medicine and metabolism unit of our university hospital [24,25]. Specifically, 98 individuals suffering lipid alterations and associated disorders (including obesity, metabolic syndrome, diabetes and cardiovascular disease) and 59 healthy volunteers free of metabolic disorders willing to participate were included in the study. Individuals with known serious diseases, including cancer, chronic lung, renal or liver disease, were excluded from this study [26]. Diabetes and metabolic syndrome were diagnosed according to the American Diabetes Association [27] and the Adult Treatment Panel III (ATPIII) criteria [28], respectively. Obesity [body mass index (BMI) $\geq$ 30 Kg/m^2], liver steatosis [fatty liver index (FLI) $\geq$ 60 [29]] and arterial hypertension [systolic blood pressure ≥ 140 and/or diastolic blood pressure ≥ 90 mm Hg) [30]] were defined following standard clinical criteria.

Written informed consent was provided by all the study participants. This study was approved by the local Ethical and Clinical Investigation Committee according to the ethical standards outlined in the Declaration of Helsinki [31].

4.2. Clinical and Standard Biochemical Determinations

Anamnesis, anthropometric and physical examination data were recorded through standardized procedures. The information collected from the medical records includes age, gender, both systolic and diastolic blood pressure, weight, height and waist circumference, among other parameters. BMI was calculated from the weight and height measurements (Kg/m^2).

Subclinical atherosclerosis was determined by performing carotid ultrasound imaging to determine the intima media thickness (IMT) of the right and left common carotid arteries by using a MyLab 60-X Vision sonographer (Esaote, Genova, Italy). The IMT value was calculated by averaging both carotid arteries. The presence of atherosclerotic plaque was defined as an IMT > 1.5 mm or focal structures into the arterial lumen that were 50% thicker than the surrounding IMT value [32].

The baseline serum samples were obtained from each participant via centrifugation from venous fasting blood samples, and aliquots were prepared for rapid storage at −80 °C in our center's BioBank for further analysis. The cellular buffy coat was obtained, and the cells were stored at −80 °C for the DNA isolations and genotyping. Standard biochemical parameters [glucose, ultrasensitive C-reactive protein (usCRP), creatinine], lipids [total cholesterol, triglycerides and high-density lipoprotein cholesterol (HDLc)], transaminases [aspartate aminotransferase (AST) and alanine aminotransferase (ALT)] and gamma-glutamyl transpeptidase (GGT) were measured using colorimetric, enzymatic and immunoturbidimetric assays, respectively (Spinreact, SA, Spain; Wako Chemicals GmbH, Germany; and Polymedco, New York, NY, USA; CV < 4%), which were adapted to the Cobas Mira Plus Autoanalyser (Roche Diagnostics, Spain). The low-density lipoprotein cholesterol (LDLc) levels were calculated by the Friedewald formula: LDLc = total cholesterol − (HDLc + [triglycerides/5]). The acute phase glycoproteins Glyc-A and Glyc-B were assessed by nuclear magnetic resonance (1H-NMR) [33]. The serum GDF15 levels were determined in duplicate using a commercial sandwich enzyme-linked immunosorbent assay kit (Biovendor, Brno, Czech Republic; CV < 5%), following the manufacturer's instructions.

Fatty liver and liver fibrosis were estimated by the FLI and Fibrosis 4 (FIB 4) algorithms, respectively [33].

4.3. SNP Selection and Genotyping

The *GDF15* SNPs were identified from the "Type 2 Diabetes Knowledge Portal" (http://type2diabetesgenetics.org (accessed on 3 September 2021)). Three selected SNPs (rs888663, rs1054564 and rs1059369) in the *GDF15* gene were chosen for genotyping according to the relatively high minor allele frequency (MAF), existence of previous reports supporting its association with metabolic disturbances or having functional evidence or being clinically relevant, by using the International HapMap database (http://hapmap.ncbi.nlm.nih.gov/ (accessed on 3 September 2021)) and the Clin Var (https://www.ncbi.nlm.nih.gov/clinvar/ (accessed on 3 September 2021)) tools.

The genomic DNA was isolated from the peripheral leukocytes from anticoagulated venous blood using the QIAamp DNA Blood Kit (Qiagen Iberia SL, Madrid, Spain) according to the manufacturer's instructions. The selected SNPs were genotyped by Polymerase Chain Reaction (PCR) in an AbiPrism 7900HT Sequence Detection System (Applied Biosystems; Beverly, CA, USA). TaqMan SNP Genotyping assays-on-demand (Life Technologies; Madrid, Spain) were used for rs888663, rs1054564 and rs1059369 variant determination.

4.4. Statistical Analysis

The Kolmogorov–Smirnov test was used to determine the normality of the continuous variables. The data are expressed as frequencies for categorical variables, the mean ± SEM for continuous variables with normal distributions or the median and interquartile range for continuous variables with non-normal distributions. The differences between the groups were established by the Chi-squared (χ^2) test for categorical variables; for continuous variables, Student's t test or Mann–Whitney U were applied for normal and non-normal distributions, respectively. When appropriate, the differences between the groups were adjusted by confounding factors through the analysis of covariance (ANCOVA). Univariate and multivariate logistic binary regression models were performed for dichotomous variables to assess the likelihood of diabetes and atherosclerotic carotid plaque based on rs1054564. The results are presented as the odds ratio (OR) and 95% confidence interval (CI). The analyses were performed with the SPSS software (IBM SPSS Statistics, version 22.0). Statistically significant differences were considered with a two-sided $p < 0.05$.

5. Conclusions

Overall, our data identified the rs1054564 variants as potential indicators for the likelihood of diabetes and subclinical atherosclerosis.

Author Contributions: Conceptualization, R.R.-C.; methodology, M.G., J.G., D.I. and N.P.; software, M.G., J.G. and R.R.-C.; validation, M.G., J.G. and R.R.-C.; formal analysis, M.G., J.G. and R.R.-C.; investigation, M.G., J.G. and R.R.-C.; resources, E.B., M.G.-A., D.I., N.P., L.M. and R.R.-C.; data curation, M.G. and J.G.; writing—original draft preparation, R.R.-C.; writing—review and editing, M.G., J.G., E.B., M.G.-A., J.R., X.C., M.V.-C. and L.M.; visualization, R.R.-C.; supervision, R.R.-C.; project administration, E.B., M.G.-A. and R.R.-C.; funding acquisition, E.B., M.G.-A. and R.R.-C. All authors have read and agreed to the published version of the manuscript.

Funding: This research was supported by CIBER (Consorcio Centro de Investigación Biomédica en Red) (CB07/08/0028), Instituto de Salud Carlos III, Ministerio de Ciencia e Innovación [PIM04], Grant PID2021-122116OB-I00 from the MICIU/AEI/10.13039/501100011033 and "ERDF, A Way of making Europe".

Institutional Review Board Statement: This study was conducted in accordance with the Declaration of Helsinki and approved by the Ethics Committee of the Institut de Investigació Sanitària Pere Virgili (protocol code: 093/2021, date of approval: 2 June 2021).

Informed Consent Statement: Informed consent was obtained from all the subjects involved in this study.

Data Availability Statement: The data presented in this study will be provided by the corresponding author after reasonable inquiry.

Acknowledgments: Emma Barroso is lecturer Serra Húnter. The authors are grateful to Roser Rosales and Yaiza Esteban for the technical assistance.

Conflicts of Interest: The authors declare no conflicts of interest.

References

1. Bootcov, M.R.; Bauskin, A.R.; Valenzuela, S.M.; Moore, A.G.; Bansal, M.; He, X.Y.; Zhang, H.P.; Donnellan, M.; Mahler, S.; Pryor, K.; et al. MIC-1, a novel macrophage inhibitory cytokine, is a divergent member of the TGF-beta superfamily. *Proc. Natl. Acad. Sci. USA* **1997**, *94*, 11514–11519. [CrossRef]
2. Breit, S.N.; Johnen, H.; Cook, A.D.; Tsai, V.W.; Mohammad, M.G.; Kuffner, T.; Zhang, H.P.; Marquis, C.P.; Jiang, L.; Lockwood, G.; et al. The TGF-beta superfamily cytokine, MIC-1/GDF15: A pleotrophic cytokine with roles in inflammation, cancer and metabolism. *Growth Factors* **2011**, *29*, 187–195. [CrossRef] [PubMed]
3. Tsai, V.W.; Lin, S.; Brown, D.A.; Salis, A.; Breit, S.N. Anorexia-cachexia and obesity treatment may be two sides of the same coin: Role of the TGF-b superfamily cytokine MIC-1/GDF15. *Int. J. Obes.* **2016**, *40*, 193–197. [CrossRef] [PubMed]
4. Assadi, A.; Zahabi, A.; Hart, R.A. GDF15, an update of the physiological and pathological roles it plays: A review. *Pflug. Arch. Eur. J. Physiol.* **2020**, *472*, 1535–1546. [CrossRef] [PubMed]
5. McGrath, R.T.; Glastras, S.J.; Scott, E.S.; Hocking, S.L.; Fulcher, G.R. Outcomes for Women with Gestational Diabetes Treated with Metformin: A Retrospective, Case-Control Study. *J. Clin. Med.* **2018**, *7*, 50. [CrossRef] [PubMed]

6. Brown, D.A.; Breit, S.N.; Buring, J.; Fairlie, W.D.; Bauskin, A.R.; Liu, T.; Ridker, P.M. Concentration in plasma of macrophage inhibitory cytokine-1 and risk of cardiovascular events in women: A nested case-control study. *Lancet* **2002**, *359*, 2159–2163. [CrossRef]
7. De Rosa, S.; Arcidiacono, B.; Chiefari, E.; Brunetti, A.; Indolfi, C.; Foti, D.P. Type 2 Diabetes Mellitus and Cardiovascular Disease: Genetic and Epigenetic Links. *Front. Endocrinol.* **2018**, *9*, 2. [CrossRef]
8. Piccirillo, F.; Mastroberardino, S.; Nusca, A.; Frau, L.; Guarino, L.; Napoli, N.; Ussia, G.P.; Grigioni, F. Novel Antidiabetic Agents and Their Effects on Lipid Profile: A Single Shot for Several Cardiovascular Targets. *Int. J. Mol. Sci.* **2023**, *24*, 10164. [CrossRef]
9. Stern, M.P. Diabetes and cardiovascular disease. The "common soil" hypothesis. *Diabetes* **1995**, *44*, 369–374. [CrossRef]
10. Fernandes Silva, L.; Vangipurapu, J.; Laakso, M. The "Common Soil Hypothesis" Revisited-Risk Factors for Type 2 Diabetes and Cardiovascular Disease. *Metabolites* **2021**, *11*, 691. [CrossRef]
11. Ferrannini, G.; Manca, M.L.; Magnoni, M.; Andreotti, F.; Andreini, D.; Latini, R.; Maseri, A.; Maggioni, A.P.; Ostroff, R.M.; Williams, S.A.; et al. Coronary Artery Disease and Type 2 Diabetes: A Proteomic Study. *Diabetes Care* **2020**, *43*, 843–851. [CrossRef] [PubMed]
12. Kempf, T.; Guba-Quint, A.; Torgerson, J.; Magnone, M.C.; Haefliger, C.; Bobadilla, M.; Wollert, K.C. Growth differentiation factor 15 predicts future insulin resistance and impaired glucose control in obese nondiabetic individuals: Results from the XENDOS trial. *Eur. J. Endocrinol.* **2012**, *167*, 671–678. [CrossRef] [PubMed]
13. Frimodt-Moller, M.; von Scholten, B.J.; Reinhard, H.; Jacobsen, P.K.; Hansen, T.W.; Persson, F.I.; Parving, H.H.; Rossing, P. Growth differentiation factor-15 and fibroblast growth factor-23 are associated with mortality in type 2 diabetes—An observational follow-up study. *PLoS ONE* **2018**, *13*, e0196634. [CrossRef] [PubMed]
14. Daniels, L.B.; Clopton, P.; Laughlin, G.A.; Maisel, A.S.; Barrett-Connor, E. Growth-differentiation factor-15 is a robust, independent predictor of 11-year mortality risk in community-dwelling older adults: The Rancho Bernardo Study. *Circulation* **2011**, *123*, 2101–2110. [CrossRef] [PubMed]
15. Lind, L.; Wallentin, L.; Kempf, T.; Tapken, H.; Quint, A.; Lindahl, B.; Olofsson, S.; Venge, P.; Larsson, A.; Hulthe, J.; et al. Growth-differentiation factor-15 is an independent marker of cardiovascular dysfunction and disease in the elderly: Results from the Prospective Investigation of the Vasculature in Uppsala Seniors (PIVUS) Study. *Eur. Heart J.* **2009**, *30*, 2346–2353. [CrossRef]
16. Rohatgi, A.; Patel, P.; Das, S.R.; Ayers, C.R.; Khera, A.; Martinez-Rumayor, A.; Berry, J.D.; McGuire, D.K.; de Lemos, J.A. Association of growth differentiation factor-15 with coronary atherosclerosis and mortality in a young, multiethnic population: Observations from the Dallas Heart Study. *Clin. Chem.* **2012**, *58*, 172–182. [CrossRef]
17. Wiklund, F.E.; Bennet, A.M.; Magnusson, P.K.; Eriksson, U.K.; Lindmark, F.; Wu, L.; Yaghoutyfam, N.; Marquis, C.P.; Stattin, P.; Pedersen, N.L.; et al. Macrophage inhibitory cytokine-1 (MIC-1/GDF15): A new marker of all-cause mortality. *Aging Cell* **2010**, *9*, 1057–1064. [CrossRef]
18. Ho, J.E.; Mahajan, A.; Chen, M.H.; Larson, M.G.; McCabe, E.L.; Ghorbani, A.; Cheng, S.; Johnson, A.D.; Lindgren, C.M.; Kempf, T.; et al. Clinical and genetic correlates of growth differentiation factor 15 in the community. *Clin. Chem.* **2012**, *58*, 1582–1591. [CrossRef]
19. Teng, M.S.; Hsu, L.A.; Juan, S.H.; Lin, W.C.; Lee, M.C.; Su, C.W.; Wu, S.; Ko, Y.L. A GDF15 3′ UTR variant, rs1054564, results in allele-specific translational repression of GDF15 by hsa-miR-1233-3p. *PLoS ONE* **2017**, *12*, e0183187. [CrossRef]
20. Jiang, J.; Thalamuthu, A.; Ho, J.E.; Mahajan, A.; Ek, W.E.; Brown, D.A.; Breit, S.N.; Wang, T.J.; Gyllensten, U.; Chen, M.H.; et al. A Meta-Analysis of Genome-Wide Association Studies of Growth Differentiation Factor-15 Concentration in Blood. *Front. Genet.* **2018**, *9*, 97. [CrossRef]
21. Hsu, L.A.; Wu, S.; Juang, J.J.; Chiang, F.T.; Teng, M.S.; Lin, J.F.; Huang, H.L.; Ko, Y.L. Growth Differentiation Factor 15 May Predict Mortality of Peripheral and Coronary Artery Diseases and Correlate with Their Risk Factors. *Mediat. Inflamm.* **2017**, *2017*, 9398401. [CrossRef] [PubMed]
22. Brown, D.A.; Lindmark, F.; Stattin, P.; Balter, K.; Adami, H.O.; Zheng, S.L.; Xu, J.; Isaacs, W.B.; Gronberg, H.; Breit, S.N.; et al. Macrophage inhibitory cytokine 1: A new prognostic marker in prostate cancer. *Clin. Cancer Res.* **2009**, *15*, 6658–6664. [CrossRef] [PubMed]
23. Liu, Q.; Qin, L.; Liang, Y.; Xu, M.; Zhang, J.; Mo, X.; Tang, X.; Lu, Y.; Wang, X.; Cao, J.; et al. Correlations between growth differentiation factor 15 (GDF-15) serum levels and gene polymorphism with type 2 diabetes mellitus. *Heliyon* **2024**, *10*, e33044. [CrossRef] [PubMed]
24. Moreno-Vedia, J.; Llop, D.; Rodriguez-Calvo, R.; Plana, N.; Amigo, N.; Rosales, R.; Esteban, Y.; Masana, L.; Ibarretxe, D.; Girona, J. Lipidomics of triglyceride-rich lipoproteins derived from hyperlipidemic patients on inflammation. *Eur. J. Clin. Investig.* **2024**, *54*, e14132. [CrossRef] [PubMed]
25. Rodriguez-Calvo, R.; Moreno-Vedia, J.; Girona, J.; Ibarretxe, D.; Martinez-Micaelo, N.; Merino, J.; Plana, N.; Masana, L. Relationship Between Fatty Acid Binding Protein 4 and Liver Fat in Individuals at Increased Cardiometabolic Risk. *Front. Physiol.* **2021**, *12*, 781789. [CrossRef]
26. Girona, J.; Rodriguez-Borjabad, C.; Ibarretxe, D.; Vallve, J.C.; Ferre, R.; Heras, M.; Rodriguez-Calvo, R.; Guaita-Esteruelas, S.; Martinez-Micaelo, N.; Plana, N.; et al. The Circulating GRP78/BiP Is a Marker of Metabolic Diseases and Atherosclerosis: Bringing Endoplasmic Reticulum Stress into the Clinical Scenario. *J. Clin. Med.* **2019**, *8*, 1793. [CrossRef]
27. American Diabetes Association. 2. Classification and Diagnosis of Diabetes: Standards of Medical Care in Diabetes-2019. *Diabetes Care* **2019**, *42*, S13–S28. [CrossRef]
28. National Cholesterol Education Program (NCEP) Expert Panel on Detection, Evaluation, and Treatment of High Blood Cholesterol in Adults (Adult Treatment Panel III). Third Report of the National Cholesterol Education Program (NCEP) Expert Panel on Detection, Evaluation, and Treatment of High Blood Cholesterol in Adults (Adult Treatment Panel III) final report. *Circulation* **2002**, *106*, 3143–3421. [CrossRef]

29. Bedogni, G.; Bellentani, S.; Miglioli, L.; Masutti, F.; Passalacqua, M.; Castiglione, A.; Tiribelli, C. The Fatty Liver Index: A simple and accurate predictor of hepatic steatosis in the general population. *BMC Gastroenterol.* **2006**, *6*, 33. [CrossRef]
30. Unger, T.; Borghi, C.; Charchar, F.; Khan, N.A.; Poulter, N.R.; Prabhakaran, D.; Ramirez, A.; Schlaich, M.; Stergiou, G.S.; Tomaszewski, M.; et al. 2020 International Society of Hypertension Global Hypertension Practice Guidelines. *Hypertension* **2020**, *75*, 1334–1357. [CrossRef]
31. World Medical Association. World Medical Association Declaration of Helsinki: Ethical principles for medical research involving human subjects. *JAMA* **2013**, *310*, 2191–2194. [CrossRef] [PubMed]
32. Touboul, P.J.; Hennerici, M.G.; Meairs, S.; Adams, H.; Amarenco, P.; Bornstein, N.; Csiba, L.; Desvarieux, M.; Ebrahim, S.; Hernandez Hernandez, R.; et al. Mannheim carotid intima-media thickness and plaque consensus (2004–2006–2011). An update on behalf of the advisory board of the 3rd, 4th and 5th watching the risk symposia, at the 13th, 15th and 20th European Stroke Conferences, Mannheim, Germany, 2004, Brussels, Belgium, 2006, and Hamburg, Germany, 2011. *Cerebrovasc. Dis.* **2012**, *34*, 290–296. [PubMed]
33. Moreno-Vedia, J.; Rosales, R.; Ozcariz, E.; Llop, D.; Lahuerta, M.; Benavent, M.; Rodriguez-Calvo, R.; Plana, N.; Pedragosa, A.; Masana, L.; et al. Triglyceride-Rich Lipoproteins and Glycoprotein A and B Assessed by ^{1}H-NMR in Metabolic-Associated Fatty Liver Disease. *Front. Endocrinol.* **2021**, *12*, 775677. [CrossRef] [PubMed]

International Journal of
Molecular Sciences

MDPI

Article

Novel Metabolites Associated with Decreased GFR in Finnish Men: A 12-Year Follow-Up of the METSIM Cohort

Lilian Fernandes Silva [1,2], Jagadish Vangipurapu [1,3], Anniina Oravilahti [1] and Markku Laakso [1,4,*]

1 Institute of Clinical Medicine, Internal Medicine, University of Eastern Finland, 70211 Kuopio, Finland; lilianf@uef.fi (L.F.S.); jagadish.vangipurapu@uef.fi (J.V.); anniina.oravilahti@uef.fi (A.O.)
2 Department of Medicine, Division of Cardiology, David Geffen School of Medicine, University of California, Los Angeles, CA 90095, USA
3 A.I. Virtanen Institute for Molecular Sciences, University of Eastern Finland, 70211 Kuopio, Finland
4 Department of Medicine, Kuopio University Hospital, 70200 Kuopio, Finland
* Correspondence: markku.laakso@uef.fi

Abstract: Identification of the individuals having impaired kidney function is essential in preventing the complications of this disease. We measured 1009 metabolites at the baseline study in 10,159 Finnish men of the METSIM cohort and associated the metabolites with an estimated glomerular filtration rate (eGFR). A total of 7090 men participated in the 12-year follow-up study. Non-targeted metabolomics profiling was performed at Metabolon, Inc. (Morrisville, NC, USA) on EDTA plasma samples obtained after overnight fasting. We applied liquid chromatography mass spectrometry (LC-MS/MS) to identify the metabolites (the Metabolon DiscoveryHD4 platform). We performed association analyses between the eGFR and metabolites using linear regression adjusted for confounding factors. We found 108 metabolites significantly associated with a decrease in eGFR, and 28 of them were novel, including 12 amino acids, 8 xenobiotics, 5 lipids, 1 nucleotide, 1 peptide, and 1 partially characterized molecule. The most significant associations were with five amino acids, N-acetylmethionine, N-acetylvaline, gamma-carboxyglutamate, 3-methylglutaryl-carnitine, and pro-line. We identified 28 novel metabolites associated with decreased eGFR in the 12-year follow-up study of the METSIM cohort. These findings provide novel insights into the role of metabolites and metabolic pathways involved in the decline of kidney function.

Keywords: glomerular filtration rate; metabolites; metabolomics; metabolic pathways

1. Introduction

Chronic kidney disease (CKD) affects approximately 10% of the Western countries' population [1]. Glomerular filtration rate (GFR) is accepted as the best marker of impaired kidney function, calculated as an estimated GFR (eGFR) [2]. Diabetes is a major risk factor for impaired kidney function [3], but also age, sex, hypertension, obesity, increased total triglycerides, and smoking are risk factors for CKD [4]. During the last few years, genome-wide association studies identified hundreds of genetic variants for kidney diseases [5–7]. Interestingly, a recent study identified genetic variants in the individuals with and without diabetes and reported that a majority of eGFR loci were similar in individuals with and without diabetes [8]. Genetic studies advanced our understanding of CKD, but they explain only a portion of the disease progression. Additionally, clinical markers often detect an impairment in kidney function only at later stages.

The first studies aiming to identify metabolites associated with eGFR had a small size and included only a low number of metabolites [9–13]. Grams et al. [14] included 587 participants in their study, and identified five metabolites (16-hydroxypalmitate, kynurenate, homovanillate sulphate, N2,N2-dimethylguanosine, and hippurate) associated with CKD. Lin et al. performed a large metabolome-wide association study, including 640 metabolites in 3906 participants of the Hispanic Community Health Study/Study of Latinos. They identified 404 eGFR-metabolite associations and found 79 novel associations [15], where amino acids and xenobiotics were the most frequent metabolites associated with eGFR. Recently, two studies reported several metabolites associated with CKD [16,17].

Identification of novel metabolites offers additional insights into the underlying biochemical processes that are lacking in genetic or clinical markers. Metabolites can serve as early biomarkers and potentially modifiable risk factors. Early identification of individuals having impaired kidney function is essential in the prevention of CKD and its complications. However, previous studies aiming to identify metabolites associated with a decrease in eGFR were cross-sectional and included a small number of participants and metabolites. Furthermore, many of these studies never addressed the long-term progression of CKD, leaving the gaps in the understanding of the changes in metabolites over time. Our population-based study included 10,159 participants having 1009 metabolites measured at baseline. A total of 7090 participated in a 12-year follow-up. Our study is the largest study identifying novel metabolites associated with a decline in eGFR during a long follow-up. Therefore, our study has a good statistical power to reveal new metabolic pathways for impaired kidney function.

2. Results

2.1. Baseline Characteristics

We included in our study 10,159 METSIM participants. Table 1 shows the baseline characteristics of the participants according to their glucose tolerance. These groups differed significantly in age, systolic blood pressure, BMI, total triglycerides, fasting glucose, HbA1c, fasting plasma insulin, eGFR, urine albumin, and high-sensitivity C-reactive protein (hs-CPR). The difference between the three groups was statistically significant but not clinically relevant eGFR (87.9 in the NGT group, 88.6 and 86.1 in the T2D group).

Table 1. Baseline Characteristics of the Participants According to Glucose Tolerance.

Measurements	NGT (n = 3034)	Pre-Diabetes (n = 5715)	T2D (n = 1410)	p
Age (years)	56.8 ± 6.9	57.4 ± 7.2	60.6 ± 6.7	1.1×10^{-63}
Systolic blood pressure (mmHg)	134.3 ± 15.9	138.7 ± 16.2	145.2 ± 18.1	2.1×10^{-93}
Body mass index (kg/m^2)	25.8 ± 3.38	27.4 ± 3.9	30.2 ± 5.2	1.1×10^{-247}
Current smoking (%)	18.0	18.4	17.2	0.606
Total triglycerides (mmol/l)	1.22 ± 0.65	1.49 ± 1.08	1.90 ± 1.21	1.2×10^{-143}
Fasting glucose (mmol/l)	5.24 ± 0.24	5.97 ± 0.37	7.51 ± 2.01	$<1 \times 10^{-250}$
HbA1C (%)	5.59 ± 0.31	5.71 ± 0.34	6.58 ± 1.13	$<1 \times 10^{-250}$
Fasting plasma insulin (mU/l)	6.25 ± 4.11	9.32 ± 6.4	19.6 ± 28.5	$<1 \times 10^{-250}$
Creatinine (umol/l)	84.6 ± 15.9	83.4 ± 12.8	84. 6 ± 22.3	0.0003
eGFR (ml/min/1.73 m^2)	87.9 ± 12.3	88.6 ± 12.2	86.1 ± 14.5	4.5×10^{-10}
Urine albumin (mg/l)	18.4 ± 110.9	20.6 ± 82.5	93.5 ± 380.1	7.2×10^{-181}
hs-CRP (mg/l)	1.82 ± 2.96	2.13 ± 4.5	3.22 ± 6.07	3.4×10^{-40}

Abbreviations: NGT, normal glucose tolerance; T2D, type 2 diabetes; HbA1C, hemoglobin A1C; eGFR, estimated glomerular filtration rate; and hs-CRP, high sensitivity C-reactive protein.

2.2. Metabolites in Participants with Decreased and Normal eGFR

We compared metabolite concentrations between the participants having eGFR < 80 and eGFR $\geq$ 80 ml/min/1.73 m2 and found statistically significant ($p < 5 \times 10^{-5}$) differences in 586 metabolites. The top significant 100 associations are shown in Table S1. The most significant differences ($p < 1.1 \times 10^{-350}$) between the two groups were in 1-methylhistidine, 1-methyl-4-imidazoleacetate, 2,3-dihydroxy-5-methylthio-4-pentenoate (DMTPA), creatinine, hydroxy asparagine, N,N,N-trimethyl-alanylproline betaine, N-acetylalanine, and pseudouridine.

2.3. Effects of Glucose Tolerance on Metabolic Profile

We analyzed the associations of eGFR with metabolites in different subgroups of glucose tolerance (n = 1 057 in each group matched for age and BMI). Participants with NGT had 379 statistically significant associations with the metabolites, participants with prediabetes had 474 significant associations, and participants with T2D had 378 significant associations. Table S1 presents the 100 most significant metabolites in the participants with T2D, prediabetes, and NGT. Independently of the glucose tolerance, all metabolites were associated with a decrease in eGFR. The Venn diagram (Figure 1) shows that the participants in the different glucose tolerance groups shared 78% of the 100 most significant metabolite associations, 11 of the metabolites were found only in the NGT group, 7 in the prediabetes group, and 12 in the T2D group.

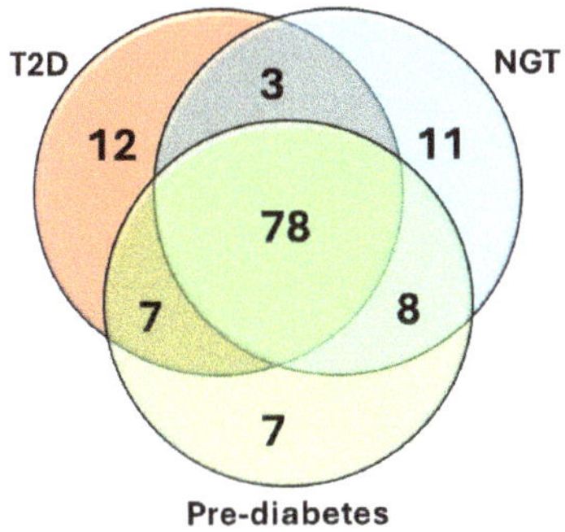

Figure 1. Venn diagram shows the top 100 most significant metabolites associated with eGFR across the participants with normal glucose tolerance (NGT), prediabetes, and type 2 diabetes (T2D).

2.4. Metabolites Associated with a Decrease in eGFR

We performed linear regression to associate 1009 metabolites with eGFR at baseline without adjustment for confounding factors, adjusted for baseline eGFR (Model 1), and adjusted for baseline eGFR, age, BMI, smoking, fasting glucose, total triglycerides, and systolic blood pressure (Model 2) (Table S2). All metabolites listed in Table S2 had $p < 5 \times 10^{-5}$ in all models. Adjustment for the baseline eGFR (Model 1) substantially decreased beta and p values. In Model 2 beta and p value further deceased but the decreases were relatively small.

We found 108 metabolites significantly associated with a decrease in eGFR (Table S2), and 28 of them were novel (Table 2). The 10 most statistically significant metabolites associated with decreased eGFR were six amino acids, creatinine, hydroxyasparagine, N,N,N-trimethyl-alanylproline betaine, N-acetylalanine, N-acetylserine, C-glycosyltryptophan, and N-formylmethionine; a nucleotide pseudouridine; xenobiotics erythritol; and carbohydrate erythronate.

Table 2. Novel Metabolites Associated with a Decrease In eGFR.

Metabolite	Sub-Class	N	Beta	*p* *	Beta	*p* **
Amino acids						
N-acetylmethionine	Methionine, cysteine, taurine metabolism	7080	−0.334	1.4×10^{-183}	−0.087	5.5×10^{-24}
N-acetylvaline	Leucine, isoleucine, valine metabolism	7082	−0.343	1.0×10^{-194}	−0.082	2.6×10^{-21}
γ-carboxyglutamate	Glutamate metabolism	6929	−0.295	1.1×10^{-138}	−0.065	2.6×10^{-14}
3-methylglutaryl-carnitine (2)	Leucine, isoleucine, valine metabolism	7001	−0.257	1.1×10^{-105}	−0.058	5.8×10^{-12}
Proline	Urea cycle; arginine proline metabolism.	7081	−0.107	1.3×10^{-19}	−0.048	3.9×10^{-9}
Pro-hydroxy-pro	Urea cycle; arginine proline metabolism	7079	−0.155	1.9×10^{-39}	−0.047	5.2×10^{-9}
4-guanidinobutanoate	Guanidino acetamido metabolism	7049	−0.158	1.7×10^{-40}	−0.049	2.3×10^{-9}
N-acetyltaurine	Methionine, cysteine, taurine metabolism	7048	−0.208	1.4×10^{-69}	−0.041	7.6×10^{-7}
Hydantoin-5-propionate	Histidine metabolism	6154	−0.211	3.6×10^{-63}	−0.043	1.1×10^{-6}
N-lactoylvaline	Lactoyl amino acid	6781	−0.182	2.5×10^{-51}	−0.043	3.1×10^{-6}
N-lactoylisoleucine	Lactoyl amino acid	5437	−0.189	4.4×10^{-45}	−0.043	1.6×10^{-5}
N-lactoylphenylalanine	Lactoyl amino acid	7033	−0.233	2.7×10^{-87}	−0.037	4.4×10^{-5}
Lipids						
11beta-hydroxy etiocholanolone glucuronide	Androgenic steroids	4891	−0.204	2.9×10^{-47}	−0.050	4.0×10^{-7}
3-decenoylcarnitine	Fatty acid metabolism	5395	−0.217	2.9×10^{-58}	−0.042	9.2×10^{-6}
Cis-3,4-methylene heptanoylglycine	Fatty acid metabolism	6825	−0.161	5.2×10^{-41}	−0.038	4.8×10^{-6}
2-methylmalonyl carnitine (C4-DC)	Fatty acid metabolism	5827	−0.235	8.0×10^{-74}	−0.042	3.1×10^{-6}
Propionylglycine	Fatty acid metabolism	3960	−0.119	4.9×10^{-14}	−0.049	1.3×10^{-5}
Nucleotide						
5-methyluridine(ribothymidine)	Pyrimidine metabolism	7082	−0.134	6.8×10^{-30}	−0.038	3.1×10^{-6}
Peptide						
Pyroglutamylvaline	Modified peptides	6398	−0.202	7.7E × 10^{-60}	−0.051	2.6×10^{-9}
Xenobiotics						
2,3-dihydroxyisovalerate	Food component/plant	6998	−0.206	3.8×10^{-68}	−0.048	6.8×10^{-9}
(S)-a-amino-omega-caprolactam	Food component/plant	7007	−0.296	1.3×10^{-141}	−0.050	1.0×10^{-8}
3-methoxycatechol sulfate (2)	Benzoate metabolism	5379	−0.185	2.0×10^{-42}	−0.044	1.9×10^{-6}
3-methyl catechol sulfate (1)	Benzoate metabolism	7065	−0.209	3.0×10^{-70}	−0.040	2.1×10^{-6}
3-methoxycatechol sulfate (1)	Benzoate metabolism	6318	−0.174	4.0×10^{-44}	−0.039	5.5×10^{-6}
2-acetamidophenol sulfate	Food component/plant	5939	−0.153	2.9×10^{-32}	−0.042	3.6×10^{-6}

Table 2. *Cont.*

Metabolite	Sub-Class	N	Beta	p *	Beta	p **
N-(2-furoyl)glycine	Food component/plant	5025	−0.235	5.0×10^{-64}	−0.042	2.4×10^{-5}
2-aminophenol sulfate	Food component/plant	7066	−0.147	2.8×10^{-35}	−0.036	1.1×10^{-5}
Other metabolites						
Glutamine_degradant	Partially characterized molecules	7060	−0.222	7.3×10^{-80}	−0.071	2.2×10^{-17}

p *: non-adjusted; p **: adjusted for eGFR at baseline, age, BMI, smoking, fasting glucose, total triglycerides, and systolic blood pressure.

Among the novel 28 metabolites decreasing eGFR, 12 were amino acids, 5 lipids, 1 nucleotide, 1 peptide, 8 xenobiotics, and 1 a partially characterized molecule. Among the amino acids, the three most significant imverse associations were with N-acetylmethionine (beta = −0.087, p = 5.5×10^{-24}), N-acetylvaline (beta = −0.082, p = 2.6×10^{-21}), and γ-carboxyglutamate (beta = −0.065, p = 2.6×10^{-14}). Among the lipids, the two most significant inverse associations were with 11beta-hydroxyetiocholanolone (beta = −0.050, p = 4.0×10^{-7}), and 2-methylmalonylcarnitive C4-DC (beta = -0.042, p = $3.1.0 \times 10^{-6}$), and among the xenobiotics for 2,3-dihydroxyisovalerate (beta = −0.048, p = 6.8×10^{-6}, and (S)-a-amino-omega-caprolactam (beta = −0.050, p = 1.0×10^{-8}).

2.5. Genetic Variants Associated with Novel Metabolites

We identified nine genetic variants significantly associated with the novel metabolites (Table 3). The most significant associations were with 5-methyluridine, glycine, proline, and N-acetylmethionine. Each of the nine genetic variants were associated with at least three different metabolites, suggesting pleiotropy of these genes. Importantly, none of these genetic variants were significantly associated with a decrease in eGFR, indicating that the effects of the metabolites on eGFR were not explained by genetic factors.

Table 3. The Variants of Nine Genes were Associated with the Novel Metabolites Related to a Decline in eGFR.

Gene-Variant	Metabolite	p
KLHDC7B-rs470118	5-methyluridine	9.9×10^{-199}
CPS1-rs715	Glycine	8.1×10^{-90}
AC007326.4-rs5992344	Proline	2.0×10^{-63}
DOCK3-rs138144932	N-acetylmethionine	1.3×10^{-44}
AOX1-rs7562507	Hydantoin-5-propionate	1.4×10^{-17}
COLEC10-rs13264172	Pro-hydroxy-pro	3.5×10^{-10}
MAGI1-rs264676	2.3-dihydroxy-5-methylthio-4-penenoate	2.9×10^{-8}
DCBLD2-rs192423025	Pyroglutamylvaline	3.4×10^{-8}
CNTNAP2-rs533473709	γ-carboxyglutamate	5.3×10^{-8}

3. Discussion

We measured 1009 metabolites with LC-MS/MS in 10,188 participants of the METSIM study. Our study reports several novel findings. We found that glucose tolerance did not have a major effect on the metabolite profile at baseline. Among the top 100 metabolites associated with eGFR, 78 were identical in participants with normal glucose tolerance, prediabetes, and diabetes (Figure 1). Our results suggest that the metabolic pathways leading to a decrease in eGFR are largely independent of glucose tolerance. This observation agrees with a previous study reporting that the majority of the eGFR loci were similar in the individuals with and without diabetes [8].

We found several statistically significant associations of the metabolites with a decrease in eGFR in the 12-year follow-up of the METSIM cohort. Of the 108 metabolites associated with a decrease in eGFR, 28 were novel (Table 2). We also replicated metabolite associations with decreased eGFR reported in previous studies [13,18–26]. The metabolic pathways independent from glucose highlights the complexity in the progression of CKD. Our findings suggest that non-glucose pathways, including amino acids, lipids, and xenobiotics, have independent effects on kidney function. However, our findings do not change the current treatment of patients with diabetes having impaired kidney function. Medication lowering hyperglycaemia remains the primary treatment in patients with diabetes and CKD.

We found three novel associations of N-acetylated amino acids (N-acetylmethionine, N-acetylvaline, and N-acetyltaurine) with a decrease in eGFR. N-acetylated amino acids are uremic toxins [27]. Aminoacylase-1 (ACY1) enzyme converts acetylated amino acids into free amino acids, and therefore the individuals having impaired activity of ACY1 or a mutation in the ACYL1 gene have increased concentrations of acetylated amino acids in blood and urine [28–33]. N-acetylated amino acids are uremic toxins that accumulate in the blood due to impaired kidney function (27). These toxins disrupt cellular processes, promote inflammation, and induce oxidative stress, worsening CKD progression (27–32). This highlights the potential of the metabolites to identify therapeutic targets for the prevention of CKD progression.

Amino acid γ-carboxyglutamate was significantly associated with a decrease in eGFR. γ-carboxyglutamate is a calcification inhibitor [34]. Atherosclerotic and vascular calcification are closely linked to the vitamin K-dependent protein matrix γ-carboxyglutamate. Vitamin K antagonists, including warfarin, are associated with increased calcification of renal and other arteries [34,35]. Coronary artery calcification was previously associated with a decline in eGFR [36]. Increased levels of γ-carboxyglutamate may represent a compensatory response to counteract vascular calcification as kidney function declines.

We report three novel associations of N-lactoyl-amino acids (N-lactoylvaline, N-lactoylisoleucine, and N-lactoylphenylalanine) with a decrease in eGFR. N-lactoylphenylalanine concentrations are increased in patients with phenylketonuria [37]. These patients have increased oxidative stress leading to tubulointerstitial disease, impaired kidney function, proteinuria, and arterial hypertension [38,39]. N-lactoylvaline and N-lactoylisoleucine were found in the urine of a patient with maple syrup urine disease [40], which is associated with nephrotic syndrome [41].

We also found that the nucleoside 5-methyluridine (ribothymidine), an endogenous methylated nucleoside, decreased eGFR. This finding was previously reported in rats with CKD [42]. Altered DNA methylation modulates the expression of pro-inflammatory and pro-fibrotic genes, stimulating renal disease progression [43]. High concentrations of homocysteine, hypoxia, and inflammation alter the epigenetic regulation of gene expression in CKD, impacting eGFR [43].

Eight of the 28 novel metabolites impairing eGFR were xenobiotics, chemical substances within an organism that are not naturally produced. Xenobiotics are food components, plant constituents, pesticides, industrial chemicals, environmental pollutants, or benzoate metabolites. An organic compound (S)-a-amino-omega-caprolactam is a uremic solute previously shown to impair kidney function [44]. 3-methyl catechol sulfate, a marker of current smoking and coffee consumption [45], decreased eGFR in our study. We also showed that genetic variants were not associated with xenobiotics, suggesting that decreased eGFR is largely regulated also by lifestyle and environmental factors.

Our findings highlight multiple metabolic pathways associated with a decrease in eGFR. We identified 28 novel metabolites among amino acids, lipids, nucleotides, peptides, and xenobiotics associated with decreased eGFR. Eight xenobiotics were associated with a decrease in eGFR, showing that non-genetic factors, including benzoate pathway, food components, and plants play a significant role in kidney dysfunction, demonstrating the influence of environmental factors on eGFR. Additionally, the effects of N-lactoyl-amino acids and 5-methyluridine show a potential for epigenetic regulation of kidney function. Overall, our novel findings provide valuable insights into the complex biochemical interactions affecting kidney function and pave the way for future studies to explore metabolic pathways on kidney function in diverse populations.

The strength of our study is that the METSIM study is the largest randomly selected population-based cohort identifying metabolites associated with a decrease in eGFR by applying the LC-MS/MS analysis method. Additionally, we followed our cohort for 12 years, and at baseline and follow-up, the metabolites identified were inversely associated with eGFR, increasing the credibility of our findings. We applied a conservative statistical significance threshold in all analyses to obtain reliable conclusions.

Our study has several limitations. The homogeneity of the study population (middle-aged and elderly Finnish men) limits the generalizability of our findings. Therefore, the replication of our findings in more diverse populations, such as women and non-European cohorts, is needed. Our study was an observational study, and therefore, we cannot establish causality between the metabolites and the decline of kidney function. Additionally, there may be unmeasured confounders, such as medication use or environmental factors, having effects on our findings. The LC-MS/MS platform provides a broad metabolite coverage, but the results are not fully generalizable to other metabolomics platforms.

In vitro and animal studies could help to establish causal relationships between the metabolites and the decline of the kidney function. Incorporating these findings into CKD risk prediction models may improve early detection and personalized treatment. Additionally, targeting specific metabolic pathways, for example, those involving uremic toxins, could reveal novel therapeutic approaches. Expanding the approach to include other omics data, such as genomics and proteomics, could further enhance our understanding of CKD progression.

4. Materials and Methods

4.1. Study Population and Laboratory Measurements

The METabolic Syndrome in Men (METSIM) study includes 10,197 men, aged from 45 to 73 years at baseline, and randomly selected from the population register of Kuopio, Eastern Finland. The METSIM study was approved by the Ethics Committee of the Kuopio University Hospital, Finland. All participants provided written informed consent.

The design and methods of the METSIM study were previously described in detail [46,47]. A total of 10,159 men were included in the current study, 3034 had normal glucose tolerance (NGT, fasting glucose < 6.1 mmol/L, 2-hour glucose < 7.8 mmol/L), 5715 prediabetes [impaired fasting glucose (6.1–6.9 mmol/L) or impaired glucose tolerance (7.8 to 11.0 mmol/L) or both], and 1410 T2D, [fasting glucose $\geq$ 7.0 mmol/L, or 2-hour glucose $\geq$ 11.1 mmol/L or glycated hemoglobin A1c (HbA1c) $\geq$ 6.5%] according to the American Diabetes Association classification [48]. BMI was calculated as weight divided by height squared. Smoking status was defined as current smoking (yes/no). All participants, excluding participants with T2D at baseline, underwent a 2-hour oral glucose tolerance test (75 g of glucose), and samples for plasma glucose and insulin were drawn at 0, 30, and 120 min. Other laboratory measurements were previously explained [46]. eGFR was calculated using the CKD-Epi equation [49].

4.2. Metabolomics

Non-targeted metabolomics profiling was performed at Metabolon, Inc. (Morrisville, NC, USA) on EDTA plasma samples obtained after overnight fasting, as previously described in detail [47,50]. We applied liquid chromatography mass spectrometry (LC-MS/MS) to identify the metabolites (the Metabolon DiscoveryHD4 platform). The LC-MS/MS platform was chosen for its high sensitivity, specificity, and broad dynamic range, making it ideal for detecting a wide variety of metabolites. Compared to other platforms, especially proton NMR, LC-MS/MS offers superior sensitivity, allowing for the identification of subtle metabolic changes, which is crucial in discovering early biomarkers for kidney function decline. Although limitations such as ion suppression and complex data processing exist for LC-MS/MS, its advantages in sensitivity and metabolite coverage makes it the best choice for this study. All samples were processed together for peak quantification and data scaling. We quantified raw mass spectrometry peaks for each metabolite using the area under the curve, and evaluated overall process variability by the median relative standard deviation for endogenous metabolites present in all 20 technical replicates in each batch. We adjusted for variation caused by day-to-day instrument tuning differences and columns used for biochemical extraction by scaling the raw peak quantifications to the median for each metabolite by the Metabolon batch.

4.3. Selection of Genetic Variants Decreasing Glomerular Filtration Rate

We identified genetic variants associated with a decrease in eGFR from previously published studies and the GWAS Catalog (The NHGRI-EBI Catalog of human genome-wide association studies (https://www.ebi.ac.uk/gwas/%E2%80%94accessed on 4 July 2024) in individuals of European ancestry. Altogether, 117 genes were found to be associated with impaired eGFR.

4.4. Statistical Analysis

We conducted statistical analyses using IBM SPSS Statistics, version 29. We log-transformed all continuous variables except for age and follow-up time to correct for their skewed distribution. We performed association analyses between the eGFR and metabolites using linear regression adjusted for confounding factors (Model 1, adjustment for eGFR at baseline, Model 2, adjustment for eGFR at baseline, age, BMI, smoking, systolic blood pressure, fasting glucose, and total triglycerides). The variables in Model 2 (age, BMI, smoking, systolic blood pressure, fasting glucose, and triglycerides) were selected because they are well-known risk factors for both CKD progression and metabolic changes. These variables were chosen to reduce bias and ensure that the associations between metabolites and eGFR are not influenced by these factors. We give the results as standardized beta coefficients and p values with the metabolite as a dependent variable. We used one-way ANOVA to assess the differences in clinical traits and metabolites between the two groups at baseline. We applied the Bonferroni correction to determine statistical significance for the metabolites identified ($p < 5.0 \times 10^{-5}$).

5. Conclusions

We measured 1009 metabolites in 10,159 Finnish men of the METSIM cohort and associated the metabolites with eGFR in the 12-year follow-up study. We found 108 metabolites significantly associated with a decrease in eGFR, and 28 of them were novel, including especially amino acids, xenobiotics, and lipids, showing that hyperglycaemia is not the only cause for impaired eGFR. Our findings provide novel insights into the role of metabolites and metabolic pathways involved in the decline of kidney function.

Supplementary Materials: The following supporting information can be downloaded at: https://www.mdpi.com/article/10.3390/ijms251810044/s1.

Author Contributions: Conceptualization: L.F.S. and M.L.; methodology: A.O., J.V., L.F.S. and M.L.; investigation: L.F.S., J.V., A.O. and M.L.; visualization: L.F.S. and A.O.; funding acquisition: M.L.; project administration: M.L.; supervision: M.L.; writing—original draft: L.F.S., J.V., A.O. and M.L. All authors have read and agreed to the published version of the manuscript.

Funding: The research leading to these results has received support from the Innovative Medicines Initiative Joint Undertaking under grant agreements no. 115372 EMIF (to M.L), and no. 115974 BEAt-DKD (to M.L.). This Joint Undertaking received support from the European Union's 7th Framework (EMIF) resp. Horizon 2020 (BEAt-DKD) research and innovation programmes and EFPIA, with JDRF (BEAt-DKD). Academy of Finland grant no. 321428 (ML). Centre of Excellence of Cardiovascular and Metabolic Diseases, the Academy of Finland grant no. 271961 (ML). Sigrid Juselius Foundation grant (ML). Finnish Foundation for Cardiovascular Research grant (ML). Kuopio University Hospital grant (ML).

Institutional Review Board Statement: The METSIM study was approved by the Ethics Committee of the Kuopio University Hospital, Finland. All participants provided written informed consent.

Informed Consent Statement: Informed consent was obtained from all subjects involved in the study.

Data Availability Statement: The data that support the findings of this study are available from the corresponding author M.L. upon reasonable request.

Conflicts of Interest: The authors declare no conflicts of interest.

References

1. Sekula, P.; Goek, O.-N.; Quaye, L.; Barrios, C.; Levey, A.S.; Römisch-Margl, W.; Menni, C.; Yet, I.; Gieger, C.; Inker, L.A.; et al. A Metabolome-Wide Association Study of Kidney Function and Disease in the General Population. *J. Am. Soc. Nephrol.* **2016**, *27*, 1175–1188. [CrossRef] [PubMed]
2. Levey, A.S.; Stevens, L.A.; Schmid, C.H.; Zhang, Y.L.; Castro, A.F., 3rd; Feldman, H.I.; Kusek, J.W.; Eggers, P.; Van Lente, F.; Greene, T.; et al. A New Equation to Estimate Glomerular Filtration Rate. *Ann. Intern. Med.* **2009**, *150*, 604–612. [CrossRef] [PubMed]
3. Baumeister, S.E.; Böger, C.A.; Krämer, B.K.; Döring, A.; Eheberg, D.; Fischer, B.; John, J.; Koenig, W.; Meisinger, C. Effect of Chronic Kidney Disease and Comorbid Conditions on Health Care Costs: A 10-Year Observational Study in a General Population. *Am. J. Nephrol.* **2009**, *31*, 222–229. [CrossRef] [PubMed]
4. Kazancioğlu, R. Risk factors for chronic kidney disease: An update. *Kidney Int. Suppl.* **2013**, *3*, 368–371. [CrossRef] [PubMed]
5. Wuttke, M.; Li, Y.; Li, M.; Sieber, K.B.; Feitosa, M.F.; Gorski, M.; Tin, A.; Wang, L.; Chu, A.Y.; Hoppmann, A.; et al. A catalog of genetic loci associated with kidney function from analyses of a million individuals. *Nat. Genet.* **2019**, *51*, 957–972. [CrossRef]
6. Wuttke, M.; König, E.; Katsara, M.-A.; Kirsten, H.; Farahani, S.K.; Teumer, A.; Li, Y.; Lang, M.; Göcmen, B.; Pattaro, C.; et al. Imputation-powered whole-exome analysis identifies genes associated with kidney function and disease in the UK Biobank. *Nat. Commun.* **2023**, *14*, 1–16. [CrossRef]
7. Gorski, M.; Rasheed, H.; Teumer, A.; Thomas, L.F.; Graham, S.E.; Sveinbjornsson, G.; Winkler, T.W.; Günther, F.; Stark, K.J.; Chai, J.-F.; et al. Genetic loci and prioritization of genes for kidney function decline derived from a meta-analysis of 62 longitudinal genome-wide association studies. *Kidney Int.* **2022**, *102*, 624–639. [CrossRef]
8. Winkler, T.W.; Rasheed, H.; Teumer, A.; Gorski, M.; Rowan, B.X.; Stanzick, K.J.; Thomas, L.F.; Tin, A.; Hoppmann, A.; Chu, A.Y.; et al. Differential and shared genetic effects on kidney function between diabetic and non-diabetic individuals. *Commun. Biol.* **2022**, *5*, 1–20. [CrossRef]
9. Titan, S.; Venturini, G.; Padilha, K.; Tavares, G.; Zatz, R.; Bensenor, I.; Lotufo, P.; Rhee, E.; Thadhani, R.; Pereira, A. Metabolites related to eGFR: Evaluation of candidate molecules for GFR estimation using untargeted metabolomics. *Clin. Chim. Acta* **2018**, *489*, 242–248. [CrossRef]
10. Lee, H.; Jang, H.B.; Yoo, M.-G.; Park, S.I.; Lee, H.-J. Amino Acid Metabolites Associated with Chronic Kidney Disease: An Eight-Year Follow-Up Korean Epidemiology Study. *Biomedicines* **2020**, *8*, 222. [CrossRef]
11. Lee, S.; Han, M.; Moon, S.; Kim, K.; An, W.J.; Ryu, H.; Oh, K.-H.; Park, S.K. Identifying Genetic Variants and Metabolites Associated with Rapid Estimated Glomerular Filtration Rate Decline in Korea Based on Genome–Metabolomic Integrative Analysis. *Metabolites* **2022**, *12*, 1139. [CrossRef] [PubMed]
12. Peng, H.; Liu, X.; Aoieong, C.; Tou, T.; Tsai, T.; Ngai, K.; I Cheang, H.; Liu, Z.; Liu, P.; Zhu, H. Identification of Metabolite Markers Associated with Kidney Function. *J. Immunol. Res.* **2022**, *2022*, 1–9. [CrossRef]
13. Peng, H.; Liu, X.; Ieong, C.A.; Tou, T.; Tsai, T.; Zhu, H.; Liu, Z.; Liu, P. A Metabolomics study of metabolites associated with the glomerular filtration rate. *BMC Nephrol.* **2023**, *24*, 1–12. [CrossRef]
14. Grams, M.E.; Tin, A.; Rebholz, C.M.; Shafi, T.; Köttgen, A.; Perrone, R.D.; Sarnak, M.J.; Inker, L.A.; Levey, A.S.; Coresh, J. Metabolomic Alterations Associated with Cause of CKD. *Clin. J. Am. Soc. Nephrol.* **2017**, *12*, 1787–1794. [CrossRef] [PubMed]
15. Lin, B.M.; Zhang, Y.; Yu, B.; Boerwinkle, E.; Thygarajan, B.; Yunes, M.; Daviglus, M.L.; Qi, Q.; Kaplan, R.; Lash, J.; et al. Metabolome-wide association study of estimated glomerular filtration rates in Hispanics. *Kidney Int.* **2021**, *101*, 144–151. [CrossRef] [PubMed]
16. Schlosser, P.; Scherer, N.; Grundner-Culemann, F.; Monteiro-Martins, S.; Haug, S.; Steinbrenner, I.; Uluvar, B.; Wuttke, M.; Cheng, Y.; Ekici, A.B.; et al. Genetic studies of paired metabolomes reveal enzymatic and transport processes at the interface of plasma and urine. *Nat. Genet.* **2023**, *55*, 995–1008. [CrossRef] [PubMed]
17. Nierenberg, J.L.; He, J.; Li, C.; Gu, X.; Shi, M.; Razavi, A.C.; Mi, X.; Li, S.; Bazzano, L.A.; Anderson, A.H.; et al. Novel associations between blood metabolites and kidney function among Bogalusa Heart Study and Multi-Ethnic Study of Atherosclerosis participants. *Metabolomics* **2019**, *15*, 149. [CrossRef]
18. Wang, F.; Sun, L.; Sun, Q.; Liang, L.; Gao, X.; Li, R.; Pan, A.; Li, H.; Deng, Y.; Hu, F.B.; et al. Associations of Plasma Amino Acid and Acylcarnitine Profiles with Incident Reduced Glomerular Filtration Rate. *Clin. J. Am. Soc. Nephrol.* **2018**, *13*, 560–568. [CrossRef]

19. Kwan, B.; Fuhrer, T.; Zhang, J.; Darshi, M.; Van Espen, B.; Montemayor, D.; de Boer, I.H.; Dobre, M.; Hsu, C.-Y.; Kelly, T.N.; et al. Metabolomic Markers of Kidney Function Decline in Patients With Diabetes: Evidence From the Chronic Renal Insufficiency Cohort (CRIC) Study. *Am. J. Kidney Dis.* **2020**, *76*, 511–520. [CrossRef]
20. Wen, D.; Zheng, Z.; Surapaneni, A.; Yu, B.; Zhou, L.; Zhou, W.; Xie, D.; Shou, H.; Avila-Pacheco, J.; Kalim, S.; et al. Metabolite profiling of CKD progression in the chronic renal insufficiency cohort study. *J. Clin. Investig.* **2022**, *7*. [CrossRef]
21. Bernard, L.; Zhou, L.; Surapaneni, A.; Chen, J.; Rebholz, C.M.; Coresh, J.; Yu, B.; Boerwinkle, E.; Schlosser, P.; Grams, M.E. Serum Metabolites and Kidney Outcomes: The Atherosclerosis Risk in Communities Study. *Kidney Med.* **2022**, *4*, 100522. [CrossRef] [PubMed]
22. Guo, X.; Peng, H.; Liu, P.; Tang, L.; Fang, J.; AoIeong, C.; Tou, T.; Tsai, T.; Liu, X. Novel Metabolites to Improve Glomerular Filtration Rate Estimation. *Kidney Blood Press. Res.* **2023**, *48*, 287–296. [CrossRef] [PubMed]
23. Au, A.Y.M.; Mantik, K.; Bahadory, F.; Stathakis, P.; Guiney, H.; Erlich, J.; Walker, R.; Poulton, R.; Horvath, A.R.; Endre, Z.H. Plasma arginine metabolites in health and chronic kidney disease. *Nephrol. Dial. Transplant.* **2023**, *38*, 2767–2775. [CrossRef]
24. Liu, J.-J.; Ching, J.; Wee, H.N.; Liu, S.; Gurung, R.L.; Lee, J.; M., Y.; Zheng, H.; Lee, L.S.; Ang, K.; et al. Plasma Tryptophan-Kynurenine Pathway Metabolites and Risk for Progression to End-Stage Kidney Disease in Patients With Type 2 Diabetes. *Diabetes Care* **2023**, *46*, 2223–2231. [CrossRef]
25. Hou, Y.; Xiao, Z.; Zhu, Y.; Li, Y.; Liu, Q.; Wang, Z. Blood metabolites and chronic kidney disease: A Mendelian randomization study. *BMC Med Genom.* **2024**, *17*, 1–14. [CrossRef]
26. Luo, Y.; Zhang, W.; Qin, G. Metabolomics in diabetic nephropathy: Unveiling novel biomarkers for diagnosis (Review). *Mol. Med. Rep.* **2024**, *30*, 1–13. [CrossRef]
27. Wishart, D.S.; Guo, A.; Oler, E.; Wang, F.; Anjum, A.; Peters, H.; Dizon, R.; Sayeeda, Z.; Tian, S.; Lee, B.L.; et al. HMDB 5.0: The Human Metabolome Database for 2022. *Nucleic Acids Res.* **2021**, *50*, D622–D631. [CrossRef]
28. Stelzer, G.; Rosen, R.; Plaschkes, I.; Zimmerman, S.; Twik, M.; Fishilevich, S.; Stein, T.I.; Nudel, R.; Lieder, I.; Mazor, Y.; et al. The GeneCards Suite: From gene data mining to disease genome sequence analysis. *Curr. Protoc. Bioinform.* **2016**, *54*, 1.30.1–13033. [CrossRef] [PubMed]
29. Jellum, E.; Horn, L.; Thoresen, O.; A Kvittingen, E.; Stokke, O. Urinary excretion of N-acetyl amino acids in patients with some inborn errors of amino acid metabolism. *Scand. J. Clin. Lab. Investigation. Suppl.* **1986**, *184*, 21–26.
30. Engelke, U.F.H.; Sass, J.O.; Van Coster, R.N.; Gerlo, E.; Olbrich, H.; Krywawych, S.; Calvin, J.; Hart, C.; Omran, H.; Wevers, R.A. NMR spectroscopy of aminoacylase 1 deficiency, a novel inborn error of metabolism. *NMR Biomed.* **2008**, *21*, 138–147. [CrossRef]
31. Okajima, K.; Inoue, M.; Morino, Y. Studies on the mechanism for renal elimination of *N*-acetylphenylalanine: Its pathophysiologic significance in phenylketonuria. *J. Lab. Clin. Med.* **1985**, *105*, 132–138. [PubMed]
32. Sass, J.O.; Mohr, V.; Olbrich, H.; Engelke, U.; Horvath, J.; Fliegauf, M.; Loges, N.T.; Schweitzer-Krantz, S.; Moebus, R.; Weiler, P.; et al. Mutations in ACY1, the Gene Encoding Aminoacylase 1, Cause a Novel Inborn Error of Metabolism. *Am. J. Hum. Genet.* **2006**, *78*, 401–409. [CrossRef]
33. Coster, V.; Gerlo, E.; Giardina, T.; Engelke, U.; Smet, J.; De Praeter, C.; Meersschaut, V.; De Meirleir, L.; Seneca, S.; Devreese, B.; et al. Aminoacylase I deficiency: A novel inborn error of metabolism. *Biochem. Biophys. Res. Commun.* **2005**, *338*, 1322–1326. [CrossRef]
34. Luo, G.; Ducy, P.; McKee, M.D.; Pinero, G.J.; Loyer, E.; Behringer, R.R.; Karsenty, G. Spontaneous calcification of arteries and cartilage in mice lacking matrix GLA protein. *Nature* **1997**, *386*, 78–81. [CrossRef]
35. Chatrou, M.L.; Winckers, K.; Hackeng, T.M.; Reutelingsperger, C.P.; Schurgers, L.J. Vascular calcification: The price to pay for anticoagulation therapy with vitamin K-antagonists. *Blood Rev.* **2012**, *26*, 155–166. [CrossRef] [PubMed]
36. Budoff, M.J.; Rader, D.J.; Reilly, M.P.; Mohler, E.R., 3rd; Lash, J.; Yang, W.; Rosen, L.; Glenn, M.; Teal, V.; Feldman, H.I.; et al. Relationship of estimated GFR and coronary artery calcification in the CRIC (Chronic Renal Insufficiency Cohort) Study. *Am. J. Kidney Dis.* **2011**, *58*, 519–526. [CrossRef]
37. Jansen, R.S.; Addie, R.; Merkx, R.; Fish, A.; Mahakena, S.; Bleijerveld, O.B.; Altelaar, M.; Ijlst, L.; Wanders, R.J.; Borst, P.; et al. *N*-lactoyl-amino acids are ubiquitous metabolites that originate from CNDP2-mediated reverse proteolysis of lactate and amino acids. *Proc. Natl. Acad. Sci. USA* **2015**, *112*, 6601–6606. [CrossRef]
38. Hennermann, J.B.; Roloff, S.; Gellermann, J.; Vollmer, I.; Windt, E.; Vetter, B.; Plöckinger, U.; Mönch, E.; Querfeld, U. Chronic kidney disease in adolescent and adult patients with phenylketonuria. *J. Inherit. Metab. Dis.* **2012**, *36*, 747–756. [CrossRef] [PubMed]
39. Burton, B.K.; Jones, K.B.; Cederbaum, S.; Rohr, F.; Waisbren, S.; Irwin, D.E.; Kim, G.; Lilienstein, J.; Alvarez, I.; Jurecki, E.; et al. Prevalence of comorbid conditions among adult patients diagnosed with phenylketonuria. *Mol. Genet. Metab.* **2018**, *125*, 228–234. [CrossRef]
40. Hagenfeldt, L.; Naglo, A. New conjugated urinary metabolites in intermediate type maple syrup urine disease. *Clin. Chim. Acta* **1987**, *169*, 77–83. [CrossRef]
41. Maceda, E.B.G.; E Abadingo, M.; Magbanua-Calalo, C.J.; A Dator, M.; Resontoc, L.P.R.; De Castro-Hamoy, L.; Abacan, M.A.R.; Chiong, M.A.D.; Estrada, S.C. Maple syrup urine disease associated with nephrotic syndrome in a Filipino child. *BMJ Case Rep.* **2021**, *14*, e242689. [CrossRef] [PubMed]
42. Zhang, Z.-H.; He, J.-Q.; Qin, W.-W.; Zhao, Y.-Y.; Tan, N.-H. Biomarkers of obstructive nephropathy using a metabolomics approach in rat. *Chem. Interactions* **2018**, *296*, 229–239. [CrossRef] [PubMed]

43. Rysz, J.; Franczyk, B.; Rysz-Górzyńska, M.; Gluba-Brzózka, A. Are alterations in DNA methylation related to CKD development? *Int. J. Mol. Sci.* **2022**, *23*, 7108. [CrossRef] [PubMed]
44. Ganesan, L.L.; O'brien, F.J.; Sirich, T.L.; Plummer, N.S.; Sheth, R.; Fajardo, C.; Brakeman, P.; Sutherland, S.M.; Meyer, T.W. Association of Plasma Uremic Solute Levels with Residual Kidney Function in Children on Peritoneal Dialysis. *Clin. J. Am. Soc. Nephrol.* **2021**, *16*, 1531–1538. [CrossRef]
45. He, W.J.; Chen, J.; Razavi, A.C.; Hu, E.A.; Grams, M.E.; Yu, B.; Parikh, C.R.; Boerwinkle, E.; Bazzano, L.; Qi, L.; et al. Metabolites Associated with Coffee Consumption and Incident Chronic Kidney Disease. *Clin. J. Am. Soc. Nephrol.* **2021**, *16*, 1620–1629. [CrossRef]
46. Laakso, M.; Kuusisto, J.; Stančáková, A.; Kuulasmaa, T.; Pajukanta, P.; Lusis, A.J.; Collins, F.S.; Mohlke, K.L.; Boehnke, M. The Metabolic Syndrome in Men study: A resource for studies of metabolic and cardiovascular diseases. *J. Lipid Res.* **2017**, *58*, 481–493. [CrossRef]
47. Yin, X.; Chan, L.S.; Bose, D.; Jackson, A.U.; VandeHaar, P.; Locke, A.E.; Fuchsberger, C.; Stringham, H.M.; Welch, R.; Yu, K.; et al. Genome-wide association studies of metabolites in Finnish men identify disease-relevant loci. *Nat. Commun.* **2022**, *13*, 1–14. [CrossRef]
48. Diagnosis and Classification of Diabetes Mellitus. *Diabetes Care* **2010**, *33*, S62–S69. [CrossRef]
49. Inker, L.A.; Eneanya, N.D.; Coresh, J.; Tighiouart, H.; Wang, D.; Sang, Y.; Crews, D.C.; Doria, A.; Estrella, M.M.; Froissart, M.; et al. New Creatinine- and Cystatin C–Based Equations to Estimate GFR without Race. *N. Engl. J. Med.* **2021**, *385*, 1737–1749. [CrossRef]
50. Silva, L.F.; Vangipurapu, J.; Kuulasmaa, T.; Laakso, M. An intronic variant in the GCKR gene is associated with multiple lipids. *Sci. Rep.* **2019**, *9*, 10240. [CrossRef]

International Journal of *Molecular Sciences*

MDPI

Article

Oxidative Stress and Histomorphometric Remodeling: Two Key Intestinal Features of Type 2 Diabetes in Goto–Kakizaki Rats

Marisa Esteves-Monteiro [1,2,3,*], Mariana Ferreira-Duarte [1,3], Cláudia Vitorino-Oliveira [4,5], José Costa-Pires [2], Sara Oliveira [6,7,8], Paulo Matafome [6,7,8,9], Manuela Morato [1,3], Patrícia Dias-Pereira [10], Vera Marisa Costa [4,5] and Margarida Duarte-Araújo [1,2,*]

1 Associated Laboratory for Green Chemistry (LAQV) Network of Chemistry and Technology (REQUIMTE), University of Porto, 4050-313 Porto, Portugal; mmorato@ff.up.pt (M.M.)
2 Department of Immuno-Physiology and Pharmacology, Institute of Biomedical Sciences Abel Salazar, University of Porto (ICBAS-UP), 4050-313 Porto, Portugal
3 Laboratory of Pharmacology, Department of Drug Sciences, Faculty of Pharmacy, University of Porto (FFUP), 4050-313 Porto, Portugal
4 Institute for Health and Bioeconomy (i4HB), Laboratory of Toxicology, Department of Biological Sciences, FFUP, 4050-313 Porto, Portugal; veramcosta@ff.up.pt (V.M.C.)
5 Research Unit on Applied Molecular Biosciences (UCIBIO), FFUP, Laboratory of Toxicology, Department of Biological Sciences, FFUP, 4050-313 Porto, Portugal
6 Coimbra Institute for Clinical and Biomedical Research (iCBR) and Institute of Physiology, Faculty of Medicine, University of Coimbra (UC), 3000-548 Coimbra, Portugal; saraoliveira116@gmail.com (S.O.)
7 Center for Innovative Biomedicine and Biotechnology (CIBB), UC, 3000-548 Coimbra, Portugal
8 Clinical Academic Center of Coimbra (CACC), 3000-548 Coimbra, Portugal
9 Coimbra Health School (ESTeSC), Polytechnic University of Coimbra, 3046-854 Coimbra, Portugal
10 Department of Pathology and Molecular Immunology, ICBAS-UP, 4050-313 Porto, Portugal; pdiaspereira@yahoo.com.br
* Correspondence: mmem89@gmail.com (M.E.-M.); mdcma@icbas.up.pt (M.D.-A.)

Abstract: Gastrointestinal complications of diabetes are often overlooked, despite affecting up to 75% of patients. This study innovatively explores local glutathione levels and morphometric changes in the gut of Goto–Kakizaki (GK) rats, a type 2 diabetes animal model. Segments of the intestine, cecum, and colon were collected for histopathological analysis and glutathione quantification. A significant increase in the total thickness of the intestinal wall of GK rats was observed, particularly in the duodenum (1089.02 ± 39.19 vs. 864.19 ± 37.17 μm), ileum (726.29 ± 24.75 vs. 498.76 ± 16.86 μm), cecum (642.24 ± 34.15 vs. 500.97 ± 28.81 μm), and distal colon (1211.81 ± 51.32 vs. 831.71 ± 53.2 μm). Additionally, diabetic rats exhibited thickening of the muscular layers in all segments, except for the duodenum, which was also the only portion where the number of smooth muscle cells did not decrease. Moreover, myenteric neuronal density was lower in GK rats, suggesting neurological loss. Total glutathione levels were lower in all intestinal segments of diabetic rats (except duodenum), and the reduced/oxidized glutathione ratio (GSH/GSSG) was significantly decreased in GK rats, indicating increased oxidative stress. These findings strongly indicate that GK rats undergo significant intestinal remodeling, notable shifts in neuronal populations, and heightened oxidative stress—factors that likely contribute to the functional gastrointestinal alterations seen in diabetic patients.

Keywords: diabetes; GK rats; gut remodeling; oxidative stress

1. Introduction

Diabetes is a highly prevalent metabolic disorder characterized by a state of hyperglycemia [1]. The most recent data from the International Diabetes Federation indicate that diabetes affected 537 million people worldwide in 2021, a number that is expected to grow to 643 million by 2030 [2]. Besides the substantial economic impact of the disease, the importance of diabetes is also related to significant mortality and morbidity rates, being considered a major public health problem [3–5]. There are two main forms of diabetes: type 1 diabetes (T1D) and type 2 diabetes (T2D). T1D is caused by an absolute insulin deficiency while T2D is a combination of insulin resistance in target organs and relative deficiency caused by dysfunctional pancreatic β-cells [6]. T2D is far more prevalent, accounting for 90 to 95% of all cases [7,8]. Around 80% of adult T2D patients are considered overweight or obese. However, 10–15% are not obese, and those patients present higher hypoglycemic events and mortality rates [9]. Given the significant importance and wide prevalence of diabetes, numerous animal models are employed in the study of diabetes-related complications [10]. The Goto–Kakizaki (GK) rat is a non-obese animal model of T2D that was developed by Goto, Kakizaki, and Masaki in 1975 [11,12]. This model was obtained by selective reproduction of non-diabetic Wistar rats with slight glucose intolerance. Consequently, the rats from posterior generations spontaneously developed T2D without becoming obese [13,14]. GK rats exhibit a reduced pancreatic β-cell number and function, moderate hyperglycemia, glucose intolerance, and peripheral insulin resistance [15].

Diabetes frequently concurs with gastrointestinal (GI) complications that are associated with significant morbidity, affecting up to 75% of patients. Currently, it is unclear whether the prevalence differs between T1D and T2D [16]. In the small intestine and colon, diabetes-related complications usually result in symptoms like chronic constipation, diarrhea, and fecal incontinence that may result in potential complications such as megacolon, pseudo-obstruction, stercoral ulcer, and perforation [17,18]. In addition, diabetes seems to worsen clinical conditions such as colorectal cancer and inflammatory bowel disease [19]. Although highly prevalent, these symptoms are often overlooked, as they do not significantly contribute to mortality in diabetic patients. However, it is crucial to acknowledge that they negatively impact health status and quality of life, making them a significant source of morbidity [20]. The relationship between diabetes and the pathogenesis of the described gut disorders is not completely understood and seems to be multifactorial [21]. Mechanical factors contribute to intestinal disorders, since it has been found that diabetes seems to cause structural remodeling that can affect histomorphometry and biomechanical properties, increase stiffness, and decrease the resting compliance and relaxation capacity of the intestinal wall [21]. A previous study by our group also showed significant histomorphometry changes and evidenced lower reactivity to angiotensin II of the ileum and colon of T1D-induced rats [22]. Also, Zhao et al. demonstrated the existence of remodeling in the esophagus and stomach of GK rats [23], while Pereira et al. showed alterations of the small intestine in the same animal model [24]. So far, only one study has shown colon remodeling in a T2D model, associating it with the formation of advanced glycation end products [25].

In diabetes, various pathways contribute to tissue damage, but a common hallmark is heightened oxidative stress, characterized by elevated levels of reactive oxygen species (ROS) [26]. Moreover, chronic hyperglycemia is linked to decreased cellular levels of glutathione (GSH) [27]. GSH is the most powerful intercellular antioxidant in an organism, undergoing oxidation to GSSG (glutathione disulphide or oxidized glutathione) after contact with electrophiles. These reactions can be catalyzed by GSH-peroxidase. GSSG can subsequently be regenerated back to GSH by GSH-reductase, using NADPH as a cofactor, or it is excluded from the cell through membrane transporters (e.g., multidrug resistance-associated proteins, MRPs). Maintaining an optimal ratio of GSH to GSSG within the cell is crucial for survival, and a decrease in this ratio may be used as a marker of oxidative stress [28]. Oxidative stress and ROS formation have already been described to be markedly increased by uncontrolled hyperglycemia [29]. Also, a decrease in GSH has been observed in the liver [30], erythrocytes [31], and colon [32] of long-term diabetic patients. But so far, there are no data regarding GSH local levels in diabetic small intestines.

Curiously, most researchers studying diabetes-related complications in the GI tract use animal models of T1D, even though T2D is the most common form [21]. Considering this, and that diabetic patients commonly present GI complications, we innovatively aimed to characterize the entire gut histomorphometry and the local glutathione system in an animal model of T2D. Examining the entire gut—from the duodenum to the distal colon—in the same animals allows for a direct comparison between segments, providing a clearer understanding of how diabetes uniquely affects each part of the gastrointestinal tract. Additionally, local glutathione levels offer a more precise picture than systemic levels because they provide insights into the specific redox environment and oxidative stress within a targeted tissue or organ—like the gut in this case. Systemic GSH levels represent an overall average throughout the body, which can mask localized changes or stresses. In contrast, studying local GSH concentrations allows us to understand how the oxidative balance is maintained or disrupted in a specific region, which is particularly relevant for organs impacted by T2D, where localized oxidative stress can contribute to disease progression. To achieve this goal, we took samples of GK rats' duodenum, middle jejunum, distal ileum, cecum, and proximal and distal colon and measured the individual layers of the intestinal wall, analyzed smooth muscle cells and myenteric neurons, and quantified GSH and GSSG levels.

2. Results

2.1. Animal Monitorization and Insulin Tolerance Test

GK rats presented elevated fasted glucose concentrations compared to controls (237.88 ± 81.05 mg/dL vs. 100 ± 1.73 mg/dL, respectively, $p < 0.05$) (time 0, Figure 1). After a 6 h fasting and insulin administration, the glycemia of the GK group increased during the first 30 min and then decreased, reaching the initial glycemic quantification at the end of the insulin tolerance test (ITT, time 120 min, Figure 1). In the control group, after insulin injection a slight decline in blood glucose values was observed. Compared to the control (CTRL) group, the GK group's blood glucose concentration was higher at all time points ($p < 0.0001$, Figure 1).

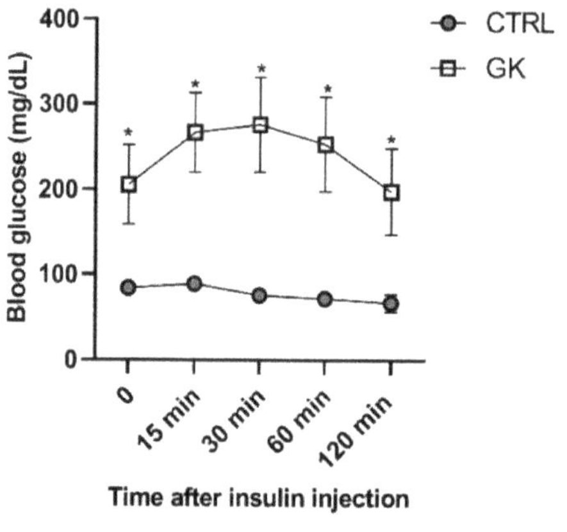

Figure 1. Blood glucose concentrations of control (CTRL, $n = 5$) and GK animals ($n = 6$) measured before (time 0) and during the insulin tolerance test—ITT. Values are presented as mean ± SEM, and a paired Student's t test was used to compare the two experimental groups (CTRL and GK). * Statistical difference, $p < 0.05$.

At the beginning of the protocol, the weight of the GK group was on average 329.17 ± 7.4 g, increasing 2 weeks later to 340.17 ± 6.95 g, representing an average weight gain of 3.24 ± 0.62%. The control rats weighed 402.20 ± 9.56 g at the beginning of the protocol and 417.40 ± 8.81 at the end, representing an average weight gain of 3.65 ± 0.81% (Figure 2). So, the initial and final weights of GK rats were both lower compared to controls ($p < 0.0001$), but the % of weight gain during the experimental period was roughly the same in the two groups ($p > 0.05$) (Figure 2). Despite maintaining the same amount of weight gain as controls, the food intake of GK rats (28.95 ± 1.40 mg/day/rat) was significantly higher than that of controls (21.50 ± 0.50 mg/day/rat) (Figure 2). Regarding water intake, it was significantly higher in diabetic rats compared to controls (Figure 2). The GK group drank 64.38 ± 5.63 mL/day/rat (n = 6), which was more than double the water that control animals drank (30.30 ± 0.40 mL/day/rat, n = 5).

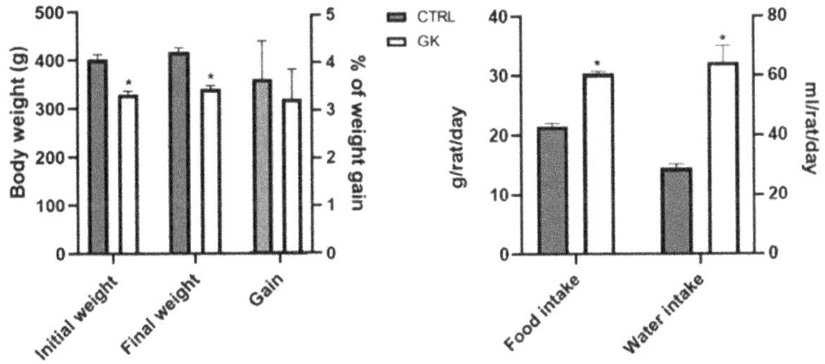

Figure 2. Evaluation during the experimental protocol of control (CTRL, n = 5) and GK diabetic rats (GK, n = 6) of: body weight; body weight gain; food intake and water intake. Values are presented as mean ± SEM and unpaired Student's t test was used to compare the two experimental groups (CTRL and GK). * Statistical difference, $p < 0.05$.

2.2. Small Intestine and Colon Microscopic Evaluation

To assess whether gut remodeling occurs and follows a proximal-to-distal progression, as previously observed in T1D rat models, we measured both the mucosal and muscle layers in a T2D rat model. The histomorphometric evaluation of the small and large intestine of GK animals showed a higher thickness of the total intestinal wall of the duodenum, ileum, cecum, and distal colon (DC) compared to controls (duodenum: 1089.02 ± 39.19 μm vs. 864.19 ± 37.17 μm; ileum: 726.29 ± 24.75 μm vs. 498.76± 16.86 μm; cecum: 642.24 ± 34.15 μm vs. 500.97 ± 28.81 μm; DC: 1211.81 ± 51.32 μm vs. 831.71 ± 53.25 μm, respectively, $p < 0.01$ for all). There was no difference between GK and control rats in a histomorphometric evaluation of the jejunum and proximal colon (PC) (jejunum: 796.16 ± 43.86 μm vs. 722.12 ± 28.75 μm and PC: 1060.18 ± 18.93 μm vs. 1029.01 ± 59.84 μm, respectively, $p > 0.05$ for both) (Figure 3).

The muscular layers of the intestinal wall of GK animals were increased in all segments except in the duodenum compared to controls (jejunum—longitudinal muscle (lm): 41.69 ± 2.80 μm vs. 25.54 ± 2.28 μm, circular muscle (cm): 91.99 ± 5.03 μm vs. 55.33 ± 3.73 μm; ileum—lm: 51.99 ± 2.90 μm vs. 27.87 ± 3.14 μm, cm: 100.11 ± 5.96 μm vs. 57.19 ± 5.38 μm; cecum—lm: 54.44± 5.33 μm vs. 36.57 ± 3.15 μm, cm: 179.36 ± 10.84 μm vs. 107.82 ± 8.09 μm; PC—lm: 77.70 ± 8.97 μm vs. 42.52 ± 1.87 μm, cm: 212.03 ± 13.73 μm vs. 146.03 ± 11.12 μm; DC—lm: 83.31 ± 6.54 μm vs. 46.04 ± 3.51 μm, cm: 283.40 ± 33. 86 μm vs. 164.43 ± 3.51 μm, respectively, $p < 0.05$ for all; duodenum—lm: 43.81 ± 2.67 μm vs. 35.12 ± 4.30 μm, cm: 99.36 ± 7.80 μm vs. 78.08 ± 9.93 μm, respectively, $p > 0.05$) (Figure 3). Submucosal values were consistent across all portions, except for the ileum in GK rats, where an increase was observed (GK: 41.73 ± 2.9 μm vs. CTRL: 28.04 ± 4.38 μm). The mucosa was only increased in the duodenum (GK: 892.48 ± 31.21 μm vs. CTRL: 710.60 ± 24.82 μm), ileum (GK: 532.46 ± 15.87 μm vs. CTRL: 385.66 ± 24.20 μm), and DC (GK:765.84 ± 16.86 μm vs. CTRL: 566.01 ± 44.33 μm) of GK rats compared with controls, while the jejunum (GK: 630.34 ± 49.26 μm vs. CTRL: 615.97 ± 30.80 μm), cecum (GK: 354.70 ± 24.00 μm vs. CTRL: 292.72 ± 30.77 μm), and PC (GK: 728.53 ± 45.79 μm vs. CTRL: 808.19 ± 51.10 μm) presented similar results in both GK and control animals (Figure 3). Additionally, in the epithelial layer, villi length and crypt depth were also increased in the duodenum and ileum of GK rats (Supplementary Materials, Figure S1).

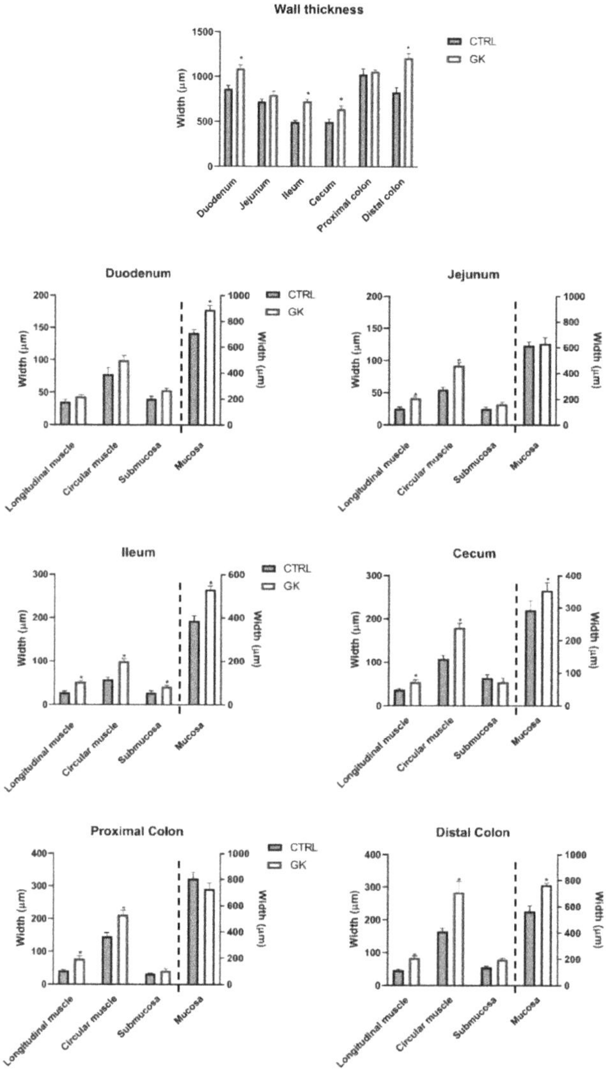

Figure 3. Morphometric evaluation of intestinal segments (duodenum, jejunum, ileum, cecum, proximal colon, and distal colon) of control (CTRL, $n = 5$) and GK diabetic rats (GK, $n = 6$): total wall thickness (μm) of each intestinal segment and thickness (μm) of the intestinal layers (longitudinal muscle, circular muscle, submucosa, and mucosa) of duodenum, jejunum, ileum, cecum, proximal colon, and distal colon). Values are presented as mean ± SEM, and a 2-way ANOVA followed by an unpaired *t* test with Welch's correction was used to compare the two experimental groups (CTRL and GK). * Statistical difference $p < 0.05$ vs. correspondent control. Unpaired *t* test with Welch's correction was used to compare the two experimental groups.

In Figure 4, representative images of both control (CTRL) and GK animals are displayed, encompassing all the studied sections. These images provide a comprehensive visual comparison, highlighting the differences in each portion analyzed.

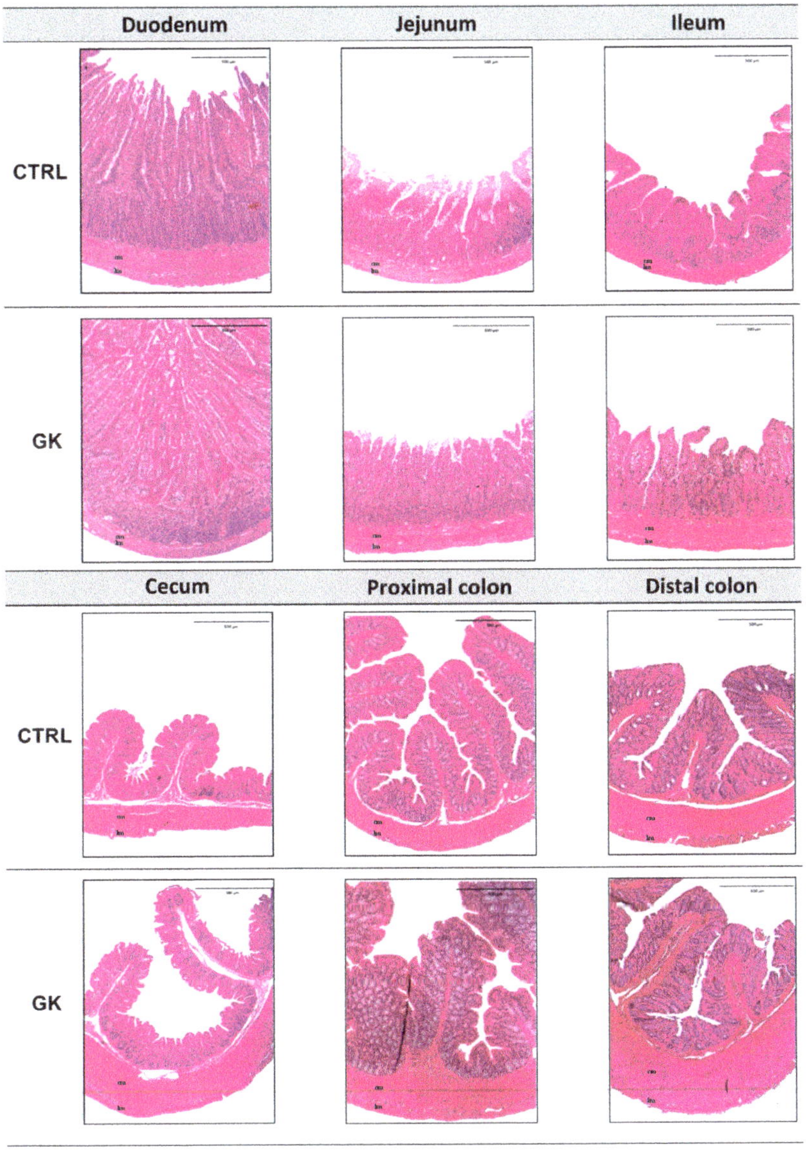

Figure 4. Representative microscopic photographs of duodenum, jejunum, ileum, cecum, proximal colon, and distal colon of control (CTRL) and GK rats (GK) stained with hematoxylin (blue) and eosin (pink), captured using 40× magnification. Longitudinal muscle (lm) and circular muscle (cm) were identified in all images.

Then, collagen deposition was measured to evaluate potential tissue remodeling and fibrosis, conditions commonly linked to chronic hyperglycemia. These factors could explain the increased thickness of the muscular layers observed in the histomorphometric analysis. Masson's trichrome and periodic acid–Schiff (PAS) stains were assessed by an experienced pathologist blinded to the experiments. Interestingly, qualitative evaluation revealed no discernible differences between the control and GK diabetic animals. This suggests the absence of collagen deposition and no meaningful disparity in the proportion of carbohydrate macromolecules, such as glycogen, between the GK and control animal groups. Representative microscopic photographs of the colon with both staining techniques are shown in Figure 5.

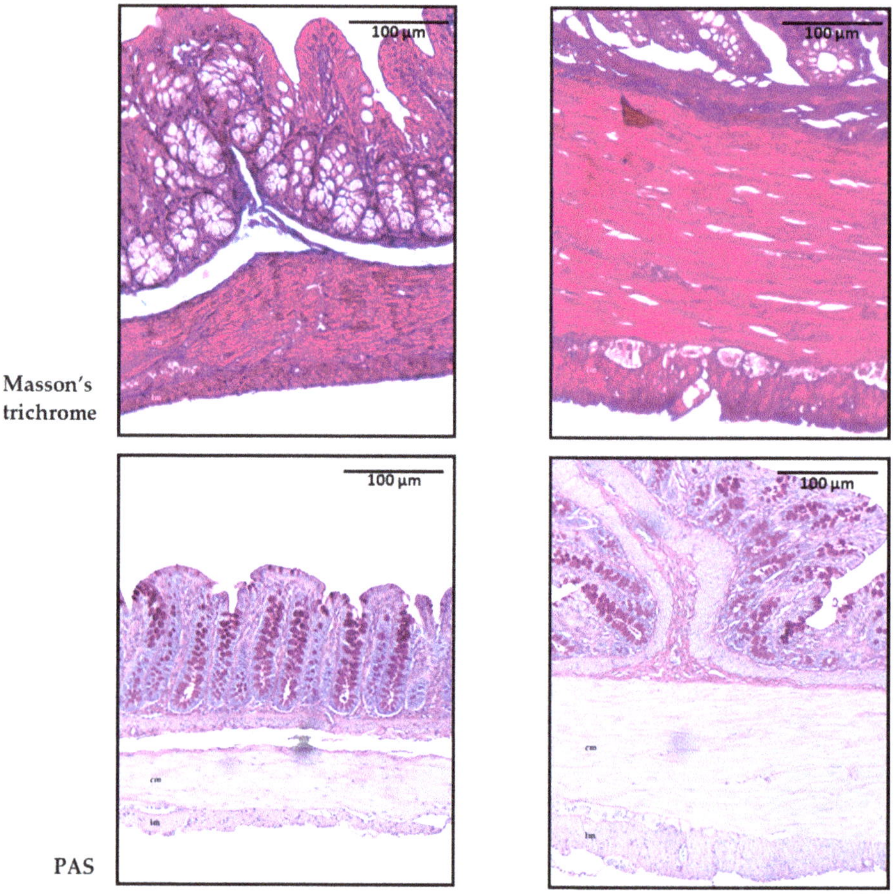

Figure 5. Representative microscopic photographs of the colon of control (CTRL) and GK rats (GK) stained with Masson's trichrome and periodic acid–Schiff (PAS), captured at 100× magnification. Longitudinal muscle (lm) and circular muscle (cm) were identified in all images.

2.3. *Smooth Muscle Cell Density in the Muscular Layers*

Smooth muscle cell density was quantified in response to the negative results from Masson's trichrome staining to determine whether an increase in density might be linked to smooth muscle hypertrophy. The number of nuclei of smooth muscle cells (SMCs) was lower in GK rats in all portions studied except for the duodenum compared to controls (jejunum: 15.42 ± 0.89 vs. 18.75 ± 0.1; ileum: 12.23 ± 0.80 vs. 15.35 ± 0.57; cecum: 9.65 ± 0.65 vs. 13.50 ± 0.67; PC: 14.90 ± 0.80 vs. 18.68 ± 0.52; DC: 10.06 ± 0.64 vs. 13.25 ± 0.51, respectively, $p < 0.02$ for all; duodenum: 17.58 ± 0.74 vs. 19.08 ± 0.31, respectively, $p > 0.05$). Representative microscopic images focusing on the muscular layers are depicted in Figure 6.

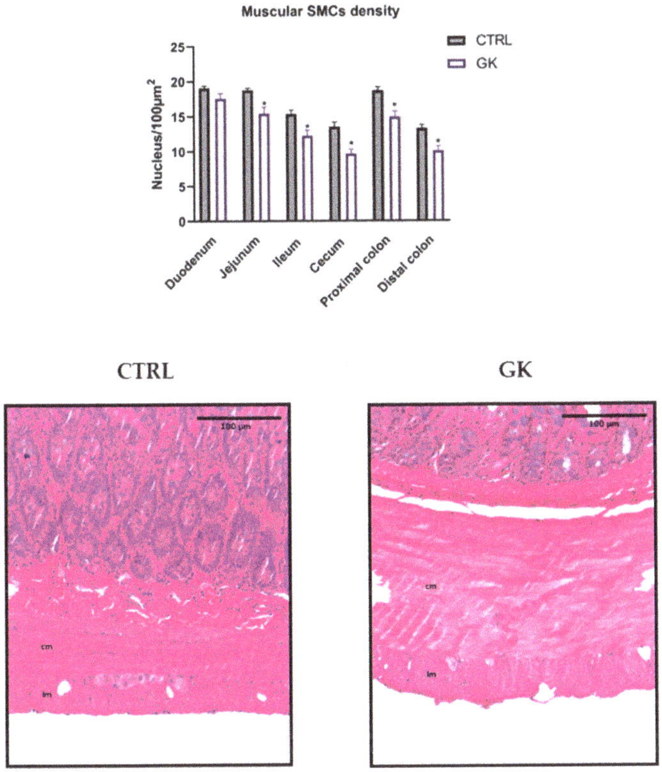

Figure 6. Morphoquantitative analyses of the density of smooth muscle cells (SMCs) in the muscular layers of duodenum, jejunum, ileum, cecum, and proximal and distal colon of control group (CTRL, $n = 5$) and GK diabetic rats (GK, $n = 6$). Data are expressed as the mean ± SEM, and comparisons between the two groups were made using Student's t test. * Statistical difference, $p < 0.05$. Representative microscopic photographs of the muscle layers of distal colon of control (CTRL) and GK rats (GK) stained with hematoxylin and eosin, captured at 100× magnification. Longitudinal muscle (lm) and circular muscle (cm) were identified in both images.

2.4. Neuronal Density in the Myenteric Plexi

Neuron density was also assessed, based on the findings of Honoré et al. (2011), which suggested that neuronal loss could contribute to increased colonic thickness, potentially due to the greater force required for motility. Both smooth muscle cell and neuron densities are critical for maintaining proper gastrointestinal motility and function, which may be compromised by prolonged diabetes.

The neuronal density in the myenteric plexus was lower in the GK group, when compared to control rats (Figure 7). The number of nuclei per mm^2 was statistically lower in diabetic animals compared to controls in all portions studied (duodenum: 444.95 ± 13.97 vs. 540.54 ± 21.47; jejunum: 461.65 ± 31.78 vs. 562.62 ± 10.86; ileum: 396.36 ± 12.73 vs. 546.63 ± 15.94; cecum: 363.81 ± 17.74 vs. 440.65 ± 24.82; PC: 382.36 ± 12.34 vs. 511.90 ± 11.85; DC: 352.65 ± 27.94 vs. 491.03 ± 21.47, respectively, $p < 0.05$ for all). Representative microscopic images focusing on the myenteric plexus are depicted in Figure 7.

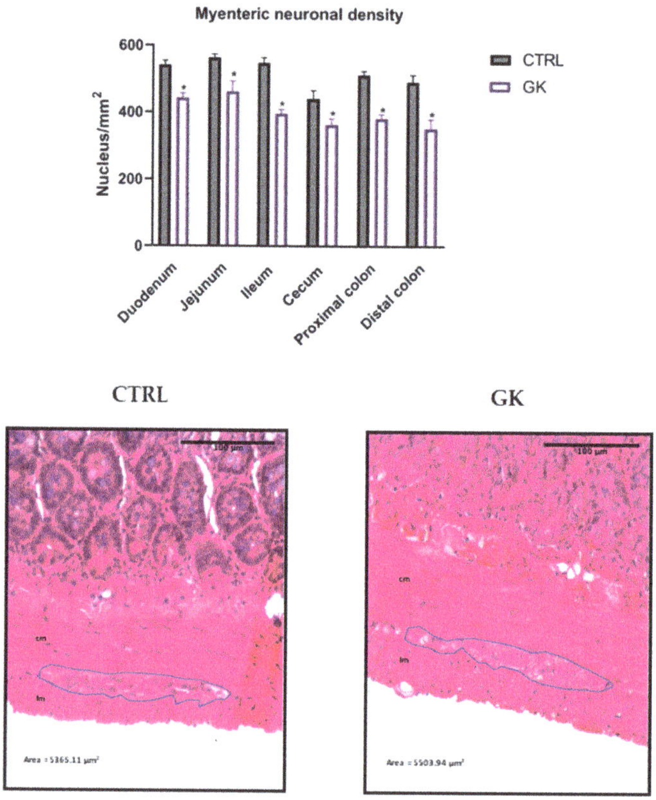

Figure 7. Morphoquantitative analyses of the neuronal density in the myenteric plexus of duodenum, jejunum, ileum, cecum, and proximal and distal colon of control group (CTRL, $n = 5$) vs. GK diabetic rats (GK, $n = 6$). Data are expressed as the mean ± SEM, and comparisons between the two groups were made using Student's t test. * Statistical difference, $p < 0.05$. Representative microscopic photographs of the myenteric plexus proximal colon of control (CTRL) and GK rats (GK) stained with hematoxylin and eosin, captured at 100× magnification. Longitudinal muscle (lm) and circular muscle (cm) were identified in both images.

2.5. Total GSH and GSSG Quantification

To investigate the potential causes of decreased neuron density observed in the myenteric plexus of GK rats, we decided to measure GSH levels as an indicator of oxidative stress, a critical factor in the development of diabetic complications. The results of the total glutathione quantification showed a decrease in tGSH in diabetic animals compared to controls in all portions studied except the duodenum (in nmol tGSH/mg protein, jejunum: 1.01 ± 0.06 vs. 2.11 ± 0,03; ileum: 0.92 ± 0.11 vs. 2.17 ± 0,15; cecum: 0.91 ± 0.01 vs. 2.24 ± 0.15; PC: 0.94 ± 0.06 vs. 2.09 ± 0.12; DC: 0.87 ± 0.02 vs. 2.44 ± 0.19, respectively, $p < 0.02$ for all; duodenum: 1.11 ± 0.20 vs. 1.17 ± 0.07, respectively, $p > 0.05$). However, the quantification of GSSG revealed comparable values between GK rats and controls across all studied portions ($p > 0.05$ for all) (Figure 8). Regarding the GSH/GSSG ratio, a decrease was observed in all portions of GK diabetic rats compared to controls (duodenum: 6.04 ± 0.24 vs. 8.28 ± 0.32; jejunum: 4.77 ± 0.31 vs. 9.39 ± 1.31; ileum: 4.62 ± 0.52 vs. 10.27 ± 1.20; cecum: 3.94 ± 0.31 vs. 10.84 ± 1.22; PC: 4.84 ± 0.53 vs. 9.15 ± 0.16; DC: 4.35 ± 0.47 vs. 8.83 ± 0.62, respectively, $p < 0.05$ for all) (Figure 8).

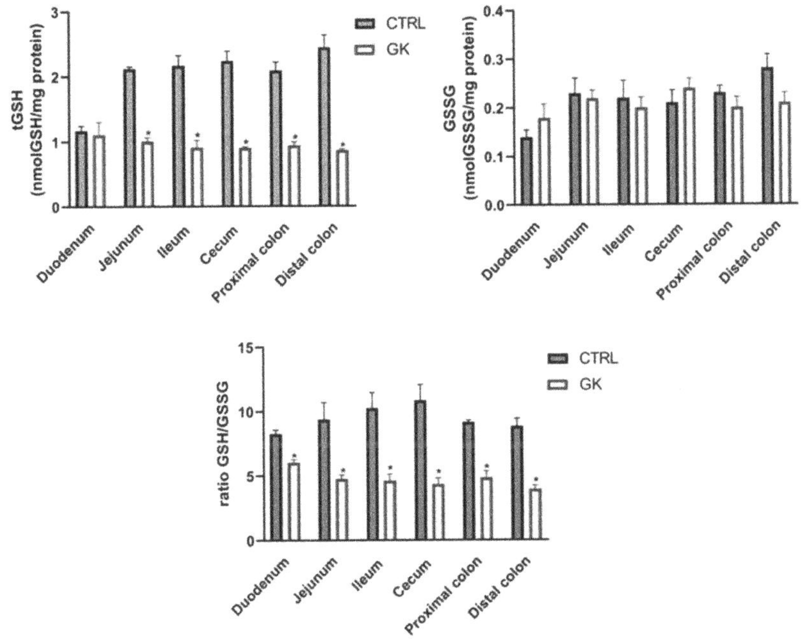

Figure 8. Glutathione evaluation of intestinal segments (duodenum, jejunum, ileum, cecum, proximal colon, and distal colon) of control (CTRL, $n = 5$) and GK diabetic rats (GK, $n = 6$): total glutathione (tGSH) quantification (nmol GSH/mg protein); oxidized glutathione (GSSG) quantification (nmol GSSG/mg protein) and ratio GSH/GSSG. Values are mean ± SEM, and an unpaired Student's t test with Welch's correction was used to compare the two experimental groups (CTRL and GK). * Statistical difference $p < 0.05$ vs. correspondent control.

3. Discussion

This study presents a novel approach to examining local glutathione levels and morphometric changes in the entire gut of GK rats. To the best of our knowledge, this is the first comprehensive study to analyze histomorphometry and quantify tGSH, GSSG, and the GSH/GSSG ratio across the entire gut, including the proximal and distal colon, all intestinal segments, and the cecum in a T2D model. The assessment of local GSH provides a more localized and precise evaluation of oxidative stress within the gastrointestinal tract. This is a novel approach in diabetic animal models, offering new insights into how different regions of the intestine respond to diabetes-related oxidative stress.

In this work, 21-week-old male GK rats exhibited reduced weight compared to their Wistar counterparts while presenting higher food intake. They also showed fasting basal hyperglycemia and impaired insulin sensitivity when compared to the control group. It has already been shown that GK animals fail to accumulate body fat despite their higher calorie consumption and that adipose tissue is a major contributor to the differential weight in these animals [33]. These changes are due to an impairment in pre-adipocyte differentiation into mature adipocytes, leading to a defect in triglyceride storage [33]. Therefore, these findings align with the expectations for the GK model, since the average body weight of a GK rat is expected to be 10–30% less than that of an age-matched control Wistar rat [34]. Basal hyperglycemia has also been documented in GK rats, often manifesting as early as 3 weeks of age [35]. At birth, the β-cell mass of a GK rat is already severely reduced compared to that of a Wistar rat [36], and in adult GK rats, the β-cell mass is usually reduced up to 60% with markedly decreased insulin secretion [37,38], which explains the early hyperglycemia. Insulin resistance, another well-documented trait of this genetic model of T2D [15,39], also aligns with the findings of our study. Although inadequate β-cell proliferation in early life is a limitation as it relates to the human condition, other characteristics are consistent with descriptions in the literature and validate GK rats as a non-obese T2D animal model [34,40].

Histomorphometric changes in the gut of other animal models of diabetes have already been described [21,22,24,41,42], but this is the first study to comprehensively examine the gut from the duodenum to the distal colon. This approach was chosen in order to ascertain whether we would observe a similar proximal-to-distal progression of the disease as previously described in models of T1D [22,43]. In the GK rat, a T2D model, we did not observe such a pattern. Gut remodeling appears to occur in both mucosa and muscle layers, in different regions of the gut. The increase in the mucosa layer has been reported, and it was suggested that it is a mechanism to augment the absorptive surface area and functional capacity of the intestine [44]. Hyperphagia occurs in almost every model of diabetes and has also been suggested as a contributor to the increase in the thickness of intestinal mucosa [43]. However, it appears that the hyperglycemic state itself is sufficient to promote significant mucosal growth independent of food intake [45]. This was reinforced by another study where insulin administration prevented a marked increase in the intestinal epithelial cell proliferation rate of type 1 diabetic rats, resulting in reduced intestinal mucosal growth compared to non-treated diabetic animals [46]. Adachi et al. showed that GK rats also exhibited intestinal hyperplasia, possibly due to the increased expression of transcription factors and proteins involved in cell regeneration, differentiation, and/or proliferation [47]. An increase in the muscle layers of the gut has also already been reported in several animal models of diabetes [21,24,41,48]. In our study, the increase in the thickness of muscular layers may be at least partially attributed to hypertrophy of SMCs since we observed a decreased density, rather than an increase. This finding was consistent across all portions examined (not reaching statistical significance in the duodenum). SMC hypertrophy has already been described by Horváth et al., who associated this alteration with contractile protein actin and myosin increases in diabetic patients [17].

The myenteric plexus is located between the circular and the longitudinal muscular layers and is the main thing responsible for GI motility control [16,49]. In contrast to the findings of Pereira et al. [24], who did not observe a significant difference in the number of myenteric neurons per unit area between GK animals and controls, our study revealed a decrease in the density of myenteric neurons in diabetic animals. It is worth noting that our animals were older compared to those in the study by Pereira et al. [24]. Therefore, the duration of diabetes may play a role in the development of these alterations. Additionally, several authors also reported changes in the number and size of myenteric neurons throughout the entire GI tract, including the stomach [43], duodenum [50], jejunum [51], ileum [49], cecum [52], and colon [53], in both type 1 (streptozotocin-induced diabetic rats) [54] and type 2 D (diabetic mice consuming a high-fat diet) [55]. It seems that the neuronal population of the submucosal plexus may be more susceptible to degenerative changes induced by diabetes than the myenteric plexus [56]. The mechanisms underlying neuronal loss encompass increased apoptosis, elevated levels of Advanced Glycation End products (AGEs) and Receptor of Advanced Glycation End products (RAGEs), reduced nerve growth factor levels, and heightened oxidative stress [51,53,57].

These changes in the morphology of the small intestine and colon result in biomechanical alterations such as a loss of matrix elasticity and contractility, impairing both contraction and relaxation responses, which are fundamental for maintaining normal GI motility [32,58]. This leads to impaired intestinal sensory function and reduced intestinal motility [41,59,60], while increased thickness of the mucosa can affect digestion and absorption [61]. The neuronal change can further lead to improper gut motility, retrograde colonic movements, altered secretions, and even increased pain stimuli [62,63]. These alterations may provide insight into the common GI symptoms observed in diabetic patients [21].

Oxidative stress results from an imbalance between the production of ROS and antioxidant defenses and has already been implicated in gastrointestinal complications of diabetes [64,65]. Given that glutathione serves as the body's primary antioxidant, playing a crucial role in combating oxidative stress [66], and that previous studies showed that hyperglycemia-related oxidative stress was a primary inducer of neurological damage [16], we chose to quantify GSH levels locally. To the best of our knowledge, this is the first study to comprehensively evaluate tGSH and GSSG levels and the GSH/GSSG ratio across all sections of the gut in diabetic animals. In this work, we observed a decrease in tGSH levels in all examined segments of the gut, except for the duodenum. Furthermore, while the levels of GSSG were comparable between diabetic and control animals, the ratio of GSH to GSSG was significantly lower in diabetic animals (including in the duodenum), indicating increased levels of oxidative stress. The reduction in the GSH/GSSG ratio can result from either a decrease in free GSH levels or an increase in GSSG levels. In this study, the observed decrease in the GSH/GSSG ratio in GK rats was primarily due to a reduction in GSH levels in GK animals compared to controls, as there were no significant differences in GSSG levels between the two experimental groups. Chandrasekharan et al. conducted the first and only quantification of GSH but only in the diabetic colon as an indicator of oxidative stress, wherein they also observed a decrease in GSH levels associated with neurological damage and motor dysfunction [32]. These are likewise consistent with findings in individuals with T2D, who have been reported to exhibit lower blood GSH values [67,68]. Also, the depletion of GSH observed in a streptozotocin-induced model of diabetes has been shown to cause cardiac damage and cardiomyocyte apoptosis [69]. In another study, a decrease in GSH levels was observed in vascular smooth muscle cells, which was attributed to the depletion of glutathione precursors, particularly cysteine, which is a rate-limiting substrate in new glutathione synthesis [70]. Sekhar et al. described that the principal cause of oxidative stress in T2D is a deficiency of glutathione, primarily stemming from reduced synthesis due to the limited availability of the precursor amino acids cysteine and glycine, and that the supplementation of these precursors through dietary means can restore the synthesis of glutathione, consequently leading to a significant reduction in oxidative stress and markers of oxidant damage [71]. Furthermore, in individuals with type 2 diabetes, increased levels of transforming growth factor beta (TGF-β) were observed in plasma samples. This cytokine is known to reduce the expression of the catalytic subunit of glutamine-cysteine ligase, which also helps to explain why GSH levels decrease in these individuals [72].

The alterations observed in the GSH and GSSG concentrations in our study led to a decrease of up to 60% in the GSH/GSSG ratio in GK rats compared to controls. This reduction closely mirrors findings reported by Calabrese et al., who observed a 68% decrease in plasma GSH levels in T2D patients compared to control subjects [73]. The decrease in the plasma GSH/GSSG ratio not only correlates with heightened oxidative stress but also appears to adversely affect glucose availability and homeostasis, thereby exacerbating the diabetic condition [74,75]. Additionally, oxidative stress is known to play a critical role in the pathogenesis of various diabetic complications, including neuropathy, nephropathy, and retinopathy [27]. Furthermore, oxidative stress has also been identified as a significant contributor to gastrointestinal dysmotility, including post-operative ileus and diabetic gastroparesis [64]. Maintaining a balanced GSH/GSSG ratio is essential for protecting cells from oxidative damage and ensuring proper metabolic functioning [76]. Therefore, our findings highlight the importance of addressing oxidative stress when studying gastrointestinal complications of diabetes.

The results of histomorphometry and oxidative stress combined reveal an interesting pattern: all sections of the intestine showed signs of oxidative stress (indicated by a decreased GSH/GSSG ratio) and neuronal damage, but muscular remodeling was not observed in every portion. In fact, the duodenum displayed both oxidative stress and neuronal damage, but the muscular layers showed no remodeling. This raises an important question: does neuronal damage from oxidative stress occur before intestinal remodeling? These findings prompt further investigation into the sequence of events leading to gastrointestinal complications in diabetes.

In conclusion, we identified significant remodeling of the intestine and colon, along with marked alterations in the neuronal population of the myenteric plexus. The critical local deficiency of GSH, a key antioxidant, emerged as a central factor driving increased oxidative stress, which likely underlies the observed structural and neuronal damage in the gut of GK rats. Furthermore, the reduced GSH/GSSG ratio further underscores the oxidative stress in the examined gut regions. These data shed some light on the complex interplay between diabetes and gastrointestinal adjustments, offering new insights that could enhance our understanding and management of diabetic complications (Figure 9).

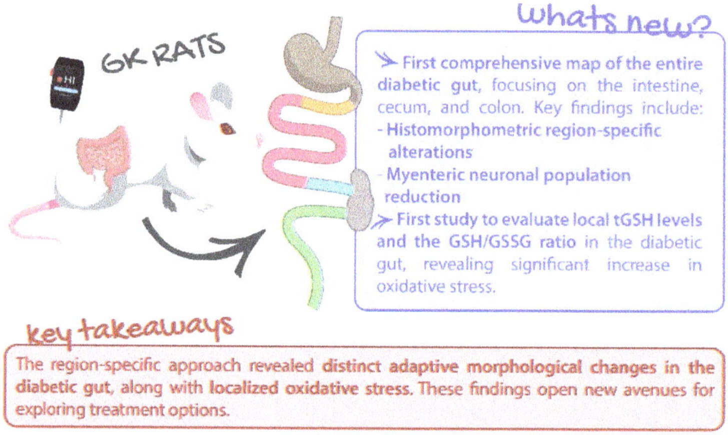

Figure 9. This study combines histomorphometry with glutathione assessments, providing a dual layer of analysis that allows for a more comprehensive understanding of tissue health and oxidative damage across different diabetic gut regions.

4. Materials and Methods

4.1. Animals

Non-obese type 2 diabetic GK male rats (n = 6), 20–21 weeks old, were obtained from the breeding colonies of the Faculty of Medicine, University of Coimbra. Wistar Han rats (n = 5) from the same colony with comparable age were used as controls. Animals were kept under standard ventilation, temperature (22.0 ± 0.1 °C), relative humidity (52.0 ± 2.0%), and light (12 h light/dark cycle) with access to autoclaved tap water and food ad libitum (standard diet A03, SAFE®, Rosenberg, Germany). All procedures involving animals were previously approved by the local animal welfare commission (ORBEA 13/18) following the European Community guidelines for the use of laboratory animals (Directive 2010/63/EU) and performed by licensed users.

4.2. In Vivo Procedures and Sample Collection

Animals' body weight, caloric intake, and blood glucose (6 h fast, blood collected from the tail vein) were monitored for 2 weeks.

Intraperitoneal insulin tolerance tests (ITTs, Humulin, Lilly®, Indianapolis, Indiana, USA, 0.25 IU/kg) were performed after a 6 h fast. Glycaemia evaluation was performed at 0, 15, 30, 60, and 120 min using a glucometer and test strips (Accu-Chek Aviva, Roche®, Basel, Switzerland) [77].

After a 6 h fast, animals were anesthetized with an intraperitoneal injection of ketamine (Nimatek, Dechra®, Northwich, England, 50 mg/kg) and xylazine (Sedaxylan, Dechra®, 6.6 mg/kg) and after blood collection were sacrificed by cervical displacement. The GI tract from the proximal part of the duodenum to the distal part of the colon was collected and weighed as previously described by our group [22].

4.3. Histological Preparation and Analyses

Samples (1 cm long) of proximal duodenum (collected 2 cm distal to pylorus), middle jejunum, distal ileum (collected 2 cm cranial to the ileocecal junction), cecum, proximal colon (PC), and distal colon (DC) were collected and fixed in 4% formalin. All samples were dehydrated in consecutive 70%, 96%, and 99% ethanol solutions and embedded in paraffin. Then, 3 μm thick cuts were made perpendicularly to the mucosa using a microtome and mounted in sterilized glass slides. Finally, the sections were rehydrated in a series of graded ethanol (99, 96, 70%), washed in water, and stained with hematoxylin and eosin (H&E). Each section was evaluated under an optical microscope (Eclipse E600Miami, Nikon Instruments®, Melville, NY, USA) and photographed in different representative regions (magnification of 40× and 100×). All stained samples were evaluated by an experienced veterinary pathologist who was blinded for the experimental groups. The thickness of the mucosa, submucosa, circular, and longitudinal muscles was then measured, by the same research team member, using the free ImageJ® software 1.54g. For each sample, the layer thickness was measured randomly in twelve different locations, and then averaged. The measurements were only carried out for images where the entire intestinal wall could be observed. To evaluate collagen deposition in the extracellular matrix, the samples were stained with Masson's trichrome, and to measure the intracellular accumulation of glycogen, the periodic acid–Schiff (PAS) reaction was performed. All histologic samples were evaluated by an experienced veterinary pathologist.

4.4. Quantitative Analysis of Smooth Muscle Cells Nuclei in the Muscular Layers

For each sample, twelve sections centered in the muscular layers were photographed (objective lens of 10×). For each section, an area of 50 μm × 200 μm (10,000 μm^2) in the center of the photo was used for nucleus quantification per unit area. Only the nuclei of the SMCs within the test area boundaries and those that touched the lines were counted. SMC nucleus density was expressed as the number of cells per mm^2 of muscular area.

4.5. Quantitative Analysis of Neuronal Nuclei in the Myenteric Plexi

For each sample, three sections stained with H&E were observed, and all myenteric plexi were photographed using 10×, 20×, and 40× objective lenses when needed. The myenteric plexi were then outlined, and their areas were measured. The neurons' nuclei within all visible sections of the myenteric plexus were counted. Myenteric neuronal density was expressed as the number of cells per mm^2 of the plexus.

4.6. Total GSH and GSSG Quantification

For total GSH (tGSH) and GSSG quantification, 1cm long samples of the proximal duodenum, middle jejunum, distal ileum, cecum, PC, and DC were collected, and 400 μL of perchloric acid 5% (*w*/*v*) was added. The tissues were homogenized and centrifugated at 16,060× *g* for 10 min at 4 °C. The pellets were then saved for protein quantification at −20 °C, and the acidic supernatant was stored at −80 °C until analysis.

The levels of tGSH and GSSG were measured using the DTNB-GSSH reductase recycling assay, following the modified Ellman's method [78]. Acidic samples were neutralized with 0.76 M potassium bicarbonate and then centrifuged (16,060× *g* for 2 min at 4 °C). The same process was applied to GSH standards ranging from 0 to 15 μM. In 96-well plates, 100 μL of sample was mixed with 65 μL of reagent solution containing NADPH (0.63 mM) and DTNB (3.96 mM) and prepared in phosphate buffer (71.5 mM Na_2HPO_4, 71.5 mM NaH_2PO_4, 0.63 mM EDTA). The mixture was incubated at 30 °C for 15 min. Subsequently, 40 μL of glutathione reductase (10 U/mL in phosphate buffer) was added, and absorbance readings were taken at 415 nm for 3 min with 10 s intervals, using a Biotek PowerWaxe X spectrophotometer (Charlotte, VT, USA). The tGSH and GSSG levels were normalized to protein levels and expressed as nmol/mg of protein. The GSH/GSSG ratio was calculated using the following formula:

$$\text{GSH/GSSG} = \frac{tGSH - 2 \times GSSG}{GSSG} \quad (1)$$

4.7. Protein Quantification

The pellets described in the previous section were dissolved in 0.5 M NaOH, and an albumin stock solution was prepared with concentrations ranging from 0.0625 mg/mL to 1 mg/mL. The pellets were homogenized, and protein levels were assessed spectrophotometrically using a microplate reader (Biotek-Powerwave HT®, Charlotte, VT, USA), following the method described by Lowry et al., with measurements taken at a wavelength of 700 nm [79].

4.8. Statistical Analysis

The GraphPad Prism 8.1.2 was used for statistical analysis of data. The unpaired Student's t test was used for comparison between 2 experimental groups (CTRL and GK), and data were expressed as mean ± SEM, where n refers to the number of experimental animals. The Shapiro–Wilk test was employed to assess the normality of the data. All datasets had $p > 0.05$ and were considered to have passed the normality test. To evaluate histological and oxidative stress data, a two-way ANOVA followed by an unpaired t test with Welch's correction was used to compare the two experimental groups. In all cases, a p value of less than 0.05 was used to identify a statistically significant difference.

Supplementary Materials: The following supporting information can be downloaded at: https://www.mdpi.com/article/10.3390/ijms252212115/s1.

Author Contributions: Conceptualization—M.E.-M. and M.D.-A.; methodology—M.E.-M., M.M., P.D.-P., V.M.C. and M.D.-A.; formal analysis—M.E.-M. and V.M.C.; investigation—M.E.-M., M.F.-D., C.V.-O., J.C.-P., S.O., P.M. and M.D.-A.; resources—P.M., M.M., P.D.-P., V.M.C. and M.D.-A.; writing/original draft preparation—M.E.-M.; writing/review and editing—M.F.-D., P.M., M.M., P.D.-P., V.M.C. and M.D.-A.; visualization—M.E.-M. and M.D.-A.; supervision—M.E.-M. and M.D.-A.; project administration—M.D.-A.; funding acquisition—P.M., M.M. and M.D.-A. All authors have read and agreed to the published version of the manuscript.

Funding: This work received financial support from Fundação para a Ciência e Tecnologia—FCT/MCTES (UIDP/50006/2020 DOI 10.54499/UIDP/50006/2020) through national funds. These funds were also in the scope of the projects: UIDP/04378/2020 and UIDB/04378/2020 of the Research Unit on Applied Molecular Biosciences (UCIBIO), LA/P/0140/2020 of the Associate Laboratory Institute for Health and Bioeconomy—i4HB, and UIDB/04046/2020 (DOI:10.54499/UIDB/04046/2020) and UIDP/04046/2020 (DOI:10.54499/UIDP/04046/2020) of the Research Unit BioISI-Biosystems & Integrative Sciences Institute. This work was also supported by an FCT PhD scholarship (2020.06502.BD, 10.54499/2020.06502.BD) to M.E.-M.

Institutional Review Board Statement: All procedures involving animals were previously approved by the Faculty of Medicine, University of Coimbra Animal Welfare Commission (ORBEA 13/18), following the European Community guidelines for the humane and responsible animal care of laboratory animals (Directive 2010/63/EU) and performed by licensed users.

Informed Consent Statement: Not applicable.

Data Availability Statement: Dataset available on request from the authors.

Acknowledgments: This work received support and help from FCT/MCTES (LA/P/0008/2020 DOI 10.54499/LA/P/0008/2020, UIDP/50006/2020 DOI 10.54499/UIDP/50006/2020 and UIDB/50006/2020 DOI 10.54499/UIDB/50006/2020). The authors thank Céu Pereira for the excellent technical support.

Conflicts of Interest: The authors declare no conflicts of interest.

References

1. World Health Organization. *Definition and Diagnosis of Diabetes Mellitus and Intermediate Hyperglycaemia*; WHO and International Diabetes Federation: Geneva, Switzerland, 2006.
2. International Diabetes Federation. *IDF Diabetes Atlas*, 10th ed.; International Diabetes Federation: Brussels, Belgium, 2021.
3. World Health Organization. *Global Report on Diabetes*; WHO Library Cataloguing: Geneva, Switzerland, 2016.
4. Sattar, N.; Eliasson, B.; Svensson, A.M.; Zethelius, B.; Miftaraj, M. Risk Factors, Mortality, and Cardiovascular Outcomes in Patients with Type 2 Diabetes. *N. Engl. J. Med.* **2018**, *379*, 633–644. [CrossRef]
5. NCD-RisC. Worldwide Trends in Diabetes since 1980: A Pooled Analysis of 751 Population-Based Studies with 4·4 Million Participants. *Lancet* **2008**, *387*, 1513–1530. [CrossRef]
6. Chatterjee, S.; Khunti, K.; Davies, M.J. Type 2 Diabetes. *Lancet* **2017**, *389*, 2239–2251. [CrossRef] [PubMed]
7. American, D.A. Diagnosis and Classification of Diabetes Mellitus. *Diabetes Care* **2014**, *37*, 81–90. [CrossRef] [PubMed]
8. Galicia-Garcia, U.; Benito-Vicente, A.; Jebari, S.; Larrea-Sebal, A.; Siddiqi, H.; Uribe, K.B.; Ostolaza, H.; Martín, C. Pathophysiology of Type 2 Diabetes Mellitus. *Int. J. Mol. Sci.* **2020**, *21*, 6275. [CrossRef]
9. Gimenes, G.M.; Santana, G.O.; Scervino, M.V.M.; Curi, R.; Pereira, J.N.B. A Short Review on the Features of the Non-Obese Diabetic Goto-Kakizaki Rat Intestine. *Braz. J. Med. Biol. Res.* **2022**, *55*, e11910. [CrossRef]
10. Maqbool, M.; Dar, M.A.; Gani, I.; Mir, S.A. Animal Models in Diabetes Mellitus: An Overview. *J. Drug Deliv. Ther.* **2019**, *9*, 472–475. [CrossRef]
11. Goto, Y.; Kakizaki, M.; Masaki, N. Production of Spontaneous Diabetic Rats by Repetition of Selective Breeding. *Tohoku J. Exp. Med.* **1976**, *119*, 85–90. [CrossRef]
12. Goto, Y.; Kakizaki, M.; Masaki, N. Spontaneous Diabetes Produced by Selective Breeding o f Normal Wistar Rats. *Proc. Jpn. Acad.* **1975**, *51*, 80–85. [CrossRef]
13. Goto, Y.; Suzuki, K.; Ono, T.; Sasaki, M.; Toyota, T. Development of Diabetes in the Non-Obese NIDDM Rat (GK Rat). In *Prediabetes*; Springer: Berlin/Heidelberg, Germany, 1988; pp. 29–31. [CrossRef]
14. Goto, Y.; Kakizaki, M. The Spontaneous-Diabetes Rat: A Model of Noninsulin Dependent Diabetes Mellitus. *Proc. Jpn. Acad.* **1981**, *57*, 381–384. [CrossRef]
15. Bisbis, S.; Bailbe, D.; Tormo, M.A.; Picarel-Blanchot, F.; Derouet, M.; Simon, J.; Portha, B. Insulin Resistance in the GK Rat: Decreased Receptor Number but Normal Kinase Activity in Liver. *Am. J. Physiol.-Endocrinol. Metab.* **1993**, *265*, E807–E813. [CrossRef] [PubMed]
16. Chandrasekharan, B.; Srinivasan, S. Diabetes and the Enteric Nervous System. *Neurogastroenterol. Motil.* **2007**, *19*, 951–960. [CrossRef] [PubMed]
17. Horváth, V.J.; Putz, Z.; Izbéki, F.; Körei, A.E. Diabetes-Related Dysfunction of the Small Intestine and the Colon: Focus on Motility. *Curr. Diabetes Rep.* **2015**, *15*, 1–8. [CrossRef] [PubMed]
18. Krishnan, B.; Babu, S.; Walker, J.; Walker, A.B.; Pappachan, J.M. Gastrointestinal Complications of Diabetes Mellitus. *World J. Diabetes* **2013**, *4*, 51–63. [CrossRef] [PubMed]
19. Piper, M.S.; Saad, R.J. Diabetes Mellitus and the Colon. *Curr. Treat. Options Gastroenterol.* **2018**, *15*, 460–474. [CrossRef]
20. Paul, J.; Veettil Hussain Shihaz, A. "Complications of Gastrointestinal System in Diabetes Mellitus"-Neglected Part of Diabetic Management. *Iran. J. Diabetes Obes.* **2020**, *12*. [CrossRef]
21. Zhao, M.; Liao, D.; Zhao, J. Diabetes-Induced Mechanophysiological Changes in the Small Intestine and Colon. *World J. Diabetes* **2017**, *8*, 9358. [CrossRef]
22. Esteves-Monteiro, M.; Menezes-Pinto, D.; Ferreira-Duarte, M.; Dias-Pereira, P.; Morato, M.; Duarte-Araújo, M. Histomorphometry Changes and Decreased Reactivity to Angiotensin II in the Ileum and Colon of Streptozotocin-Induced Diabetic Rats. *Int. J. Mol. Sci.* **2022**, *23*, 13233. [CrossRef]
23. Zhao, J.; Gregersen, H. Diabetes-Induced Mechanophysiological Changes in the Esophagus. *Ann. N. Y. Acad. Sci.* **2016**, *1380*, 139–154. [CrossRef]
24. Pereira, J.N.B.; Murata, G.M.; Sato, F.T.; Marosti, A.R.; Carvalho, C.R.d.O.; Curi, R. Small Intestine Remodeling in Male Goto–Kakizaki Rats. *Physiol. Rep.* **2021**, *9*, e14755. [CrossRef]
25. Chen, P.-M.; Gregersen, H.; Zhao, J.-B. Advanced Glycation End-Product Expression Is Upregulated in the Gastrointestinal Tract of Type 2 Diabetic Rats. *World J Diabetes* **2015**, *6*, 662. [CrossRef] [PubMed]
26. Singh, V.P.; Bali, A.; Singh, N.; Jaggi, A.S. Advanced Glycation End Products and Diabetic Complications. *Korean J. Physiol. Pharmacol.* **2014**, *18*, 1. [CrossRef] [PubMed]
27. Giacco, F.; Brownlee, M. Oxidative Stress and Diabetic Complications. *Circ. Res.* **2010**, *107*, 1058–1070. [CrossRef] [PubMed]
28. Zitka, O.; Sablikova, S.; Gumulec, J.; Masaryk, M.; Adam, V.; Hubalek, J.; Trankova, L.; Kruseova, J.; Eckschlager, T.; Kizek, R. Redox Status Expressed as GSH: GSSG Ratio as a Marker for Oxidative Stress in Paediatric Tumour Patients. *Oncol. Lett.* **2012**, *4*, 1247–1253. [CrossRef] [PubMed]
29. Kuyvenhoven, J.P.; Meinders, A.E. Oxidative Stress and Diabetes Mellitus Pathogenesis of Long-Term Complications. *Eur. J. Intern. Med.* **1999**, *10*, 9–19. [CrossRef]
30. Mclennan, S.V.; Heffernan, S.; Wright, L.; Rae, C.; Fisher, E.; Yue, D.K.; Turtle, J.R. Changes in Hepatic Glutathione Metabolism in Diabetes. *Diabetes* **1991**, *40*, 344–348. [CrossRef]

31. Nwose, E.U.; Jelinek, H.F.; Richards, R.S.; Kerr, P.G. Changes in the Erythrocyte Glutathione Concentration in the Course of Diabetes Mellitus. *Redox Rep.* **2006**, *11*, 99–104. [CrossRef]
32. Chandrasekharan, B.; Anitha, M.; Blatt, R.; Shahnavaz, N.; Kooby, D.; Staley, C.; Mwangi, S.; Jones, D.P.; Sitaraman, S.V.; Srinivasan, S. Colonic Motor Dysfunction in Human Diabetes Is Associated with Enteric Neuronal Loss and Increased Oxidative Stress. *Neurogastroenterol. Motil.* **2011**, *23*, 131-e26. [CrossRef]
33. Xue, B.; Sukumaran, S.; Nie, J.; Jusko, W.J.; DuBois, D.C.; Almon, R.R. Adipose Tissue Deficiency and Chronic Inflammation in Diabetic Goto-Kakizaki Rats. *PLoS ONE* **2011**, *6*, e17386. [CrossRef]
34. Akash, M.; Rehman, K.; Chen, S. Goto-Kakizaki Rats: Its Suitability as Non-Obese Diabetic Animal Model for Spontaneous Type 2 Diabetes Mellitus. *Curr. Diabetes Rev.* **2013**, *9*, 387–396. [CrossRef]
35. Picarel-Blanchot, F.; Berthelier, C.; Bailbe, D.; Portha, B. Impaired Insulin Secretion and Excessive Hepatic Glucose Production Are Both Early Events in the Diabetic GK Rat. *Am. J. Physiol.-Endocrinol. Metab.* **1996**, *271*, E755–E762. [CrossRef] [PubMed]
36. Miralles, F.; Portha, B. Early Development of Beta-Cells Is Impaired in the GK Rat Model of Type 2 Diabetes. *Diabetes* **2001**, *50* (Suppl. S1), S84. [CrossRef] [PubMed]
37. Portha, B.; Giroix, M.H.; Serradas, P.; Gangnerau, M.N.; Movassat, J.; Rajas, F.; Bailbe, D.; Plachot, C.; Mithieux, G.; Marie, J.C. Beta-Cell Function and Viability in the Spontaneously Diabetic GK Rat: Information from the GK/Par Colony. *Diabetes* **2001**, *50* (Suppl. S1), S89. [CrossRef] [PubMed]
38. Movassat, J.; Saulnier, C.; Serradas, P.; Portha, B. Impaired Development of Pancreatic Beta-Cell Mass Is a Primary Event during the Progression to Diabetes in the GK Rat. *Diabetologia* **1997**, *40*, 916–925. [CrossRef]
39. Guest, P.C. Characterization of the Goto-Kakizaki (GK) Rat Model of Type 2 Diabetes. In *Methods in Molecular Biology*; Humana Press Inc.: Tova, NJ, USA, 2019; Volume 1916, pp. 203–211. [CrossRef]
40. Kuwabara, W.M.T.; Panveloski-Costa, A.C.; Yokota, C.N.F.; Pereira, J.N.B.; Filho, J.M.; Torres, R.P.; Hirabara, S.M.; Curi, R.; Alba-Loureiro, T.C. Comparison of Goto-Kakizaki Rats and High Fat Diet-Induced Obese Rats: Are They Reliable Models to Study Type 2 Diabetes Mellitus? *PLoS ONE* **2017**, *12*, e0189622. [CrossRef]
41. Zhao, J.; Chen, P.; Gregersen, H. Morpho-Mechanical Intestinal Remodeling in Type 2 Diabetic GK Rats-Is It Related to Advanced Glycation End Product Formation? *J. Biomech.* **2013**, *46*, 1128–1134. [CrossRef]
42. Sha, H.; Tong, X.; Zhao, J. Abnormal Expressions of AGEs, TGF- β 1, BDNF and Their Receptors in Diabetic Rat Colon—Associations with Colonic Morphometric and Biomechanical Remodeling. *Sci. Rep.* **2018**, *8*, 9437. [CrossRef]
43. Fregonesi, C.; Miranda-neto, M.; Molinari, S.L.; Zanoni, J.N. Quantitative Study If the Myenteric Plexus of the Stomach of Rats with Streptozotocin-Induced Diabetes. *Arq. De Neuro-Psiquiatr.* **2001**, *59*, 50–53. [CrossRef]
44. Dailey, M.J. Nutrient-Induced Intestinal Adaption and Its Effect in Obesity. *Physiol. Behav.* **2014**, *136*, 74–78. [CrossRef]
45. Miller, D.L.; Hanson, W.; Schedl, H.P.; Osborne, J.W. Proliferation Rate and Transit Time of Mucosal Cells in Small Intestine of the Diabetic Rat. *Gastroenterology* **1977**, *73*, 1326–1332. [CrossRef]
46. Sukhotnik, I.; Shamir, R.; Bashenko, Y.; Mogilner, J.G.; Chemodanov, E.; Shaoul, R.; Coran, A.G.; Shehadeh, N. Effect of Oral Insulin on Diabetes-Induced Intestinal Mucosal Growth in Rats. *Dig. Dis. Sci.* **2011**, *56*, 2566–2574. [CrossRef] [PubMed]
47. Adachi, T.; Mori, C.; Sakurai, K.; Shihara, N.; Tsuda, K.; Yasuda, K. Morphological Changes and Increased Sucrase and Isomaltase Activity in Small Intestines of Insulin-Deficient and Type 2 Diabetic Rats. *Endocr. J.* **2003**, *50*, 271–279. [CrossRef] [PubMed]
48. Siegman, M.J.; Eto, M.; Butler, T.M. Remodeling of the rat distal colon in diabetes: Function and ultrastructure. *Am. J. Physiol. Cell Physiol.* **2016**, *310*, C151–C160. [CrossRef] [PubMed]
49. Brasileiro, A.D.; Garcia, L.P.; Carvalho, S.D.; Rocha, L.B.; Pedrosa, A.L.; Vieira, A.S.; José, V.; Rogelis, A.; Rodrigues, A. Effects of Diabetes Mellitus on Myenteric Neuronal Density and Sodium Channel Expression in the Rat Ileum. *Brain Res.* **2019**, *1708*, 1–9. [CrossRef] [PubMed]
50. Sanders, K.M.; Ward, M.; Koh, S.D. Interstitial Cells: Regulators of Smooth Muscle Function. *Physiol. Rev.* **2014**, *94*, 859–907. [CrossRef]
51. Honoré, S.M.; Zelarayan, L.C.; Genta, S.B.; Sánchez, S.S. Neuronal Loss and Abnormal BMP / Smad Signaling in the Myenteric Plexus of Diabetic Rats. *Auton. Neurosci.* **2011**, *164*, 51–61. [CrossRef]
52. Zanoni, J.; Miranda, N.; Bazzote, R. Morphological and Quantitative Analysis of the Neurons of the Myenteric Plexus of the Cecum of Streptozotocin-Induced Diabetic Rats. *Arq. Neuropsiquiatr.* **1998**, *55*, 696–702. [CrossRef]
53. Furlan, M.M.D.P.; Molinari, S.L.; Neto, M.H.d.M. Morphoquantitative Effects of Acute Diabetes on the Myenteric Neurons of the Proximal Colon on Adult Rats. *Arq. de Neuro-Psiquiatr.* **2002**, *60*, 576–581. [CrossRef]
54. Voukali, E.; Shotton, H.R.; Lincoln, J. Selective Responses of Myenteric Neurons to Oxidative Stress and Diabetic Stimuli. *Neurogastroenterol. Motil.* **2011**, *23*, 964. [CrossRef]
55. Larsson, S.; Voss, U. Neuroprotective Effects of Vitamin D on High Fat Diet- and Palmitic Acid-Induced Enteric Neuronal Loss in Mice. *BMC Gastroenterol.* **2018**, *18*, 175. [CrossRef]
56. Ferreira, P.E.B.; Beraldi, E.J.; Borges, S.C.; Natali, M.R.M.; Buttow, N.C. Resveratrol Promotes Neuroprotection and Attenuates Oxidative and Nitrosative Stress in the Small Intestine in Diabetic Rats. *Biomed. Pharmacother.* **2018**, *105*, 724–733. [CrossRef] [PubMed]
57. He, C.; Soffer, E.D.Y.E.; Ferris, C.D.; Walsh, R.M.; Szurszewski, J.H.; Farrugia, G. Loss of Interstitial Cells of Cajal and Inhibitory Innervation in Insulin-Dependent Diabetes. *Gastroenterology* **2001**, *121*, 427–434. [CrossRef] [PubMed]

58. Talubmook, C.; Forrest, A.S.; Parsons, M.E. Alterations in Basal and Evoked Contractile Responses of Iintestinal Tissue from Streptozotocin-Induced Diabetic Rats. *Br. J. Pharmacol. (Proc. Suppl.)* **2002**, *136*, 42.
59. Chen, P.; Zhao, J.; Gregersen, H. Up-Regulated Expression of Advanced Glycation End-Products and Their Receptor in the Small Intestine and Colon of Diabetic Rats. *Dig. Dis. Sci.* **2011**, *57*, 48–57. [CrossRef]
60. Zhao, J.; Yang, J.; Gregersen, H. Biomechanical and Morphometric Intestinal Remodelling during Experimental Diabetes in Rats. *Diabetologia* **2003**, *46*, 1688–1697. [CrossRef]
61. Lerkdumnernkit, N.; Sricharoenvej, S.; Lanlua, P.; Niyomchan, A.; Baimai, S.; Chookliang, A.; Plaengrit, K.; Pianrumluk, S.; Manoonpol, C. The Effects of Early Diabetes on Duodenal Alterations in the Rats. *Int. J. Morphol.* **2022**, *40*, 389–395. [CrossRef]
62. Klinge, M.W.; Haase, A.; Mark, E.B.; Sutter, N.; Fynne, L.V.; Drewes, A.M.; Schlageter, V.; Lund, S.; Borghammer, P.; Krogh, K. Colonic Motility in Patients with Type 1 Diabetes and Gastrointestinal Symptoms. *Neurogastroenterol. Motil.* **2020**, *32*, e13948. [CrossRef]
63. Bulc, M.; Całka, J.; Palus, K. Effect of Streptozotocin-Inducted Diabetes on the Pathophysiology of Enteric Neurons in the Small Intestine Based on the Porcine Diabetes Model. *Int. J. Mol. Sci.* **2020**, *21*, 2047. [CrossRef]
64. Kashyap, P.; Farrugia, G. Oxidative Stress: Key Player in Gastrointestinal Complications of Diabetes. *Neurogastroenterol. Motil.* **2011**, *23*, 111–114. [CrossRef]
65. Rajini, P.S. Exacerbation of Intestinal Brush Border Enzyme Activities and Oxidative Stress in Streptozotocin-Induced Diabetic Rats by Monocrotophos. *Chem. Biol. Interact.* **2014**, *211*, 11–19. [CrossRef]
66. Maritim, A.C.; Sanders, R.A.; Watkins, J.B. Diabetes, Oxidative Stress, and Antioxidants: A Review. *J. Biochem. Mol. Toxicol.* **2003**, *17*, 24–38. [CrossRef] [PubMed]
67. Kalkan, I.H.; Suher, M. The Relationship between the Level of Glutathione, Impairment of Glucose Metabolism and Complications of Diabetes Mellitus. *Pak. J. Med. Sci.* **2013**, *29*, 938–942. [CrossRef] [PubMed]
68. Seghrouchni, I.; Drai, J.; Bannier, E.; Rivière, J.; Calmard, P.; Garcia, I.; Orgiazzi, J.; Revol, A. Oxidative Stress Parameters in Type I, Type II and Insulin-Treated Type 2 Diabetes Mellitus; Insulin Treatment Efficiency. *Clin. Chim. Acta* **2002**, *321*, 89–96. [CrossRef] [PubMed]
69. Ghosh, S.; Pulinilkunnil, T.; Yuen, G.; Kewalramani, G.; An, D.; Qi, D.; Abrahani, A.; Rodrigues, B. Cardiomyocyte Apoptosis Induced by Short-Term Diabetes Requires Mitochondrial GSH Depletion. *Am. J. Physiol. -Heart Circ. Physiol.* **2005**, *289*, H768–H776. [CrossRef]
70. Powell, L.A.; Nally, S.M.; McMaster, D.; Catherwood, M.A.; Trimble, E.R. Restoration of Glutathione Levels in Vascular Smooth Muscle Cells Exposed to High Glucose Conditions. *Free. Radic. Biol. Med.* **2001**, *31*, 1149–1155. [CrossRef]
71. Sekhar, R.V.; McKay, S.V.; Patel, S.G.; Guthikonda, A.P.; Reddy, V.T.; Balasubramanyam, A.; Jahoor, F. Glutathione Synthesis Is Diminished in Patients with Uncontrolled Diabetes and Restored by Dietary Supplementation with Cysteine and Glycine. *Diabetes Care* **2011**, *34*, 162–167. [CrossRef]
72. Lagman, M.; Ly, J.; Saing, T.; Kaur Singh, M.; Vera Tudela, E.; Morris, D.; Chi, P.-T.; Ochoa, C.; Sathananthan, A.; Venketaraman, V. Investigating the Causes for Decreased Levels of Glutathione in Individuals with Type II Diabetes. *PLoS ONE* **2015**, *10*, e0118436. [CrossRef]
73. Calabrese, V.; Cornelius, C.; Leso, V.; Trovato-Salinaro, A.; Ventimiglia, B.; Cavallaro, M.; Scuto, M.; Rizza, S.; Zanoli, L.; Neri, S.; et al. Oxidative Stress, Glutathione Status, Sirtuin and Cellular Stress Response in Type 2 Diabetes. *Biochim. Et Biophys. Acta (BBA) -Mol. Basis Dis.* **2012**, *1822*, 729–736. [CrossRef]
74. Paolisso, G.; Di Maro, G.; Pizza, G.; D'Amore, A.; Sgambato, S.; Tesauro, P.; Varricchio, M.; D'Onofrio, F. Plasma GSH/GSSG Affects Glucose Homeostasis in Healthy Subjects and Non-Insulin-Dependent Diabetics. *Am. J. Physiol. -Endocrinol. Metab.* **1992**, *263*, E435–E440. [CrossRef]
75. De Mattia, G.; Bravi, M.C.; Laurenti, O.; Cassone-Faldetta, M.; Armiento, A.; Ferri, C.; Balsano, F. Influence of Reduced Glutathione Infusion on Glucose Metabolism in Patients with Non—Insulin-Dependent Diabetes Mellitus. *Metabolism* **1998**, *47*, 993–997. [CrossRef]
76. Hatem, E.; Berthonaud, V.; Dardalhon, M.; Lagniel, G.; Baudouin-Cornu, P.; Huang, M.-E.; Labarre, J.; Chédin, S. Glutathione Is Essential to Preserve Nuclear Function and Cell Survival under Oxidative Stress. *Free. Radic. Biol. Med.* **2014**, *67*, 103–114. [CrossRef] [PubMed]
77. Oliveira, S.; Monteiro-Alfredo, T.; Henriques, R.; Ribeiro, C.F.; Seiça, R.; Cruz, T.; Cabral, C.; Fernandes, R.; Piedade, F.; Robalo, M.P.; et al. Improvement of Glycaemia and Endothelial Function by a New Low-Dose Curcuminoid in an Animal Model of Type 2 Diabetes. *Int. J. Mol. Sci.* **2022**, *23*, 5652. [CrossRef] [PubMed]
78. Bastos, M.L.; Carvalho, M.; Milhazes, N.; Remio, F.; Borges, F.; Fernandes, E.; Amado, F.; Monks, T.J.; Carvalho, F. Hepatotoxicity of 3,4-Methylenedioxyamphetamine and?-Methyldopamine in Isolated Rat Hepatocytes: Formation of Glutathione Conjugates. *Arch. Toxicol.* **2004**, *78*, 16–24. [CrossRef] [PubMed]
79. Lowry, O.H.; Rosebrough, N.J.; Farr, A.L.; Randall, R.J. Protein Measurement with the Folin Phenol Reagent. *J. Biol. Chem.* **1951**, *193*, 265–275. [CrossRef] [PubMed]

International Journal of
Molecular Sciences

MDPI

Article

New Application of an Old Drug: Anti-Diabetic Properties of Phloroglucinol

Krzysztof Drygalski [1,*], Mateusz Maciejczyk [2], Urszula Miksza [3], Andrzej Ustymowicz [4], Joanna Godzień [5], Angelika Buczyńska [5], Andrzej Chomentowski [6], Iga Walczak [7], Karolina Pietrowska [5], Julia Siemińska [5], Cezary Pawlukianiec [8], Przemysław Czajkowski [3], Joanna Fiedorczuk [3], Monika Moroz [3], Beata Modzelewska [6], Anna Zalewska [9], Barbara Kutryb-Zając [7], Tomasz Kleszczewski [6], Michał Ciborowski [5], Hady Razak Hady [10], Marc Foretz [11] and Edyta Adamska-Patruno [3]

1 Department of Hypertension and Diabetology, Medical University of Gdansk, 80-214 Gdansk, Poland
2 Department of Hygiene, Epidemiology and Ergonomics, Medical University of Bialystok, 15-089 Bialystok, Poland; mat.maciejczyk@gmail.com
3 Clinical Research Support Centre, Medical University of Bialystok, 15-089 Bialystok, Poland; urszula.miksza@umb.edu.pl (U.M.); przemyslaw.czajkowski@umb.edu.pl (P.C.); joanna.fiedorczuk@umb.edu.pl (J.F.); monika.moroz@umb.edu.pl (M.M.); edyta.adamska-patruno@umb.edu.pl (E.A.-P.)
4 Department of Radiology, Medical University of Bialystok, 15-089 Bialystok, Poland
5 Clinical Research Centre, Medical University of Bialystok, 15-089 Bialystok, Poland; angelika.buczynska@umb.edu.pl (A.B.); karolina.pietrowska@umb.edu.pl (K.P.); julia.sieminska@umb.edu.pl (J.S.); michal.ciborowski@umb.edu.pl (M.C.)
6 Department of Biophysics, Medical University of Bialystok, 15-089 Bialystok, Poland; andrzej.chomentowski@sd.umb.edu.pl (A.C.); beata.modzelewska@umb.edu.pl (B.M.); tomasz.kleszczewski@umb.edu.pl (T.K.)
7 Department of Biochemistry, Medical University of Gdansk, 80-214 Gdansk, Poland; igawalczak@gumed.edu.pl (I.W.); b.kutryb-zajac@gumed.edu.pl (B.K.-Z.)
8 Students Scientific Club "Biochemistry of Civilization Diseases", Department of Hygiene, Epidemiology and Ergonomics, Medical University of Bialystok, 15-089 Bialystok, Poland
9 Experimental Dentistry Laboratory, Medical University of Bialystok, 15-089 Bialystok, Poland; stom.dos@umb.edu.pl
10 Clinical Department of General and Endocrine Surgery, Medical University of Bialystok, 15-089 Bialystok, Poland
11 Institut Cochin, Université Paris Cité, CNRS, INSERM, F-75014 Paris, France; marc.foretz@inserm.fr
* Correspondence: drygalskikrzysztof@gmail.com or krzysztof.drygalski@gumed.edu.pl

Abstract: Phloroglucinol (PHG), an analgesic and spasmolytic drug, shows promise in preventing high-fat-diet (HFD)-induced non-alcoholic fatty liver disease (NAFLD) and insulin resistance. In Wistar rats, 10 weeks of PHG treatment did not prevent HFD-induced weight gain but significantly mitigated fasting hyperglycemia, impaired insulin responses, and liver steatosis. This protective effect was not linked to hepatic lipogenesis or AMP-activated protein kinase (AMPK) activation. Instead, PHG improved mitochondrial function by reducing oxidative stress, enhancing ATP production, and increasing anti-oxidant enzyme activity. PHG also relaxed gastric smooth muscles via potassium channel activation and nitric oxide (NO) signaling, potentially delaying gastric emptying. A pilot intervention in pre-diabetic men confirmed PHG's efficacy in improving postprandial glycemic control and altering lipid metabolism. These findings suggest PHG as a potential therapeutic for NAFLD and insulin resistance, acting through mechanisms involving mitochondrial protection, anti-oxidant activity, and gastric motility modulation. Further clinical evaluation is warranted to explore PHG's full therapeutic potential.

Keywords: phloroglucinol; NAFLD; diabetes; oxidative stress; insulin resistance; liver steatosis; anti-spasmodic; lipid metabolism

1. Introduction

The increasing worldwide rates of overweight and obesity constitute a growing medical issue and a major public health burden [1]. The unprecedented access to high-caloric foods combined with a simultaneous reduction in physical activity and energy expenditure is leading to a long-lasting positive energy balance, resulting in weight gain. An increased body weight and high-fat diet lead to fat accumulation in peripheral tissues, such as the liver, causing non-alcoholic fatty liver disease (NAFLD). Hepatocytes' steatosis makes them less responsive to insulin and stimulates insulin resistance, as well as oxidative stress and mitochondrial damage [2–5]. To compensate insulin resistance in peripheral insulin-dependent tissues, the pancreas produces more insulin to maintain normal blood glucose levels. Over time, this compensatory mechanism may become insufficient, leading to elevated blood sugar levels and the eventual development of type 2 diabetes (T2DM). On the other hand, as an anabolic, insulin promotes the storage of fat, and when its levels are persistently high due to insulin resistance, it can lead to increased fat accumulation, particularly in visceral adipose tissue. This creates a vicious cycle, as excess body fat, especially visceral fat, further exacerbates insulin resistance, resulting in difficulty in losing weight or weight gain [6].

Additionally, insulin resistance hampers the body's ability to use glucose effectively for energy, often resulting in fatigue, postprandial hypoglycemia, and reduced physical activity, which further contributes to weight management issues. Addressing insulin resistance through lifestyle changes such as dietary interventions, increased physical activity, and weight management is essential for preventing its adverse health outcomes but is often fruitless [7,8]. That is why patients with insulin resistance or who are pre-diabetic seek pharmacological help.

The most widely used anti-diabetic drug that effectively reduces insulin resistance is metformin. By activating AMP-activated protein kinase (AMPK), metformin enhances cellular glucose uptake and utilization, particularly in muscle tissue, while simultaneously inhibiting hepatic glucose production [9]. This dual action helps lower blood glucose levels and improves insulin sensitivity. Additionally, metformin's influence on AMPK contributes to reduced fat accumulation and improved lipid metabolism, which further mitigates insulin resistance. Its efficacy, safety profile, and additional benefits, such as weight stabilization and potential cardiovascular protection, make metformin a cornerstone in the management of insulin resistance and T2DM [10–12].

Using a drug repurposing approach, we show that phloroglucinol has anti-diabetic and anti-oxidative properties and could be used in the treatment of non-alcoholic fatty liver disease (NAFLD).

However, in the last decade, numerous new compounds have been tested to treat insulin resistance and liver steatosis. Beginning with the resveratrol boom in the early XXI century, multiple polyphenolic substances, such as curcumin or enterolactone, have shown anti-diabetic, anti-oxidant, or anti-inflammatory properties in preclinical studies [13–17]. However, despite promising results in some cases, all these attempts failed to be introduced into clinical practice due to the poor pharmacokinetics of polyphenols and very low availability in humans [18]. That is why, in our project, we focused on phloroglucinol (PHG) (benzene-1,3,5-triol), an organic compound from the phenol group similar to resveratrol but characterized with good availability in vivo and anti-oxidative, anti-inflammatory and anti-glycating properties [19,20]. Furthermore, being used for a long time as a generic drug, PHG is well tolerated and affordable, which makes it a promising candidate for drug repurposing in the treatment of NAFLD and insulin resistance.

In the following study, we describe the anti-diabetic properties of PHG, an antispasmodic and analgesic drug also found in *Ecklonia cava* extracts. It reduced insulin resistance, liver steatosis, and mitochondrial oxidative stress in rat models of NAFLD, as well as ameliorated postprandial glucose tolerance and lipid metabolism in pre-diabetic men. As an example of translatory medicine, our study shows that drug repurposing may provide an efficient approach to discovering new therapeutic uses for existing medications.

2. Results

2.1. PHG Protects Wistar Rats from High-Fat-Diet-Induced NAFLD and Insulin Resistance

Ten weeks of a high-fat diet (HFD) led to a significant increase in mean caloric intake, resulting in higher body weight, liver steatosis, and insulin resistance in Wistar rats. Despite PHG treatment not preventing HFD-induced weight gain, it was sufficient to prevent its metabolic consequences, such as increased fasting glucose or impaired response to insulin in an insulin tolerance test (ITT) (Figure 1A,G,I,J). Moreover, PHG also ameliorated glycemic control in control diet rats, as seen in the change in the oral glucose tolerance test (OGTT) AUC (Figure 1E). These anti-diabetic effects of PHG treatment in the HFD group could be associated with a significant decrease in the histological grade of liver steatosis (Figure 1N,O).

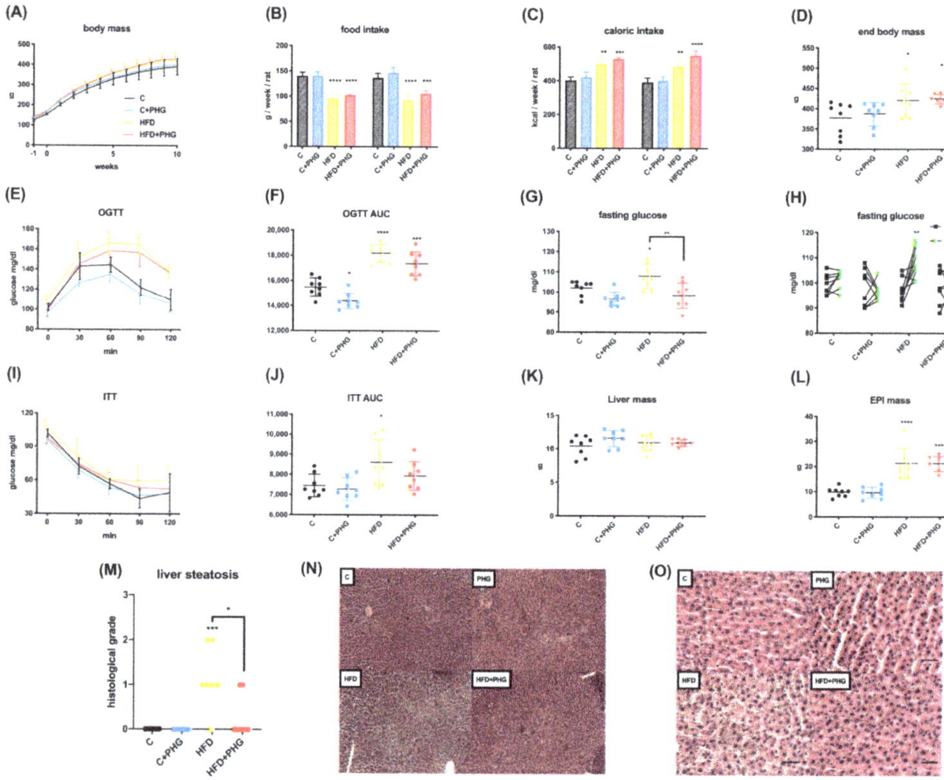

Figure 1. The effect of PHG treatment on insulin resistance in the NAFLD model. The effect of ten−week HFD feeding of Wistar rats on body weight (**A**). Mean food (**B**) and caloric (**C**) intake. End body weight (**D**). The effect of ten-week HFD and PHG treatment on oral glucose tolerance test (OGTT) (**E**), OGTT AUC (**F**), end-day fasting glucose (**G**), change in fasting glucose during the experiment (**H**), insulin tolerance test (ITT) (**I**), ITT AUC (**J**), liver mass (**K**), epididymal fat mass (**L**), histological liver steatosis (**M–O**). n = 32; * $p < 0.05$, ** $p < 0.01$, *** $p < 0.001$, **** $p < 0.0001$. If not indicated otherwise * refers to control vs. particular group. For H * refers to day 0 vs. end day. Scale bar refers to 50 μm.

2.2. The Effect of PHG Is Not Related to Hepatic Lipogenesis or AMPK Mediated

One of the possible explanations for the results observed in Wistar rats could be due to AMPK activation, resulting in decreased lipogenesis in the liver. To assay that potential, the effect of PHG on lipogenesis was assessed via the incorporation of 14C-acetate tracer into the lipids in control and AMPK KO mice hepatocytes during a 3 h treatment. In both the control and AMPK-deficient hepatocytes, PHG had no significant effect on lipid synthesis up to 100 µM. At very high concentrations (300 µM and 1 mM), lipid synthesis was inhibited by barely 20%, similarly in control and AMPK KO hepatocytes (Figure 2A). In contrast, the treatment of hepatocytes with AICAR at 200 µM, a well-known AMPK activator, inhibited lipogenesis by >95% in control, and this effect was fully blunted in AMPK KO hepatocytes, demonstrating that AICAR inhibits lipogenesis in an AMPK-dependent manner. Western blot analysis showed that treatment of control hepatocytes with PHG for 3 h did not increase AMPK phosphorylation on Thr172 or the phosphorylation of its target ACC on Ser79, even at concentrations as high as 1 mM. In contrast, AICAR at 200 µM induced robust phosphorylation of AMPK and ACC (Figure 2B). In conclusion, it seems clear that in hepatocytes, PHG does not affect lipid synthesis (<20% at high concentrations >300 µM), and it is not an AMPK activator, even at concentrations as high as 1 mM. Thus, the observed decrease in liver steatosis in rats in response to PHG does not seem to be mediated by lipogenesis inhibition or by AMPK activation in hepatocytes.

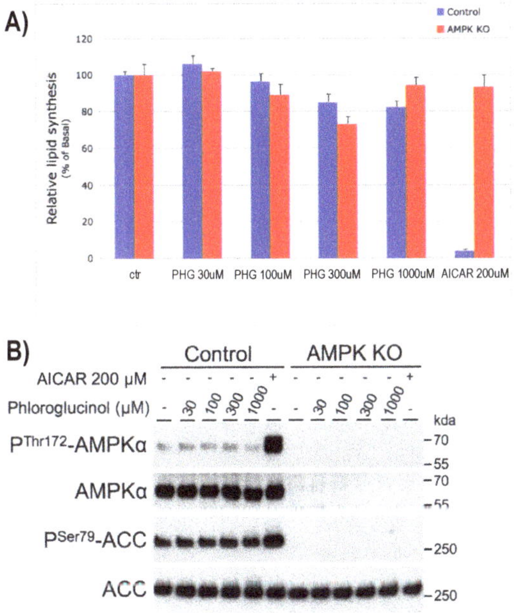

Figure 2. Effect of phloroglucinol on lipid synthesis in control and AMPK-deficient primary mouse hepatocytes. The effect of PHG and AICAR on de novo lipid synthesis (**A**) and phosphorylation of AMPKα at Thr172, and ACC at Ser79. A representative picture of the Western blotting membrane is shown (**B**). n = 3.

2.3. PHG Protects Mitochondria from Oxidative Stress

Since the effect of PHG on liver steatosis and insulin resistance was not AMPK mediated or related to lipid synthesis, we focused on its potential effect on mitochondrial function, as they are responsible for lipid oxidation. To assess that, we used two cell culture models: HepG2 and differentiated 3T3-L1, which were exposed to a 0.75 mM palmitic acid (PA) medium for 24 h to induce steatosis and oxidative stress. The exposure to PA significantly inhibited the activity of both complexes I and II + III, which was partly reversed in the presence of PHG. The reduced activity of respiratory complexes was accompanied by increased ROS production and an ADP/ATP ratio in the group exposed to PA (Figure 3). In summary, these results indicate PA-induced impairment of mitochondrial function, manifested by reduced energy production and increased electron leakage, leading to intensified free radical formation.

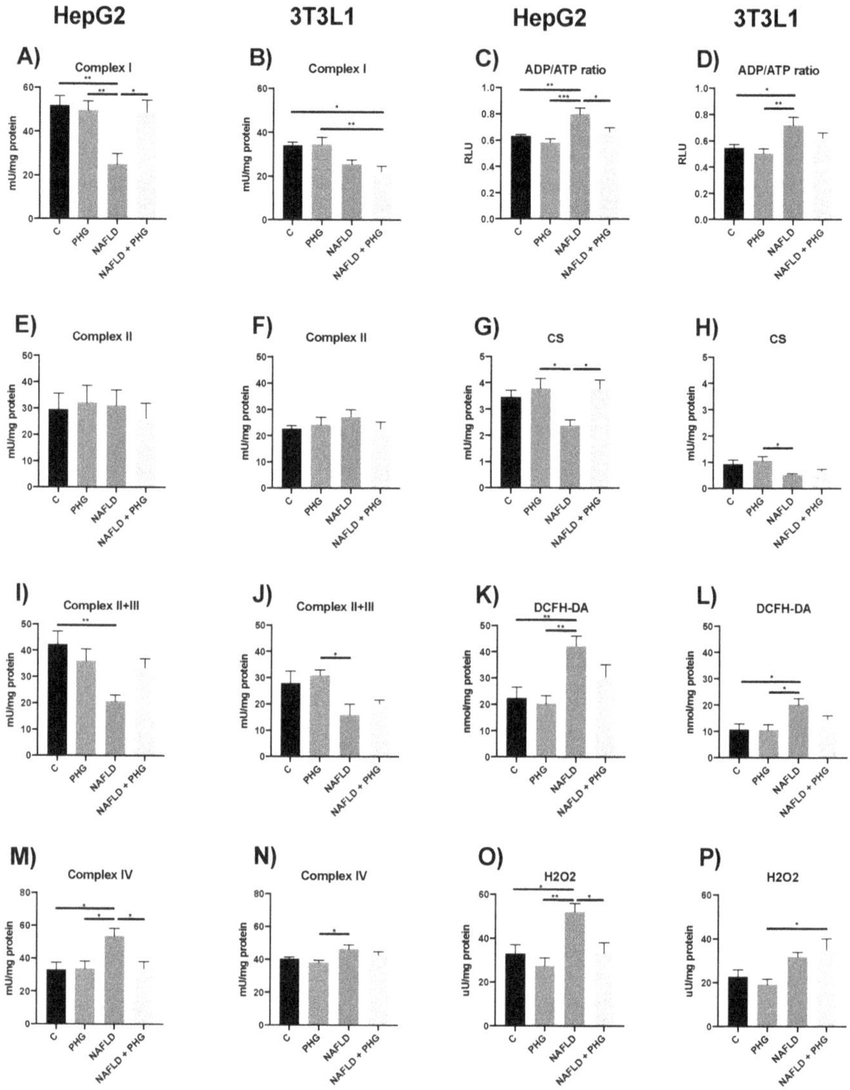

Figure 3. The effect of PHG on mitochondrial function. The effect of PHG on mitochondrial complex I (**A**,**B**), complex II (**E**,**F**), complexes II+III (**I**,**J**) and complex IV (**M**,**N**) activity, ATP/ADP ratio (**C**,**D**), CS (**G**,**H**), DCFH-DA (**K**,**L**) and H_2O_2 production (**O**,**P**) in mitochondria isolated from HepG2 and 3T3L1 model of NAFLD respectively. n = 3. * $p < 0.05$, ** $p < 0.01$, *** $p < 0.001$.

Unsurprisingly, the increased ROS production resulted in elevated mitochondrial damage, which could have been prevented when PHG was added to the culture media. The drug normalized the concentration of both reduced and oxidized glutathione and restored a control redox ratio. Furthermore, PHG increased the activity of enzymatic anti-oxidants: GSH-Px and SOD in mitochondria, which was inhibited by PA (Figure 4). Taken together, our findings suggest that PHG normalizes mitochondrial function and protects them from oxidative damage. Its effect seems to be multifactorial, affecting cellular steatosis, mitochondrial complexes' activity, and enzymatic anti-oxidants. As a downstream effect of PHG protective activity, we observed a decrease in TNF-a and IL-1 concentrations in media and the normalization of key regulators of apoptosis: Bax and Bcl-2 in both analyzed cell lines (Figure 5).

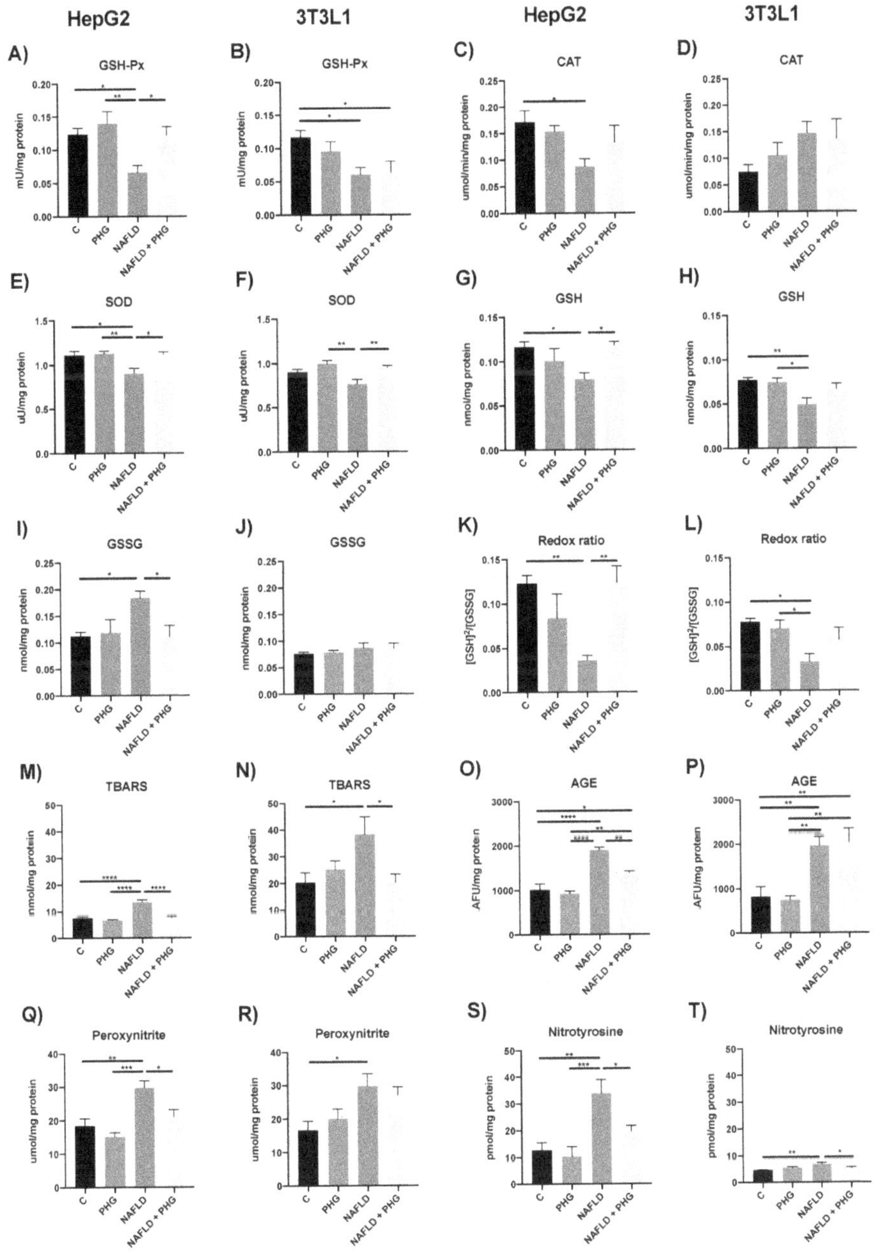

Figure 4. The effect of PHG on mitochondrial enzymatic anti-oxidants, redox ratio and oxidative damage. The effect of PHG on mitochondrial activity of GSH-Px (**A**,**B**), CAT (**C**,**D**), SOD (**E**,**F**), concertation of GSH (**G**,**H**) and GSSG (**I**,**J**), redox ratio (**K**,**L**), TBARS (**M**,**N**), AGEs (**O**,**P**), peroxynitrite (**Q**,**R**) and nitrotyrosine production (**S**,**T**) in mitochondria isolated from HepG2 and 3T3L1 model of NAFLD respectively. n = 3. * $p < 0.05$, ** $p < 0.01$, *** $p < 0.001$, **** $p < 0.0001$.

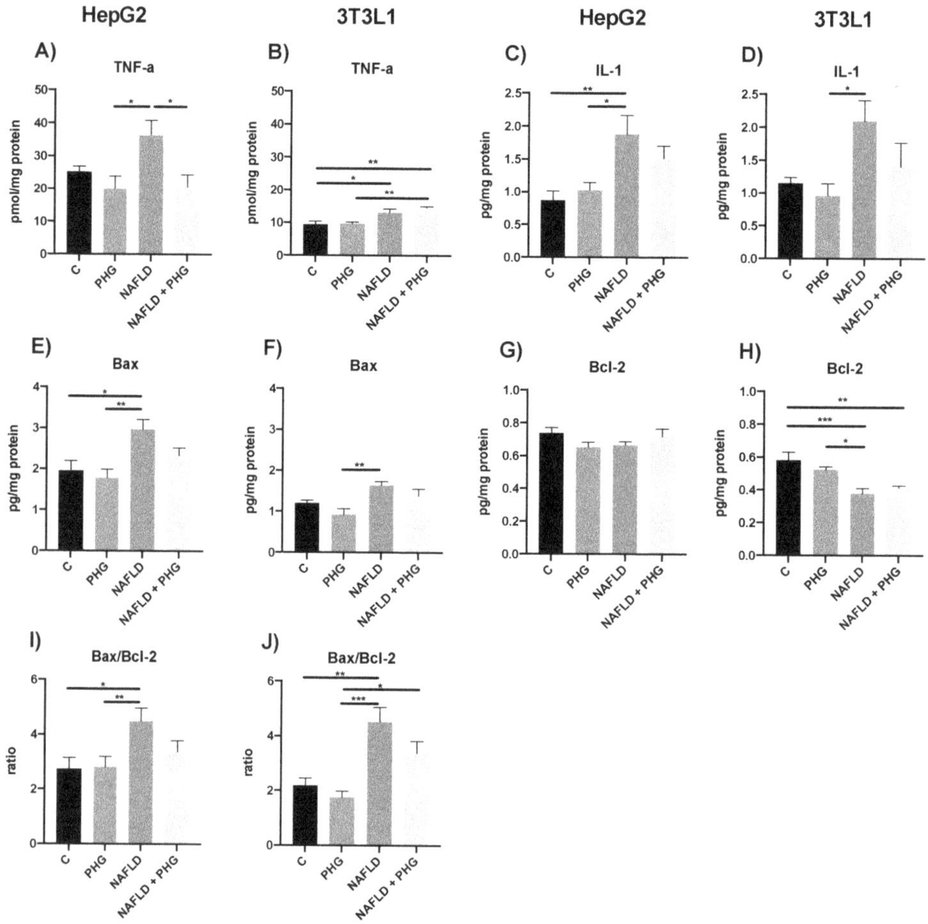

Figure 5. The effect of PHG on inflammation and apoptosis. The effect of PHG on media TNFα (**A**,**B**) and Il-1 (**C**,**D**) concentration, as well as mitochondrial expression of Bax (**E**,**F**), Bcl-2 (**G**,**H**) and Bax/Bcl-2 ratio (**I**,**J**) in mitochondria isolated from HepG2 and 3T3L1 model of NAFLD respectively. n = 3. * $p < 0.05$, ** $p < 0.01$, *** $p < 0.001$.

2.4. PHG Relaxes Gastric Smooth Muscles

Since PHG is registered as an analgesic and spasmolytic drug, another possible explanation for its effects on glycaemic control could be gastric muscle relaxation, resulting in delayed gastric emptying and food passages. Interestingly, despite PHG being used as a spasmolytic drug for decades, the molecular mechanism of its action has not been clearly explained so far. Thus, to answer whether PHG affects gastric smooth muscle contractility and to look for a molecular explanation of its effects, we established an ex vivo gastric smooth muscle bath culture. In the muscle strips that were isolated from the upper half of the human stomach, carbachol, in the concentration of 10^{-6} mol/L, promoted a noticeable and stable muscle contractility (Figure 6A). Typical tracings show the response of gastric smooth muscle strips when cumulatively applied to PHG (Figure 6B).

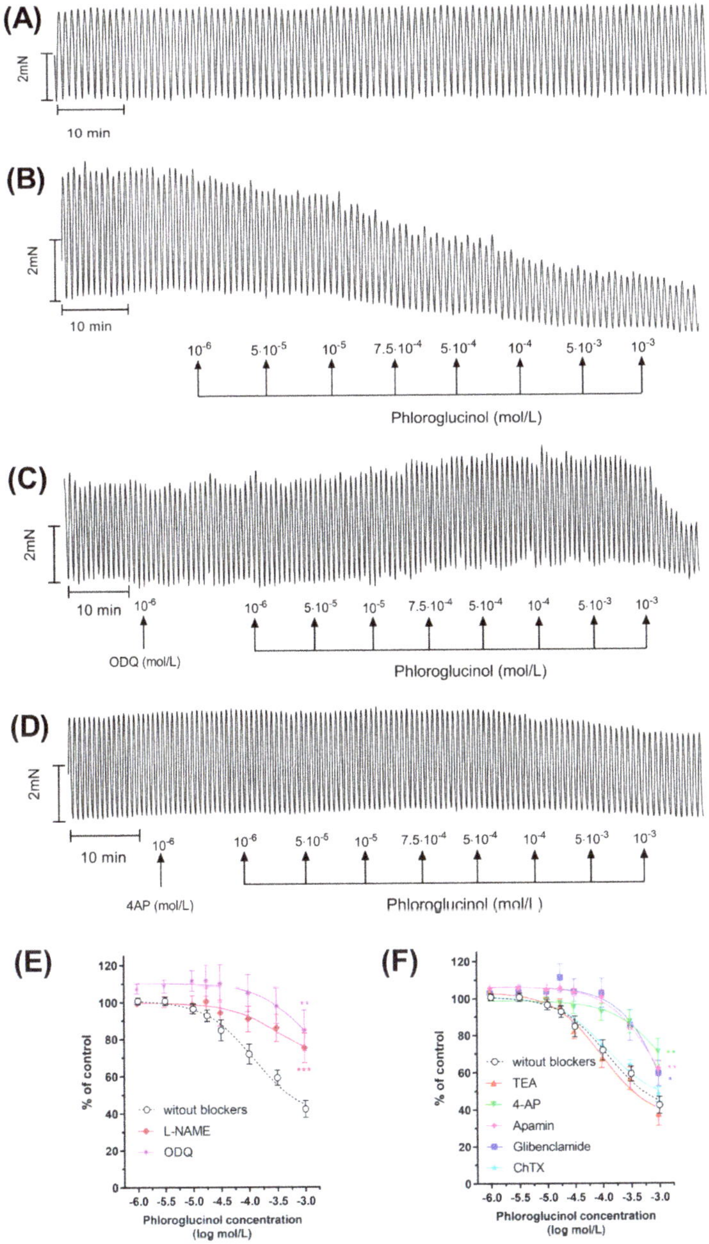

Figure 6. PHG relaxes gastric smooth muscles. A typical recording of carbachol-induced contractile activity of the human gastric strips (**A**) and the effect of cumulatively administered PHG (range 10^{-6}–10^{-3} mol/L) (**B**). Representative recording for the blocking effect of ODQ (**C**) and 4-AP (**D**) pre-treatment on PHG-induced muscle relaxation. Effects of PHG after preincubation with L-NAME, ODQ (**E**); and TEA, 4-AP, ChTX, glibenclamide, or apamin (**F**) on the gastric strips, as measured by AUC. Each point represents the mean ± SEM of values obtained from individual gastric strips (n = 10) from ten different patients. Contractions of the gastric strips before phloroglucinol were treated as controls. * $p < 0.05$, ** $p < 0.01$, *** $p < 0.001$ versus PHG alone.

Pre-treatment with apamin, a specific blocker of the SK_{Ca} channel, altered PHG-induced relaxation significantly at all concentrations of the drug. Similarly, preincubation with a K_{ATP} channel blocker, glibenclamide, or a K_V channel blocker, 4-AP, caused considerable inhibition of muscle contractions (Figure 6D). A statistically significant shift to the right from the concentration–response curve for PHG observed after preincubation with both apamin, glibenclamide, or 4-AP showed a noncompetitive antagonism of tissue response to high efficacy (receptor reserve present). However, PHG-induced relaxation was not inhibited by non-selective K+ channel inhibitors TEA and ChTX, inhibitors of BK_{Ca} and IK_{Ca}, respectively (Figure 6F).

Since NO, as one of the most significant NANC neurotransmitters in the GI tract, controls smooth muscle relaxation by cGMP-dependent or -independent mechanisms, we decided to test its relationship with PHG. The data presented clearly show that after preincubation with a guanylate cyclase blocker—ODQ—a statistically significant decrease in AUC was observed at all PHG concentrations (Figure 6C). In addition, PHG-induced relaxation was blocked by NO inhibition with L-NAME. Both effects were accompanied by a statistically significant shift to the right of the concentration–response curves for PHG, indicating that gastric strip relaxation caused by it is also related to NO production (Figure 6E). All Log EC50 and Emax data are summarized in Table 1. These results indicate that PHG relaxes gastric smooth muscles by activating certain types of potassium channels, SK_{Ca}, K_{ATP}, and K_V channels, as well as through NO signaling.

Table 1. Log EC50 and Emax for PHG on carbachol-induced contractility of the human gastric muscles. The values are mean ± SEM of n = 10 individual gastric strips from different patients. * $p < 0.05$, ** $p < 0.01$, *** $p < 0.001$, **** $p < 0.0001$ versus phloroglucinol alone.

	$logEC_{50}$	p-Value	E_{max}	p-Value
Phloroglucinol alone	−3.92 ± 0.14		42.27 ± 4.54	
Phloroglucinol after preincubation with				
L-NAME	−3.50 ± 0.60 *	0.021	75.15 ± 7.92 ***	p < 0.001
ODQ	−2.98 ± 1.17 *	0.0106	84.64 ± 11.12 **	0.0082
4-AP	−2.94 ± 0.57 ****	p < 0.0001	70.95 ± 7.06 **	0.0049
ChTX	−3.92 ± 0.11	0.472	48.11 ± 3.84	0.5887
Apamin	−3.00 ± 0.22 ****	p < 0.0001	62.11 ± 3.64 **	0.0077
Glibenclamide	−2.65 ± 0.82 ****	p < 0.0001	59.47 ± 6.87 *	0.0408
TEA	−4.01 ± 0.14	0.1001	37.49 ± 6.25	0.516

2.5. *PHG Ameliorates Postprandial Glucose Tolerance and Calorimetry*

Because acute exposure to PHG induced gastric muscle relaxation, we wanted to verify if this effect can affect postprandial glycemic control, portal flow, and calorimetry in humans. To measure that, we recruited 15 pre-diabetic male volunteers to conduct a pilot, double-blinded, single-dosage, crossover intervention. We conducted a challenge test with a high-carbohydrate meal intake, and surprisingly, even after a single dosage of PHG, we observed a significantly lower postprandial peak of glycemia and insulin concentrations (Figure 7B,F). However, it has to be noted that due to high variability between participants, statistically significant changes in postprandial insulin concentrations were observed only if results were expressed as a % change from the baseline (fasting) insulin concentration for each individual. Nevertheless, considering the small group size and acute exposure, it seems promising to assess the anti-diabetic effects of chronic PHG treatment in future clinical trials in pre-diabetic or T2DM patients. Since it is unlikely that a single dosage affected insulin resistance, the observed effect on postprandial glycemic control might support our hypothesis on delayed gastric emptying and food passages. PHG did not affect the portal vein diameter or portal flow (Figure 7E,I). On the other hand, pre-treatment with PHG led to slightly increased postprandial energy expenditure and oxygen consumption with a pronounced yet not significant trend toward lipid oxidation (Figure 7J–R).

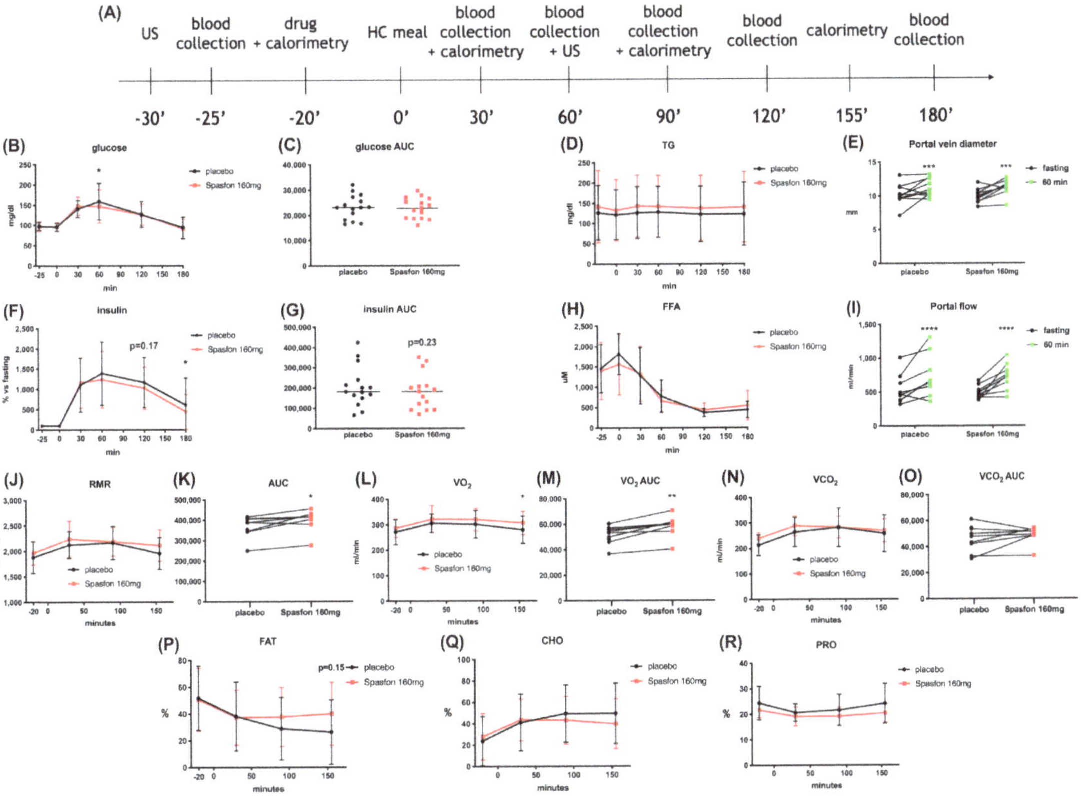

Figure 7. The effect of single dose Spasfon 160 mg on postprandial glycemic control and calorimetry in pre-diabetic male volunteers. Meal test and IMP intervention visit design (**A**). Postprandial glucose level (**B**), glucose AUC (**C**), triglyceride (**D**), insulin (**F**), insulin AUC (**G**), free fatty acid level (**H**). Postprandial change in portal vein diameter (**E**) and portal flow (**I**). Postprandial energy expenditure (resting metabolic rate (RMR)) (**J**), RMR AUC (**K**), oxygen consumption (**L**), oxygen consumption AUC (**M**), carbon dioxide production (**N**), carbon dioxide production AUC (**O**), percentage of fat (**P**), carbohydrates (**Q**) and protein (**R**) in substrates utilization. For biochemical assays n = 15, for calorimetry and US examination n = 10. * $p < 0.05$, ** $p < 0.01$, *** $p < 0.001$, **** $p < 0.0001$ placebo vs. Spasfon 160 mg. For portal diameter and flow * refers to fasting vs. postprandial.

2.6. PHG Affects Postprandial Lipid Metabolism

To screen for possible metabolic effects of PHG, we applied untargeted metabolomics in plasma samples collected during the high-carbohydrate meal test. Pathway analysis was performed for 29 unique annotated metabolites. The analysis revealed that the treatment highly affected the glycerophospholipid, sphingolipid, and linoleic acid metabolism (Figure 8A). Almost all metabolites differentiating between placebos and the drug in postprandial metabolomics were classified as lipid: 64% belonged to the glycerophospholipids class, 10% were sphingolipids, 10% were fatty acyls, and 2% were sterol lipids as well as glycerophospholipids (Figure 8B). Within fatty acyls, carnitines (CAR 6:0, CAR 12:1, and CAR 14:1) decreased after the drug administration, while aminopentanoic acid increased. Lysophosphatidylcholines (LPC 18:0, LPC 18:1, LPC 18:2 and LPC 18:3) increased after drug administration similarly to ether-lysophosphatidylcholines (LPC O-16:0, LPC O-18:2/LPC P-18:1 and LPC O-18:2/LPC P-18:1). In contrast, phosphatidylcholines and ether-phosphatidylcholines behaved differently: most phosphatidylcholines (PC 36:4, PC 37:4, PC 38:5, PC 40:7) decreased, while the majority of ether-phosphatidylcholines (PC O-34:2/PC P-34:1, PC O-38:67/PC P-38:6 and PC O-21:0) increased after the drug administration. Glycerophosphatidylethanolamines behaved similarly to the glycerophosphocholines: lysophosphatidylethanolamines (LPE 18:1 and LPE 18:3) increased after the drug administration, while diacyl glycerophosphoethanolamines (PE 38:1 and PE 38:4) decreased. Interestingly, all discriminating LPCs and LPEs were annotated as sn-1 forms. Within sphingolipids both sphingomyelins (SM(d33:1), SM(d34:0), SM(d38:2)), and galactosylceramide (d34:1) decreased after drug administration. Finally, only one sterol lipid was found to significantly differ between groups, cholesterol, slightly decreasing in the plasma of patients receiving the drug. Taken together, the metabolomics results suggest that PHG may affect the activity of phospholipase A1 (PLA_1) since it decreased PCs and increased LPCs.

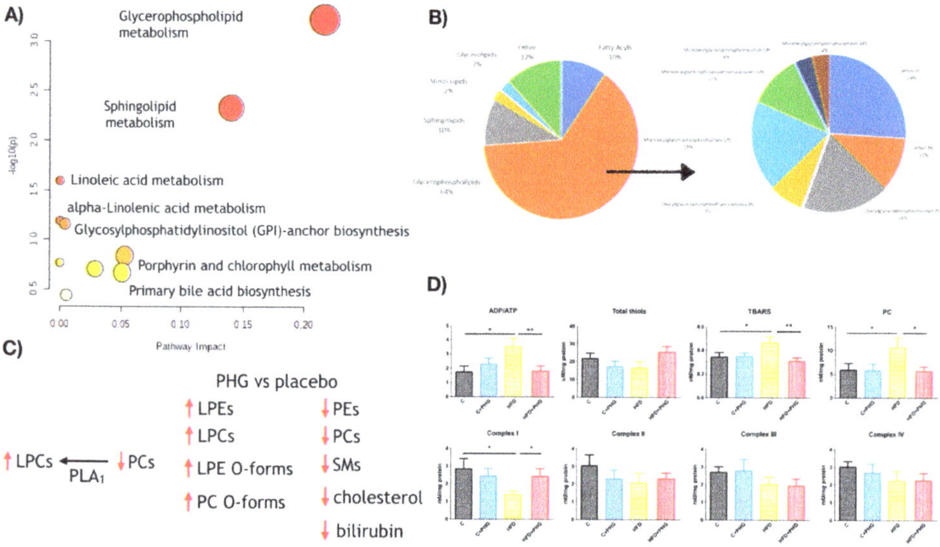

Figure 8. The effect of PHG on lipid metabolism. Pathway analysis of the effects of PHG on postprandial metabolomics (**A**) and the classification of significantly changed metabolites (**B**). Main changes in serum metabolites induced by a single-dosage PHG administration in pre-diabetic men (**C**). The effects of ten-week PHG treatment on mitochondrial activity and oxidative damage in the liver of high-fat-fed Wistar rats (**D**). For human metabolomics data n = 10, for Wistar NAFLD model n = 32; * $p < 0.05$, ** $p < 0.01$.

Similarly, increased ether forms of PCs and LPCs may confirm its anti-oxidative effects in vivo since these modified phospholipids served as cellular anti-oxidants (Figure 8C). The metabolomic results and postprandial calorimetry seem to be in line with the results observed in the rat model of NAFLD. Where chronic PHG treatment restored mitochondrial complexes' activities and ATP production in liver mitochondria what was disrupted by HFD-feeding. However, when it was combined with the anti-oxidative properties of PHG, it led to a significant decrease in mitochondrial oxidative damage (Figure 8D).

3. Discussion

Drug repurposing offers a promising strategy for discovering new therapeutic applications, leveraging the benefits of reduced development time, cost, and established safety profiles [21,22]. In this study, we explored the repurposing of PHG, a generic drug registered as an analgesic and spasmolytic, for its effects on insulin resistance and liver steatosis in a Wistar rat model of non-alcoholic fatty liver disease (NAFLD) and pre-diabetic men. This translational research approach revealed significant findings on the protective effects of PHG against high-fat-diet (HFD)-induced metabolic disorders.

Our study demonstrated that PHG effectively prevents the development of HFD-induced liver steatosis and insulin resistance in Wistar rats. In contrast to previously published results in C57BL/6J mice, we found that this protective effect of PHG is not associated with AMP-activated protein kinase (AMPK) activation or lipid synthesis pathways in the liver [23]. Instead, PHG exhibits strong anti-oxidative, anti-glycative, and anti-inflammatory properties, which protect mitochondrial function by reducing oxidative stress, restoring ATP production, and maintaining redox balance. This is in line with our previous reports on anti-glycative and anti-oxidant properties of PHG in vitro [19,20]. In this study, we showed that PHG also mitigates oxidative damage at the mitochondrial level. The effect of PHG is more pronounced in insulin-resistant hepatocytes than in adipocytes. Additionally, PHG was found to relax gastric smooth muscles through multiple pathways, including SK_{Ca}, K_{ATP}, and K_V channels, as well as through NO synthesis and signaling. Our observations on gastric track motility should not be surprising since PHG is recommended as an antispasmodic for smooth muscles in the treatment of abdominal pain [24–26]. Although it has long been used in clinical practice as an antispasmodic for painful conditions of the urogenital and gastrointestinal (GI) tract, early in vivo studies in anesthetized rats demonstrated that PHG did not affect contractions of the duodenum, ileum, and colon [27]. This may suggest that its action is limited to gastric smooth muscles. Thus, its spasmolytic activity may delay gastric emptying, potentially contributing to a slower food passage and improved postprandial glycemic control. However, to better understand the effect of PHG on GI tract motility and food digestion, it would be necessary to assay the profile of postprandial incretin hormones, which was not possible in our study. On the other hand, in irritable bowel syndrome (IBS), PHG has been shown to reduce glycerol-induced abdominal pain and inhibit colonic phasic contractions without affecting the colonic tone [28].

In a single-dosage crossover intervention study in pre-diabetic men, PHG significantly improved postprandial glycemic control and increased energy expenditure. Untargeted metabolomics pathway analysis revealed that PHG treatment significantly impacts glycerophospholipid, sphingolipid, and linoleic acid metabolism. These changes suggest that PHG may affect the activity of phospholipase A1 (PLA1), as evidenced by decreased PCs and increased LPCs. The rise in ether forms of PCs and LPCs further supports the anti-oxidant effects of PHG in vivo, as these modified phospholipids act as cellular anti-oxidants. Similarly, in a recent study assessing the oxidative stress in retinal pigment epithelium cells, PHG attenuated DNA damage and apoptosis through the inhibition of mitochondrial ROS production [29].

The metabolomics findings align with our observations from the rat NAFLD model, where chronic PHG treatment restored mitochondrial complex activity and ATP production in liver mitochondria disrupted by HFD feeding. This restoration, combined with PHG's anti-oxidative properties, led to a significant reduction in mitochondrial oxidative damage.

PHG's ability to prevent liver steatosis and insulin resistance, coupled with its impact on mitochondrial function, underscores its potential as a therapeutic agent for metabolic disorders. The relaxation of gastric smooth muscles and the subsequent delay in gastric emptying likely contribute to improved metabolic outcomes. The enhanced postprandial glycemic control and energy expenditure observed in pre-diabetic men further highlight the translational potential of PHG.

Despite the promising results, the precise molecular mechanisms underlying PHG's metabolic benefits warrant further investigation. While our study did not find a link between PHG's effects and AMPK activation, future research should explore other potential targets and pathways that mediate its actions. Additionally, the impact of PHG on other metabolic tissues, such as muscle and adipose tissue, should be examined to fully understand its systemic effects. Nevertheless, our findings indicate that PHG has a multifactorial role in mitigating metabolic dysfunctions associated with NAFLD and insulin resistance. By protecting mitochondrial integrity, enhancing anti-oxidative defenses, and modulating lipid metabolism, PHG emerges as a candidate for drug repurposing in the treatment of T2DM and NAFLD. Further functional and clinical evaluations are needed to confirm its efficacy in broader patient populations.

In conclusion, the repurposing of PHG exemplifies how existing drugs can be leveraged for new therapeutic uses, potentially offering a cost-effective and expedited route to addressing pressing health issues such as NAFLD and insulin resistance. The insights gained from this study pave the way for future investigations into PHG's mechanisms of action and its potential clinical applications.

Nevertheless, our study has some limitations. We have not assessed the effect of long-term PHG treatment on lipid profiles, hepatic lipid transporters, and lipid metabolism, nor have we assessed the drug's effect on postprandial incretin signaling. However, our work provides a good starting point for further clinical investigations that could answer at least some of these concerns and verify whether PHG can be used in the treatment of T2DM.

4. Materials and Methods

4.1. Animals

Five-week-old Wistar male rats (120–140 g) were purchased from and housed in the Experimental Medicine Center of the Medical University of Bialystok, Bialystok, Poland. All animal experiments and procedures on Wistar rats were performed according to Polish animal care regulations upon acceptance of the Local Ethical Committee in Olsztyn, Poland (22/WPN/2021). The rats were housed under ambient temperatures: 20–24 °C; 12 h of daylight/darkness cycles; 55% ± 10% humidity, two rats per cage. After 5 days of acclimatization, the rats were randomly divided into 4 groups of 8 animals each. The rats were closely monitored for behavioral changes or other adverse events to ensure animal safety. On completing the experiment, the rats were euthanized using CO_2 with the volatile method, and organs of interest were swiftly retrieved for H&E staining, mitochondria isolation, and molecular assays.

4.2. High-Fat Diet (HFD) Treatment

To induce insulin resistance and liver steatosis, the rats in experimental groups were fed with a high-fat diet (Research Diets, New Brunswick, NJ, USA, D12492) composed of 59.8% fats, 20.1% proteins and 20.1% carbohydrates for 10 weeks, while control groups were fed a standard chow diet (Agropol, Motycz, Poland) composed of 10.3% fats, 24.2% proteins and 65.5% carbohydrates for 10 weeks. All animals had ad libitum access to chow and tap water. The control group (C) received a control chow diet and 0,9% NaCl intragastric injection daily, and the PHG group received a control chow diet and 100 mg/kg of PHG intragastric injection daily, the HFD group received a high-fat diet and 0,9% NaCl intragastric injection daily while HFD + PHG group received high-fat diet and 100 mg/kg of PHG intragastric injection daily. The PHG dosage was chosen based on a literature search [23].

4.3. AMPK Knock-Out Hepatocytes and Lipid Synthesis Assay

The effects of phloroglucinol on lipid synthesis and AMPK activation were assessed in control and AMPK KO hepatocytes isolated from AMPK$\alpha1^{lox/lox}$, $\alpha2^{lox/lox}$ (control) and AMPK$\alpha1^{lox/lox}$, $\alpha2^{lox/lox}$-Alfp-Cre (liver AMPK KO) mice as described previously [30]. The effect of phloroglucinol (at 30, 100, 300, and 1000 µM) on lipogenesis flux was assessed in triplicate via the incorporation of [1-^{14}C]-acetate tracer into lipids in control and AMPK KO hepatocytes during a 3 h treatment and compared with AICAR at 200 µM, a well-known AMPK activator, as described previously [31]. AMPK phosphorylation on Thr172, as well as its target, ACC, and phosphorylation on Ser79 were analyzed using standard Western blotting [30].

4.4. Oral Glucose Tolerance Test (OGTT) and Insulin Tolerance Test (ITT)

Fasting glucose was measured on day 0 and the last day of the experiment. An oral glucose tolerance test (OGTT) was performed on the last day of the experiment upon overnight fasting. All rats received glucose (2 g/kg of body weight) in 0.9% saline solution via intragastric tube. Blood glucose level was measured before glucose administration and 30, 60, 90, and 120 min after. Insulin tolerance test (ITT) was performed 6 days before the end of the experiment upon overnight fasting. All rats received insulin (1 IU/kg of body weight i.p. NovoRapid (Novo Nordisk, Bagsværd, Denmark). Blood glucose level was measured before insulin administration and 30, 60, 90, and 120 min after. All glucose level measurements were performed in tail vein blood using an AccuCheck glucometer (Roche, Basel, Switzerland).

4.5. Liver Histology

Freshly dissected liver lobes were fixed in 4% formaldehyde (SigmaAldrich, Warsaw, Poland) overnight and replaced with 70% and 95% ethanol before paraffin embedding and tissue cutting. The slides were stained with hematoxylin and eosin (H&E) and imaged under the light microscope. A total of 2–3 liver sections per animal were assessed in a random order by a blinded pathomorphologist from the Department of Medical Pathomorphology of the Medical University of Bialystok, Poland.

4.6. Cell Culture and Mitochondria Isolation

The experimental procedures were conducted on differentiated 3T3-L1, and HepG2 cells seeded on 6-well plates. At the beginning of the experiment, HepG2 cells had a confluence of 90%. 3T3-L1 were differentiated for 8 days, as previously described [32]. To induce the NALFD model, cells were exposed to 0.75 mM of palmitic acid (PA) for 24 h. The PA was dissolved in a growth medium (DMEM, 10%FBS, 1% penicillin/streptomycin) as previously described [15,33]. Cells were divided into 4 experimental groups: control, 100 μM PHG, 0.75 mM PA, 0.75 mM PA$^+$ 100μM PHG. After the experiment, cells were homogenized (1:10, *w/v*) with a Teflon-on-glass electric homogenizer in an ice-cold mitochondria isolation buffer (250 mM sucrose, 5 mM Tris-HCl, and 2 mM ethylene glycol bis(2-aminoethyl)tetraacetic acid) [34,35]. To prevent proteolysis, protease and phosphatase inhibitors (Roche Diagnostics GmbH, Germany) were added. The homogenate was centrifuged (500× *g*, 10 min, 4 °C), and the resulting supernatant was centrifuged twice at 8000× *g* for 10 min at 4 °C [36]. The mitochondria pellet was resuspended in an isolation buffer and immediately processed [37].

4.7. Muscle Relaxation Assay

4.7.1. Sample Processing

The study was conducted under the Helsinki Declaration principles, the International Conference on Harmonisation Guideline for Good Clinical Practice, the laws and regulations of Poland, and with the approval from the Ethical Committee of Medical University of Białystok, Poland (No. R-I-002/304/2018). Tissues were obtained from patients undergoing sleeve gastrectomy due to morbid obesity (n = 10 aged 25–51; BMI 47.31 ± 1.11). The experiment did not affect the course of the operation. Samples were taken from the upper half of the stomach with larger curvature removed during the surgical procedure [38,39]. All patients were carefully informed about the aim and nature of the study before surgery and signed written consent.

4.7.2. Experimental Protocol

After the equilibration period, contractile activity was stimulated using carbachol (10^{-6} mol/L). The contractile activity of strips incubated only with carbachol was considered a control after reaching the plateau. To examine concentration–response relationships, PHG was added cumulatively to the organ chambers (range 10^{-6}–10^{-3} mol/L) at 10-min intervals, and the effects were recorded. To verify the role of NOS in the effect of resveratrol 10^{-6} mol/L of N^G-Methyl-L-arginine (N-Nitroarginine methyl ester, L-NAME), a NO synthase (NOS) blocker and 10^{-6} mol/L ODQ (soluble guanylate cyclase blocker, sGC) were used. Concentration–response curves to PHG were also constructed in the absence and presence of 10^{-3} mol/L tetraethylammonium chloride (TEA), a non-selective K^+ channel blocker; 10^{-9} mol/L charybdotoxin (ChTX), an inhibitor of high conductance Ca^{2+}-dependent (BK_{Ca}), intermediate (IK_{Ca}) conductance calcium-activated potassium channels (K_{Ca}), and slowly inactivating voltage-gated K_{Ca} channels; 10^{-3} mol/L 4-Aminopiridine (4AP)—a voltage-gated K^+ (K_v) channel blocker; 10^{-6} mol/L apamin, a small conductance K_{Ca} (SK_{Ca}) channel blocker or 10^{-6} mol/L glibenclamide, a K^+ATP-dependent (K_{ATP}) channel blocker. As far as possible, experiments were performed with strips from the same stomach sample and were studied in parallel. Appropriate controls were run under similar experimental conditions obtained from the same patient.

4.7.3. Measurement of Contraction Parameters

Gastric muscle activity was recorded by an isometric force transducer with digital output (BIO-SYS-TECH, Białystok, Poland) and calculated with the DASYLab software unit (version 9.0; Laboratory Data Acquisition System, SuperLogics, Waltham, MA, USA). The area under the curve (AUC) reflected the total change over time, representing the contractile activity of stomach muscle responses before and after the administration of the given drug. The AUC was measured as the area under all recorded contractions over a 10 min interval before the addition of each agonist or antagonist, respectively [40,41]. Values from three to four strips from each sample were averaged at each time point for each drug dose. The AUC of contractions of each muscle strip over a 10 min interval before the addition of PHG was treated as a control (100%).

4.8. Single Dose Crossed Intervention

The interventional human study was conducted under the Helsinki Declaration principles, the International Conference on Harmonisation Guideline for Good Clinical Practice, the laws and regulations of Poland, and with the approval from the Ethical Committee of Medical University of Bialystok, Bialystok, Poland (No.APK.002.375.2021). Briefly, 15 male volunteers with impaired fasting glucose (IGF) or impaired glucose tolerance (IGT) were recruited into the randomized, placebo-controlled, double-blinded, single-dose crossover intervention. Taking into consideration the fact that the investigated parameters may be characterized by sexual dimorphism and other factors may have an impact on their concentrations depending on the phase of the menstrual cycle; therefore, only male participants were enrolled in the study [42]. The study included 3 visits: a screening visit and two visits with an intervention. During the screening visit, the inclusion and exclusion criteria were evaluated based on medical history, physical examination, vital signs measurements as well as on laboratory results, including fasting hematology, lipid profile, C-reactive protein (CRP), creatine, aspartate transaminase, alanine transaminase and oral glucose tolerance test (OGTT) conducted accordingly with the American Diabetes Association (ADA) recommendations [43]. OGTT was performed only in individuals without previously known diabetes history; otherwise, only fasting glucose concentration was evaluated. Volunteers enrolled in the study who met inclusion criteria and did not meet exclusion criteria were instructed to maintain their regular lifestyle throughout the whole study participation and to avoid intensive physical activity, intake of alcohol, and food rich in polyphenols (tea, coffee, chocolate, etc.). The main study consisted of two visits during which participants received an investigational medicinal product (IMP) or placebo in a random order, and meal tests were performed. The interventional visits were conducted using a crossover method. Therefore, IMP and placebo interventions were performed in the same individuals, with at least a two-week wash-out period between interventions.

4.8.1. IMP/Placebo

Individuals received both IMP and placebo capsules at different visits in random order. The capsules containing IMP and placebo were prepared in the laboratory by a trained pharmacist just before dispensing to the participant during a meal challenge test. In the gelatin capsule, 160 mg of Spasfon Lyoc (Teva, Tel Awiw, Izrael), containing 160 mg of phloroglucinol dihydrate or a placebo made of starch, was placed. Placebo capsules were visually identical to IMP capsules and were dispensed in identical packaging. Consequently, investigators, site staff, and participants remained blinded throughout the study.

4.8.2. Biochemical Analysis

The blood for measurements of blood glucose, triglycerides, free fatty acids, insulin, and metabolite levels was collected 30, 60, 120, and 180 min after meal intake. The samples were prepared following the manufacturer's laboratory kit instructions. The plasma glucose concentrations were evaluated using the hexokinase enzymatic method (using Cobas c111, Roche Diagnostics Co., Ltd., Basel, Switzerland). Serum insulin concentrations were assessed using an immunoradiometric assay (INS-Irma, DIA Source S.A., Ottignies-Louvain-la-Neuve, Belgium; using Wallac Wizard 1470 Automatic Gamma Counter, PerkinElmer Life Sciences, Turku, Finland). The hsCRP, creatinine, ALT, AST, lipid profile including total cholesterol, LDL-C, HDL-C, and TG concentrations were assessed by the enzymatic colorimetric assays using commercially available laboratory kits (Cobas c111, Roche Diagnostic Co., Ltd., Basel, Switzerland). Serum free fatty acid level was assessed by the enzymatic colorimetric assay (Zenbio, Durham, NC, USA).

4.9. Untargeted Metabolomics

4.9.1. Chemical and Reagents

Ultrapure water was used to prepare all the aqueous solutions and was obtained "in-house" from a Milli-Q Integral 3 system (Millipore, SAS, Molsheim, France). Zomepirac sodium salt, formic acid, LC-MS-grade methanol and acetonitrile, and LC-grade ethanol were purchased from Sigma-Aldrich Chemie GmbH (Steinheim, Germany).

4.9.2. Sample Preparation

Plasma samples were prepared using the previously described method [44]. On the day of analysis, the samples were thawed on ice. For protein precipitation and metabolite extraction, one plasma sample volume was mixed with three volumes of ice-cold methanol/ethanol (1:1) containing one ppm of zomepirac (internal standard IS). After extraction, samples were stored on ice for 10 min and centrifuged at 21,000× g for 20 min at 4 °C. The supernatant was filtered into a glass HPLC vial through a 0.22 µm nylon filter (ThermoFisher Scientific, Waltham, MA, USA). Quality control samples (QCs) were prepared by mixing equal volumes of all raw samples. QCs were treated like the rest of the samples and injected at the beginning of the batch (10 injections) to equilibrate the system and every ten samples to monitor further the analysis's stability [45].

4.9.3. HPLC-MS Analysis

Plasma profiling was performed using 6546 iFunnel ESI-Q-TOF (Agilent Technologies, Santa Clara, CA, USA) coupled with a 1290 Infinity UHPLC system (Agilent Technologies, Santa Clara, CA, USA) with a degasser, binary pump and thermostated autosampler as described in Supplementary Materials Methods.

4.9.4. Metabolite Annotation

The annotation process covered two stages: first, a tentative annotation assignment based on the MS1 data and, second, spectral matching and structural elucidation based on MS/MS data.

Initial tentative identification of the features discriminating between groups based on accurate mass matching was performed using a CEU Mass Mediator (CMM, Madrid,

Spain) [46]. CMM uses ionization, adduct formation, and elution order information to rank tentative candidates retrieved from several databases. This led to the tentative assignment of experimental masses to the candidate hits retrieved from the database, which covered accurate mass matching, isotopic distribution determination, and checking of the possible ions and adducts.

To confirm the annotation of the compounds, LC-MS/MS data independent analysis (DIA) was performed. Fragmentation spectra were searched against the Metlin database. In addition, spectra that were not available in the Metlin were manually inspected, and lipid structural elucidation was performed. This was performed using known lipid fragmentation patterns [47,48].

4.9.5. Metabolic Pathway Analysis

Pathway analysis was performed using MetaboAnalyst 5.0 (http://www.metaboanalyst.ca/ accessed on 20 April 2024). Only annotated metabolites significantly discriminating between groups were used for this analysis. The Kyoto Encyclopedia of Genes and Genomes (KEGG) based Homo sapiens library was selected for analysis with a hypergeometric test in over-representation analysis and relative-betweenness centrality in pathway typology analysis (to estimate node importance). Pathway significance was determined from pathway enrichment analysis and based on values for each compound in the dataset.

4.10. Molecular Assays

4.10.1. Mitochondrial Activity

A colorimetric assay was used to measure the activity of Complex I (EC 1.6.5.3). 2,6-dichloroindophenol (DCIP) was reduced by electrons accepted from decyl ubiquinol (coenzyme Q_1), which was reduced after oxidation of NADH by Complex I [34]. The method described by Rustin et al., which is based on colorimetric measurement of succinate-ubiquinone reductase and succinate-cytochrome c reductase activities, was utilized in the analysis of the activities of Complex II and Complex II + III [49]. The activity of cytochrome c oxidase (COX, complex IV) was analyzed colorimetrically at 550 nm wavelength by measuring the oxidation of reduced cytochrome [50]. Citrate synthase (CS) activity was measured colorimetrically using the reaction with 5-thio-2-nitrobenzoic acid generated from 5,5′-dithiobis-2-nitrobenzoic acid during CS synthesis [51]. The ADP/ATP ratio was measured using a bioluminescent method in which luciferase catalysis is the conversion of ATP and luciferin to light. Abcam ADP/ATP Ratio Assay Kit ab65313 was utilized with adherence to the manufacturer's instructions.

4.10.2. Glutathione Metabolism

The content of oxidized (GSSG) and reduced (GSH) and total glutathione was assessed colorimetrically at a 412 nm wavelength based on the enzymatic reaction between NADPH, 5,5-dithiobis-(2-nitrobenzoic acid) (DTNB), and glutathione reductase GR [52]. For the GSSG determination, samples were thawed and neutralized to pH 6–7 with 1 M chlorhydrol triethanolamine (TEA). Subsequently, samples were incubated with 2-vinylpyridine (to inhibit glutathione oxidation). The concentration of GSH was calculated by comparing the difference between total glutathione and disulfide glutathione levels. The redox ratio was calculated according to the formula $[GSH]^2/[GSSG]$ [53,54]. The spectrophotometric method described by Paglia and Valentine was used to assess serum glutathione peroxidase (GPx) activity. In this method, organic peroxides are reduced in the presence of NADPH. One unit of GPx activity was defined as the amount of the enzyme needed to catalyze the oxidation of 1 μmol of NADPH for 1 min, and the absorbance was assessed at 340 nm.

4.10.3. Nitrosative Stress

Commercial enzyme-linked immunosorbent assay (ELISA) (Nitrotyrosine ELISA; Immundiagnostik AG, Bensheim, Germany) was used to measure the level of nitrotyrosine. All steps were performed in accordance with the manufacturer's instructions. The peroxynitrite content was measured using a colorimetric method involving peroxynitrite-mediated nitration of phenol to nitrophenol. The resulting color product exhibited a maximum absorption at the wavelength of 405 nm.

4.10.4. Oxidative Stress

The formation of hydrogen peroxide (H_2O_2) was quantified by measuring the increase in fluorescence after the reaction of Amplex Red with H_2O_2 in the presence of horseradish peroxidase. The observation was performed at 530/590 nm wavelength [55]. The rate of H_2O_2 formation was calculated using a standard curve of H_2O_2 stabilized solution. Spectrofluorometric analysis was performed to measure the concentration of advanced glycation end product (AGE) [56,57]. Samples were diluted 1:50 (*v*/*v*) in PBS (0.02 M, pH 7.0), and the intensity of fluorescence was assessed at 440/370 nm in a 96-well microplate spectrophotometer [58]. The results were expressed as fluorescence units (AFU)/mg protein. The activity of both catalase (CAT) and superoxide dismutase (SOD) was determined colorimetrically at 340 nm wavelength by measuring hydrogen peroxide decomposition and inhibition of oxidation of epinephrine to adrenochrome, respectively [59,60]. One unit of CAT activity was defined as an amount of the enzyme that degrades 1 µmol of hydrogen peroxide per minute. One unit of SOD activity was defined as the amount inhibiting the oxidation of epinephrine by 50%.

4.10.5. ELISA

Bax and Bcl-2 expression were assessed using ELISA kits from Cell Biolabs Inc. (San Diego, CA, USA). IL-1 and TNF-α concentrations in media were measured using ELISA kits from EIAab (Wuhan, China). Samples were standardized to assess the total protein concentration using the BCA method. All the assays were performed in triplicate according to manufacturers' instructions, and the results were averaged.

4.11. Data Analysis

Statistical analysis was analyzed using GraphPad Prism 8.3.0 for MacOS (GraphPad Software, La Jolla, CA, USA). All series of data were checked for consistency with a Gaussian distribution following the D'Agostino–Pearson normality test. For muscle contractility results dose-response was determined using one-way ANOVA or the Kruskal–Wallis test, where appropriate. Statistically significant differences between means were determined by Tukey's post-hoc or a nonparametric Mann–Whitney test, where appropriate. For the rest of the results, the statistical significance was assessed by one-way ANOVA with Tukey's post-hoc test for multiple comparisons. Values were considered to be statistically significant at $p < 0.05$.

5. Conclusions

Our study demonstrates that PHG, a known analgesic and spasmolytic drug, has potent protective effects against HFD-induced NAFLD and insulin resistance. PHG's efficacy in reducing liver steatosis and improving insulin sensitivity in Wistar rats is attributed to its anti-oxidant properties and mitochondrial protection rather than modulation of hepatic lipogenesis or AMPK activation.

PHG improves mitochondrial function by reducing oxidative stress, restoring ATP production, and enhancing anti-oxidant defenses. Additionally, it relaxes gastric smooth muscles, potentially delaying gastric emptying and thus contributing to better postprandial glycemic control. This mechanism was further supported by a pilot intervention in prediabetic men, where PHG administration improved postprandial glucose and insulin profiles and altered lipid metabolism favorably.

The translational potential of PHG as a therapeutic agent for metabolic disorders is significant, given its multifactorial effects on oxidative stress, mitochondrial function, and lipid metabolism. However, further research is necessary to elucidate the precise molecular mechanisms underlying PHG's metabolic benefits and to evaluate its long-term effects and efficacy in larger clinical populations.

In conclusion, PHG emerges as a promising candidate for repurposing in the treatment of NAFLD and insulin resistance. Its established safety profile and demonstrated metabolic benefits provide a strong foundation for future clinical trials aimed at exploring its full therapeutic potential in metabolic disorders.

Supplementary Materials: The following supporting information can be downloaded at: https://www.mdpi.com/article/10.3390/ijms251910291/s1.

Author Contributions: K.D. was responsible for whole study design, funding acquisition, cell cultures, animal procedures, study supervision, statistical analysis and manuscript writing. M.M. (Mateusz Maciejczyk), C.P. and A.Z. were responsible for molecular and oxidative stress assays. U.M., J.F. and M.M. (Monika Moroz) were responsible for patients recruitment and experimental visits procedures. A.U. performed all the ultrasound examinations. J.G., K.P., J.S. and M.C. were responsible for metabolomic assays. A.B. performed all the biochemical analysis in patients' samples. A.C., B.M. and T.K. performed muscle activity assays. I.W. and B.K.-Z. performed mitochondrial function assays. P.C. was responsible for the investigational medicinal product (IMP) and placebo preparation, pharmacological supervision and blinding. H.R.H. performed all the surgeries, recruited patients and collected stomachs for muscle contractility tests. M.F. performed lipid synthesis assays and provided AMPK knock-out model. E.A.-P. was responsible for funding acquisition, single dose crossed intervention design and supervision. K.D., M.M. (Mateusz Maciejczyk), J.G., B.M., B.K.-Z., M.F. and E.A.-P. performed final correction and article review. All authors have read and agreed to the published version of the manuscript.

Funding: This research was funded by [National Research Center, Poland] grant numbers [2018/31/N/NZ7/03242; 2020/04/X/NZ7/01134] and The APC was funded by [Medical University of Gdańsk].

Institutional Review Board Statement: The study was conducted in accordance with the Declaration of Helsinki, and approved by the Institutional Ethics Committee of Medical University o Bialystok, Poland (No.APK.002.375.2021) 2021 and (No. R-I-002/304/2018) 2018. The animal study protocol was approved by the Local Ethical Committee in Olsztyn, Poland (22/WPN/2021) 2021 for studies involving animals.

Informed Consent Statement: Informed consent was obtained from all subjects involved in the study.

Data Availability Statement: The data presented in this study are available on request from the corresponding author.

Acknowledgments: Krzysztof Drygalski was supported by the Foundation for Polish Science (FNP). The authors would like to thank J. Reszeć (Medical University of Bialystok) for providing histological grading. The author would like to thank J. Roleder (Heliodor Swiecicki Clinical Hospital in Poznan, Poland) for her invaluable, perpetually present support and help with animal dissection.

Conflicts of Interest: The authors declare no conflicts of interest.

References

1. Ward, Z.J.; Bleich, S.N.; Cradock, A.L.; Barrett, J.L.; Giles, C.M.; Flax, C.; Long, M.W.; Gortmaker, S.L. Projected U.S. State-Level Prevalence of Adult Obesity and Severe Obesity. *N. Engl. J. Med.* **2019**, *381*, 2440–2450. [CrossRef] [PubMed]
2. Mantena, S.K.; King, A.L.; Andringa, K.K.; Eccleston, H.B.; Bailey, S.M. Mitochondrial dysfunction and oxidative stress in the pathogenesis of alcohol- and obesity-induced fatty liver diseases. *Free Radic. Biol. Med.* **2008**, *44*, 1259–1272. [CrossRef] [PubMed]
3. Ansari, S.; Haboubi, H.; Haboubi, N. Adult obesity complications: Challenges and clinical impact. *Ther. Adv. Endocrinol. Metab.* **2020**, *11*, 2042018820934955. [CrossRef]
4. Pérez-Carreras, M.; Del Hoyo, P.; Martín, M.A.; Rubio, J.C.; Martín, A.; Castellano, G.; Colina, F.; Arenas, J.; Solis-Herruzo, J.A. Defective hepatic mitochondrial respiratory chain in patients with nonalcoholic steatohepatitis. *Hepatology* **2003**, *38*, 999–1007. [CrossRef]
5. Chavin, K.D.; Yang, S.Q.; Lin, H.Z.; Chatham, J.; Chacko, V.P.; Hock, J.B.; Walajtys-Rode, E.; Rashid, A.; Chen, C.H.; Huang, C.C.; et al. Obesity induces expression of uncoupling protein-2 in hepatocytes and promotes liver ATP depletion. *J. Biol. Chem.* **1999**, *274*, 5692–5700. [CrossRef]
6. Sanyal, A.J.; Van Natta, M.L.; Clark, J.; Neuschwander-Tetri, B.A.; Diehl, A.; Dasarathy, S.; Loomba, R.; Chalasani, N.; Kowdley, K.; Hameed, B.; et al. Prospective Study of Outcomes in Adults with Nonalcoholic Fatty Liver Disease. *N. Engl. J. Med.* **2021**, *385*, 1559–1569. [CrossRef] [PubMed]

7. Sumithran, P.; Prendergast, L.A.; Delbridge, E.; Purcell, K.; Shulkes, A.; Kriketos, A.; Proietto, J. Long-Term Persistence of Hormonal Adaptations to Weight Loss. *N. Engl. J. Med.* **2011**, *365*, 1597–1604. [CrossRef]
8. King, N.A.; Hopkins, M.; Caudwell, P.; Stubbs, R.J.; Blundell, J.E. Beneficial effects of exercise: Shifting the focus from body weight to other markers of health. *Br. J. Sports Med.* **2009**, *43*, 924–927. [CrossRef]
9. Foretz, M.; Guigas, B.; Viollet, B. Understanding the glucoregulatory mechanisms of metformin in type 2 diabetes mellitus. *Nat. Rev. Endocrinol.* **2019**, *15*, 569–589. [CrossRef]
10. Perazza, F.; Leoni, L.; Colosimo, S.; Musio, A.; Bocedi, G.; D'Avino, M.; Agnelli, G.; Nicastri, A.; Rossetti, C.; Sacilotto, F.; et al. Metformin and the Liver: Unlocking the Full Therapeutic Potential. *Metabolites* **2024**, *14*, 186. [CrossRef]
11. GRADE Study Research Group. Glycemia Reduction in Type 2 Diabetes—Microvascular and Cardiovascular Outcomes. *N. Engl. J. Med.* **2022**, *387*, 1075–1088. [CrossRef]
12. GRADE Study Research Group. Glycemia Reduction in Type 2 Diabetes—Glycemic Outcomes. *N. Engl. J. Med.* **2022**, *387*, 1063–1074. [CrossRef] [PubMed]
13. Guarente, L. Sirtuins, Aging, and Medicine. *N. Engl. J. Med.* **2011**, *364*, 2235–2244. [CrossRef] [PubMed]
14. Drygalski, K.; Berk, K.; Charytoniuk, T.; Iłowska, N.; Łukaszuk, B.; Chabowski, A.; Konstantynowicz-Nowicka, K. Does the enterolactone (ENL) affect fatty acid transporters and lipid metabolism in liver? *Nutr. Metab.* **2017**, *14*, 1–10. [CrossRef] [PubMed]
15. Berk, K.; Drygalski, K.; Harasim-Symbor, E.; Charytoniuk, T.; Iłowska, N.; Łukaszuk, B.; Chabowski, A.; Konstantynowicz-Nowicka, K. The effect of enterolactone on liver lipid precursors of inflammation. *Life Sci.* **2019**, *221*, 341–347. [CrossRef]
16. Ngu, M.H.; Norhayati, M.N.; Rosnani, Z.; Zulkifli, M.M. Curcumin as adjuvant treatment in patients with non-alcoholic fatty liver (NAFLD) disease: A systematic review and meta-analysis. *Complement. Ther. Med.* **2022**, *68*, 102843. [CrossRef]
17. He, Y.; Wang, H.; Lin, S.; Chen, T.; Chang, D.; Sun, Y.; Wang, C.; Liu, Y.; Lu, Y.; Song, J.; et al. Advanced effect of curcumin and resveratrol on mitigating hepatic steatosis in metabolic associated fatty liver disease via the PI3K/AKT/mTOR and HIF-1/VEGF cascade. *Biomed. Pharmacother.* **2023**, *165*, 115279. [CrossRef]
18. Soleas, G.J.; Diamandis, E.P.; Goldberg, D.M. Resveratrol: A molecule whose time has come? And gone? *Clin. Biochem.* **1997**, *30*, 91–113. [CrossRef]
19. Drygalski, K.; Siewko, K.; Chomentowski, A.; Odrzygóźdź, C.; Zalewska, A.; Krętowski, A.; Maclejczyk, M. Phloroglucinol Strengthens the Antioxidant Barrier and Reduces Oxidative/Nitrosative Stress in Nonalcoholic Fatty Liver Disease (NAFLD). *Oxid. Med. Cell. Longev.* **2021**, *2021*, 8872702. [CrossRef]
20. Drygalski, K.; Fereniec, E.; Zalewska, A.; Krętowski, A.; Żendzian-Piotrowska, M.; Maciejczyk, M. Phloroglucinol prevents albumin glycation as well as diminishes ROS production, glycooxidative damage, nitrosative stress and inflammation in hepatocytes treated with high glucose. *Biomed. Pharmacother.* **2021**, *142*, 111958. [CrossRef]
21. Pushpakom, S.; Iorio, F.; Eyers, P.A.; Escott, K.J.; Hopper, S.; Wells, A.; Doig, A.; Guilliams, T.; Latimer, J.; McNamee, C.; et al. Drug repurposing: Progress, challenges and recommendations. *Nat. Rev. Drug Discov.* **2019**, *18*, 41–58. [CrossRef] [PubMed]
22. Wu, P.; Feng, Q.P.; Kerchberger, V.E.; Nelson, S.D.; Chen, Q.; Li, B.; Edwards, T.L.; Cox, N.J.; Phillips, E.J.; Stein, C.M.; et al. Integrating gene expression and clinical data to identify drug repurposing candidates for hyperlipidemia and hypertension. *Nat. Commun.* **2022**, *13*, 1–12. [CrossRef] [PubMed]
23. Yoon, J.-Y.; Choi, H.; Jun, H.-S. The Effect of Phloroglucinol, A Component of Ecklonia cava Extract, on Hepatic Glucose Production. *Mar. Drugs* **2017**, *15*, 106. [CrossRef] [PubMed]
24. Blanchard, C.; Pouchain, D.; Vanderkam, P.; Perault-Pochat, M.C.; Boussageon, R.; Vaillant-Roussel, H. Efficacy of phloroglucinol for treatment of abdominal pain: A systematic review of literature and meta-analysis of randomised controlled trials versus placebo. *Eur. J. Clin. Pharmacol.* **2018**, *74*, 541–548. [CrossRef]
25. Annaházi, A.; Róka, R.; Rosztóczy, A.; Wittmann, T. Role of antispasmodics in the treatment of irritable bowel syndrome. *World J. Gastroenterol.* **2014**, *20*, 6031. [CrossRef]
26. Shin, S.Y.; Cha, B.K.; Kim, W.S.; Park, J.Y.; Kim, J.W.; Choi, C.H. The effect of phloroglucinol in patients with diarrhea-predominant irritable bowel syndrome: A randomized, double-blind, placebo-controlled trial. *J. Neurogastroenterol. Motil.* **2020**, *26*, 117–127. [CrossRef]
27. Subissi, A.; Brunori, P.; Bachi, M. Effects of spasmolytics on K+-induced contraction of rat intestine in vivo. *Eur. J. Pharmacol.* **1983**, *96*, 295–301. [CrossRef]
28. Chassany, O.; Bonaz, B.; Bruley Des Varannes, S.; Bueno, L.; Cargill, G.; Coffin, B.; Ducrotté, P.; Grangé, V. Acute exacerbation of pain in irritable bowel syndrome: Efficacy of phloroglucinol/trimethylphloroglucinol. A randomized, double-blind, placebo-controlled study. *Aliment. Pharmacol. Ther.* **2007**, *25*, 1115–1123. [CrossRef]
29. Park, C.; Cha, H.J.; Kim, M.Y.; Bang, E.J.; Moon, S.K.; Yun, S.J.; Kim, W.J.; Noh, J.S.; Kim, G.Y.; Cho, S.; et al. Phloroglucinol Attenuates DNA Damage and Apoptosis Induced by Oxidative Stress in Human Retinal Pigment Epithelium ARPE-19 Cells by Blocking the Production of Mitochondrial ROS. *Antioxidants* **2022**, *11*, 2353. [CrossRef]
30. Boudaba, N.; Marion, A.; Huet, C.; Pierre, R.; Viollet, B.; Foretz, M. AMPK Re-Activation Suppresses Hepatic Steatosis but Its Downregulation Does Not Promote Fatty Liver Development. *EBioMedicine* **2018**, *28*, 194–209. [CrossRef]
31. Foretz, M.; Viollet, B. Measurement of AMPK-Induced Inhibition of Lipid Synthesis Flux in Cultured Cells. *Methods Mol. Biol.* **2018**, *1732*, 363–371. [PubMed]
32. Blouin, C.M.; Le Lay, S.; Lasnier, F.; Dugail, I.; Hajduch, E. Regulated Association of Caveolins to Lipid Droplets during Differentiation of 3T3-L1 Adipocytes. *Biochem. Biophys. Res. Commun.* **2008**, *376*, 331–335. [CrossRef]

33. Charytoniuk, T.; Harasim-Symbor, E.; Polak, A.; Drygalski, K.; Berk, K.; Chabowski, A.; Konstantynowicz-Nowicka, K. Influence of Resveratrol on Sphingolipid Metabolism in Hepatocellular Carcinoma Cells in Lipid Overload State. *Anticancer. Agents Med. Chem.* **2019**, *19*, 121–129. [CrossRef] [PubMed]
34. Janssen, A.J.M.; Trijbels, F.J.M.; Sengers, R.C.A.; Smeitink, J.A.M.; Van Den Heuvel, L.P.; Wintjes, L.T.M.; Stoltenborg-Hogenkamp, B.J.M.; Rodenburg, R.J.T. Spectrophotometric Assay for Complex I of the Respiratory Chain in Tissue Samples and Cultured Fibroblasts. *Clin. Chem.* **2007**, *53*, 729–734. [CrossRef] [PubMed]
35. Zalewska, A.; Ziembicka, D.; Zendzian-Piotrowska, M.; Maclejczyk, M. The impact of high-fat diet on mitochondrial function, free radical production, and nitrosative stress in the salivary glands of wistar rats. *Oxid. Med. Cell. Longev.* **2019**, *2019*, 2606120. [CrossRef]
36. Mantena, S.K.; Vaughn, D.P.; Andringa, K.K.; Eccleston, H.B.; King, A.L.; Abrams, G.A.; Doeller, J.E.; Kraus, D.W.; Darley-Usmar, V.M.; Bailey, S.M. High fat diet induces dysregulation of hepatic oxygen gradients and mitochondrial function in vivo. *Biochem. J.* **2009**, *417*, 183–193. [CrossRef]
37. Zalewska, A.; Zięba, S.; Kostecka-Sochoń, P.; Kossakowska, A.; Żendzian-Piotrowska, M.; Matczuk, J.; Maciejczyk, M. NAC Supplementation of Hyperglycemic Rats Prevents the Development of Insulin Resistance and Improves Antioxidant Status but only Alleviates General and Salivary Gland Oxidative Stress. *Oxid. Med. Cell. Longev.* **2020**, *2020*, 8831855. [CrossRef]
38. Wojciak, P.A.; Pawłuszewicz, P.; Diemieszczyk, I.; Komorowska-Wojtunik, E.; Czerniawski, M.; Krętowski, A.; Błachnio-Zabielska, A.; Dadan, J.; Ładny, J.R.; Hady, H.R. Laparoscopic sleeve gastrectomy: A study of efficiency in treatment of metabolic syndrome components, comorbidities and influence on certain biochemical markers. *Videosurg. Other Miniinvasive Tech.* **2020**, *15*, 136–147. [CrossRef]
39. Diemieszczyk, I.; Woźniewska, P.; Gołaszewski, P.; Drygalski, K.; Nadolny, K.; Ładny, J.R.; Hady, H.R. Does weight loss after laparoscopic sleeve gastrectomy contribute to reduction in blood pressure? *Polish Arch. Intern. Med.* **2021**, *131*, 693–700. [CrossRef]
40. Modzelewska, B.; Kleszczewski, T.; Kostrzewska, A. The effect of a selective inhibition of potassium channels on the relaxation induced by nitric oxide in the human pregnant myometrium. *Cell. Mol. Biol. Lett.* **2003**, *8*, 69–75.
41. Gagnon, R.C.; Peterson, J.J. Estimation of confidence intervals for area under the curve from destructively obtained pharmacokinetic data. *J. Pharmacokinet. Biopharm.* **1998**, *26*, 87–102. [CrossRef] [PubMed]
42. Ter Horst, K.W.; Gilijamse, P.W.; De Weijer, B.A.; Kilicarslan, M.; Ackermans, M.T.; Nederveen, A.J.; Nieuwdorp, M.; Romijn, J.A.; Serlie, M.J. Sexual dimorphism in hepatic, adipose tissue, and peripheral tissue insulin sensitivity in obese humans. *Front. Endocrinol.* **2015**, *6*, 167090. [CrossRef]
43. Elsayed, N.A.; Aleppo, G.; Aroda, V.R.; Bannuru, R.R.; Brown, F.M.; Bruemmer, D.; Collins, B.S.; Hilliard, M.E.; Isaacs, D.; Johnson, E.L.; et al. 2. Classification and Diagnosis of Diabetes: Standards of Care in Diabetes—2023. *Diabetes Care* **2023**, *46*, S19–S40. [CrossRef]
44. Daniluk, U.; Daniluk, J.; Kucharski, R.; Kowalczyk, T.; Pietrowska, K.; Samczuk, P.; Filimoniuk, A.; Kretowski, A.; Lebensztejn, D.; Ciborowski, M. Untargeted Metabolomics and Inflammatory Markers Profiling in Children with Crohn's Disease and Ulcerative Colitis—A Preliminary Study. *Inflamm. Bowel Dis.* **2019**, *25*, 1120–1128. [CrossRef] [PubMed]
45. Godzien, J.; Alonso-Herranz, V.; Barbas, C.; Armitage, E.G. Controlling the quality of metabolomics data: New strategies to get the best out of the QC sample. *Metabolomics* **2015**, *11*, 518–528. [CrossRef]
46. Gil-De-La-Fuente, A.; Godzien, J.; Saugar, S.; Garcia-Carmona, R.; Badran, H.; Wishart, D.S.; Barbas, C.; Otero, A. CEU Mass Mediator 3.0: A Metabolite Annotation Tool. *J. Proteome Res.* **2019**, *18*, 797–802. [CrossRef]
47. Godzien, J.; Ciborowski, M.; Martínez-Alcázar, M.P.; Samczuk, P.; Kretowski, A.; Barbas, C. Rapid and Reliable Identification of Phospholipids for Untargeted Metabolomics with LC-ESI-QTOF-MS/MS. *J. Proteome Res.* **2015**, *14*, 3204–3216. [CrossRef]
48. Lange, M.; Angelidou, G.; Ni, Z.; Criscuolo, A.; Schiller, J.; Blüher, M.; Fedorova, M. AdipoAtlas: A Reference Lipidome for Human White Adipose Tissue. *bioRxiv* **2021**. [CrossRef]
49. Rustin, P.; Chretien, D.; Bourgeron, T.; Gérard, B.; Rötig, A.; Saudubray, J.M.; Munnich, A. Biochemical and molecular investigations in respiratory chain deficiencies. *Clin. Chim. Acta* **1994**, *228*, 35–51. [CrossRef]
50. Wharton, D.C.; Tzagoloff, A. [45] Cytochrome oxidase from beef heart mitochondria. *Methods Enzymol.* **1967**, *10*, 245–250.
51. Srere, P.A. [1] Citrate synthase: [EC 4.1.3.7. Citrate oxaloacetate-lyase (CoA-acetylating)]. *Methods Enzymol.* **1969**, *13*, 3–11.
52. Griffith, O.W. Determination of glutathione and glutathione disulfide using glutathione reductase and 2-vinylpyridine. *Anal. Biochem.* **1980**, *106*, 207–212. [CrossRef] [PubMed]
53. Choromańska, B.; Myśliwiec, P.; Łuba, M.; Wojskowicz, P.; Dadan, J.; Myśliwiec, H.; Choromańska, K.; Zalewska, A.; Maciejczyk, M. A Longitudinal Study of the Antioxidant Barrier and Oxidative Stress in Morbidly Obese Patients after Bariatric Surgery. Does the Metabolic Syndrome Affect the Redox Homeostasis of Obese People? *J. Clin. Med.* **2022**, *9*, 976. [CrossRef] [PubMed]
54. Paglia, D.E.; Valentine, W.N. Studies on the quantitative and qualitative characterization of erythrocyte glutathione peroxidase. *J. Lab. Clin. Med.* **1967**, *70*, 158–169. [PubMed]
55. Muller, F.L.; Liu, Y.; Van Remmen, H. Complex III releases superoxide to both sides of the inner mitochondrial membrane. *J. Biol. Chem.* **2004**, *279*, 49064–49073. [CrossRef]
56. Kalousová, M.; Škrha, J.; Zima, T. Advanced Glycation End-Products and Advanced Oxidation Protein Products in Patients with Diabetes Mellitus. *Physiol. Res.* **2002**, *51*, 597–604. [CrossRef]
57. Münch, G.; Keis, R.; Weβels, A.; Riederer, P.; Bahner, U.; Heidland, A.; Niwa, T.; Lemke, H.D.; Schinzel, R. Determination of advanced glycation end products in serum by fluorescence spectroscopy and competitive ELISA. *Eur. J. Clin. Chem. Clin. Biochem.* **1997**, *35*, 669–678. [CrossRef]

58. Mil, K.M.; Gryciuk, M.E.; Pawlukianiec, C.; Żendzian-Piotrowska, M.; Ładny, J.R.; Zalewska, A.; Maciejczyk, M. Pleiotropic Properties of Valsartan: Do They Result from the Antiglycooxidant Activity? Literature Review and in Vitro Study. *Oxid. Med. Cell. Longev.* **2021**, *2021*, 5575545. [CrossRef]
59. Bondy, S.C.; Guo, S.X. Effect of ethanol treatment on indices of cumulative oxidative stress. *Eur. J. Pharmacol. Environ. Toxicol. Pharmacol.* **1994**, *270*, 349–355. [CrossRef]
60. Maciejczyk, M.; Matczuk, J.; Żendzian-Piotrowska, M.; Niklińska, W.; Fejfer, K.; Szarmach, I.; Ładny, J.R.; Zieniewska, I.; Zalewska, A. Eight-Week Consumption of High-Sucrose Diet Has a Pro-Oxidant Effect and Alters the Function of the Salivary Glands of Rats. *Nutrients* **2018**, *10*, 5575545. [CrossRef]

International Journal of
Molecular Sciences

MDPI

Case Report

Carnitine Deficiency Caused by Salcaprozic Acid Sodium Contained in Oral Semaglutide in a Patient with Multiple Acyl-CoA Dehydrogenase Deficiency

Yasuko Mikami-Saito [1], Masamitsu Maekawa [2,3], Masahiro Watanabe [3], Shinichiro Hosaka [4], Kei Takahashi [4], Eriko Totsune [1], Natsuko Arai-Ichinoi [1], Atsuo Kikuchi [1], Shigeo Kure [1], Hideki Katagiri [4] and Yoichi Wada [1,*]

1 Department of Pediatrics, Tohoku University Graduate School of Medicine, 1-1 Seiryo-machi, Aoba-ku, Sendai 980-8574, Japan
2 Department of Pharmaceutical Sciences, Tohoku University Hospital, 1-1 Seiryo-machi, Aoba-ku, Sendai 980-8574, Japan
3 Graduate School of Pharmaceutical Sciences, Tohoku University, 1-1 Seiryo-machi, Aoba-ku, Sendai 980-8574, Japan
4 Department of Diabetes, Metabolism and Endocrinology, Tohoku University Graduate School of Medicine, 2-1 Seiryo-machi, Aoba-ku, Sendai 980-8575, Japan
* Correspondence: yoichi.wada.e1@tohoku.ac.jp; Tel.: +81-22-717-7287

Abstract: Carnitine plays an essential role in maintaining energy homeostasis and metabolic flexibility. Various medications, such as pivalate-conjugated antibiotics, valproic acid, and anticancer agents, can induce carnitine deficiency, inhibit the utilization of fatty acid, and contribute to the development of hypoglycemia. No studies have linked oral semaglutide to carnitine deficiency. Herein, we report the case of a 34-year-old male patient with multiple acyl-CoA dehydrogenase deficiency who developed carnitine deficiency attributable to salcaprozic acid sodium (SNAC) in oral semaglutide. The patient was diagnosed with type 2 diabetes mellitus at 32 years of age and was treated with semaglutide injections. Hypoglycemic symptoms appeared after switching to oral semaglutide, and the mean levels of blood-free carnitine significantly decreased. Liquid chromatography–tandem mass spectrometry analysis revealed a peak corresponding to the SNAC–carnitine complex (m/z 423.24) in the urine exclusively during the oral administration of semaglutide. The MS/MS spectra at m/z 423.24 contained peaks consistent with those of the SNAC and carnitine product ions. Our results suggest that through complexation with carnitine, SNAC may induce carnitine deficiency. Healthcare providers should monitor for carnitine deficiency when administering SNAC-containing medications to at-risk individuals. Furthermore, this case can raise more significant concerns about the potential impact of pharmaceutical excipients like SNAC on metabolic pathways.

Keywords: carnitine deficiency; salcaprozic acid sodium; multiple acyl-CoA dehydrogenase deficiency; type 2 diabetes; oral semaglutide

1. Introduction

Carnitine (3-hydroxy-4-N-trimethylammoniobutanoate) is an essential nutrient that plays critical roles in facilitating energy production through the fatty acids oxidation, regulating the acyl-CoA/CoA ratio within mitochondria, and excreting toxic acyl groups in the urine as carnitine esters [1–4]. Approximately 75% of body carnitine is obtained from dietary sources, with the remaining being synthesized endogenously in the liver and kidneys [5]. Skeletal and cardiac muscle contain 99% of body carnitine, with only 0.5% present in extracellular compartments, maintaining blood-free carnitine levels at 25–50 μmol/L [6–8]. More than 90% of urine-excreted carnitine is reabsorbed by the kidneys when blood-free carnitine levels are within the normal range [9]. By regulating carnitine excretion, the kidneys maintain blood carnitine concentration within physiological limits [10]. Body carnitine homeostasis relies on the coordination of dietary intake, endogenous synthesis, renal excretion, and reabsorption. Carnitine deficiency results from the disruption of any of these metabolic processes and is defined as an insufficient concentration of carnitine in tissues and blood, thereby impairing proper organ function. The underlying causes classify the primary or secondary types of carnitine deficiency, and among them, secondary carnitine deficiency can be induced by medications, including pivalate-conjugated antibiotics, valproic acid, and various anticancer drugs [6,11–13]. Pivalate-conjugated antibiotics contain pivalic acid to enhance intestinal absorption; however, pivalate binds to free carnitine in the blood, increasing urinary carnitine excretion and leading to carnitine deficiency [11]. Although the effect of pivalate on carnitine homeostasis might be negligible in healthy adults, it can contribute to carnitine deficiency in vulnerable populations, such as infants, individuals with chronic renal or hepatic insufficiency, and those with inherited metabolic disorders, leading to hypoglycemia and acute encephalopathy [14–18].

Multiple acyl-CoA dehydrogenase deficiency (MADD, MIM#231680), also known as glutaric acidemia type II, is a fatty acid oxidation disorder (FAOD) caused by defects in either electron transfer flavoprotein (ETF) or ETF dehydrogenase [19]. ETF is a heterodimeric protein in the mitochondrial matrix that accepts electrons from acyl-CoA dehydrogenase involved in fatty acid β-oxidation, as well as from various other dehydrogenases participating in amino acid, choline, and other metabolic pathways [20]. The electrons are subsequently transferred to ETF dehydrogenase, situated in the inner mitochondrial membrane, and then to the respiratory chain [21]. Consequently, MADD impairs fatty acid oxidation, resulting in acylcarnitine accumulation. Accumulated acylcarnitine can inhibit the organic cation/carnitine transporter 2-mediated transport and increase the urine excretion of carnitine, resulting in carnitine deficiency [6,22]. Treatment for individuals with MADD includes frequent high-carbohydrate, low-protein, and low-fat diet meals, along with riboflavin, carnitine, and coenzyme Q10 supplementation [23].

Salcaprozic acid sodium (SNAC), a synthetic N-acetylated amino-acid derivative of salicylic acid, functions as an intestinal permeation enhancer similar to pivalic acid [24,25]. SNAC, which exhibits weakly acidic properties with amphiphilicity and surface activity, was discovered through screening for substances capable of chaperoning drugs with low mucosal permeability across the intestine [26]. SNAC was incorporated into an oral formulation of vitamin B_{12}, which was approved as a medical food in 2014 [27]. In 2019, the first oral glucagon-like peptide-1 receptor agonist, semaglutide formulated with SNAC, was developed and approved by the Food and Drug Administration [28]. The pH buffering effect of SNAC in the stomach prevents semaglutide degradation, enhances its solubility, and promotes its monomerization, thus increasing its permeability through the gastric mucosa [29]. Oral semaglutide demonstrates safety and tolerability identical to those of subcutaneous semaglutide, and no studies have linked it to carnitine deficiency [28,30].

In this study, we report a case of MADD presenting with carnitine deficiency attributed to oral semaglutide and identify SNAC as a potential inducer of carnitine deficiency.

2. Results

2.1. Case Description

A 34-year-old man with MADD first presented with consciousness disturbance, hypoglycemia, acidosis, and hyperammonemia when he developed acute gastroenteritis at the age of 1 year. Urine organic acid analysis revealed elevated levels of ethylmalonic acid and glutanic acid. Genetic testing identified heterozygous pathogenic variants in the *ETFA* gene (NM_000126.4): c.[764G>T];[478delG], confirming MADD [31]. The patient was treated with riboflavin, levocarnitine, and allopurinol. Despite experiencing occasional hypoglycemic episodes during infancy, the patient had an uneventful course after adolescence.

At the age of 32 years, the patient was diagnosed with type 2 diabetes mellitus and began treatment with a once-daily liraglutide injection [32]. Treatment was changed to weekly semaglutide at the age of 33 years and 2 months to improve convenience. Owing to the temporary discontinuation of manufacturing and export of semaglutide injectable formulations, the patient transitioned to daily oral semaglutide at the age of 33 years and 8 months. Four months after this transition, the patient presented with headaches and vomiting. Blood glucose and free carnitine levels were 3.3 mmol/L and 19.7 μmol/L, respectively (Table 1). The mean free carnitine level during oral semaglutide administration was significantly lower than that during semaglutide injection (29.3 vs. 63.8 μmol/L, p = 0.012) (Figure S1). We observed no changes in medication adherence, dietary habits, body weight, or hepatic and renal functions. After increasing the levocarnitine dose from 1500 to 3000 mg and returning to semaglutide injection, the hypoglycemic symptoms resolved, and free blood carnitine levels increased (Table 1).

Table 1. Clinical data before and after carnitine deficiency: body weight, acylcarnitine profile, hypoglycemic symptoms, and medications.

Months Relative to the Episode	−8	−6	−4	−2	0	2	4	5	6	8
Body Weight (kg)	81.5	81.7	81.3	80.5	80	81.3	77.9	76.8	75	78.2
Acylcarnitines in dried blood spots (μmol/L)										
C0	74.92	68.68	63.8	48.69	19.7	34.45	14.29	69.87	83.73	66.02
C2	7.63	7.14	7.09	6.3	3.33	4.48	4.3	6.87	10.56	13.42
C4	0.75	0.69	0.91	0.62	0.35	0.45	0.38	0.94	1.05	1.11
C6	0.49	0.32	0.7	0.66	0.18	0.26	0.17	0.36	0.92	1.19
C8	0.87	0.3	1.19	1.31	0.32	0.44	0.33	0.5	1.44	2.35
C10	0.37	0.12	0.45	0.47	0.17	0.2	0.14	0.19	0.62	1.17
C12	0.07	0.04	0.08	0.07	0.05	0.05	0.04	0.06	0.11	0.14
C14:1	0.09	0.04	0.08	0.06	0.06	0.07	0.06	0.07	0.14	0.14
C16	0.46	0.33	0.5	0.42	0.38	0.48	0.36	0.47	0.55	0.61
Hypoglycemic symptoms	−	−	−	−	+	+	+	−	−	−
Treatment										
Semaglutide injection (mg/week)	0.5	0.5	−	−	−	−	−	0.25	0.25	0.25
Oral semaglutide (mg/day)	−	−	3	7	7	7	7	−	−	−
Levocarnitine (mg/day)	1500	1500	1500	1500	1500	1500	1500	3000	3000	3000

C0, free carnitine; −, absent; +, present.

2.2. MS Analysis Results

The period during which the patient developed carnitine deficiency coincided with the oral administration of semaglutide (Table 1), leading us to hypothesize that oral semaglutide may cause carnitine deficiency. Given the molecular structure of oral semaglutide, we considered the possibility of dehydration condensation between SNAC and free carnitine in the blood (Figure 1). We performed liquid chromatography–tandem mass spectrometry (LC-MS/MS) analysis on plasma, serum, and urine samples to investigate this hypothesis. LC-MS/MS analysis revealed a peak corresponding to the molecular weight of the SNAC–carnitine complex (m/z 423.24) in urine during oral semaglutide administration (Figure 2a, Table S1). Although the m/z 423.24 peak eluted at 4.9 and 6.8 min (Figure 2b), both peaks exhibited identical MS/MS spectra (Table S2). The MS/MS spectra at m/z 423.24 contained fragment ions corresponding to carnitine and SNAC product ions (Figure 2c, Table S3). We were unable to identify the corresponding peak for the SNAC–carnitine complex in serum or plasma samples.

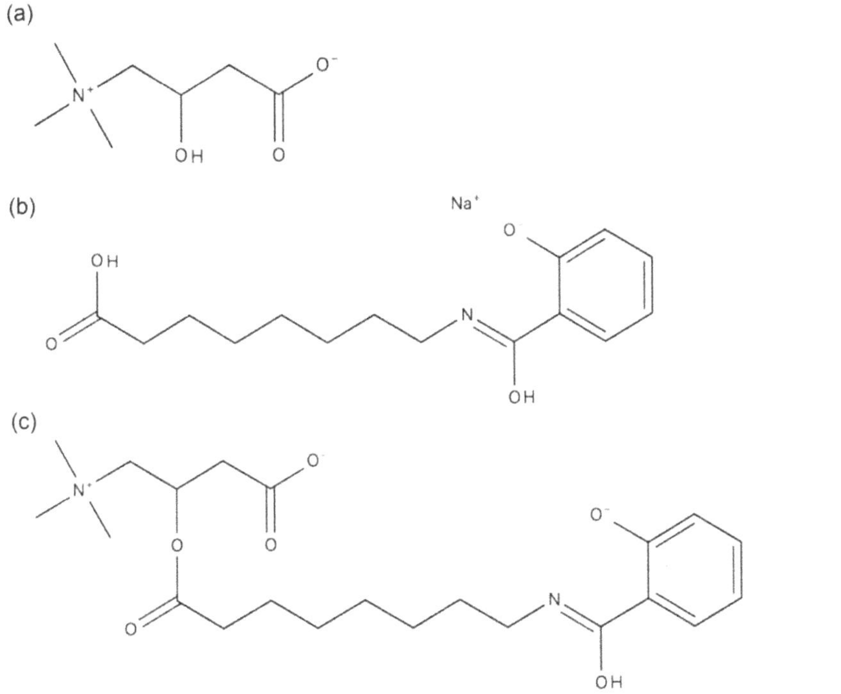

Figure 1. Chemical structures of (**a**) carnitine, (**b**) salcaprozic acid sodium (SNAC), and (**c**) SNAC–carnitine complex.

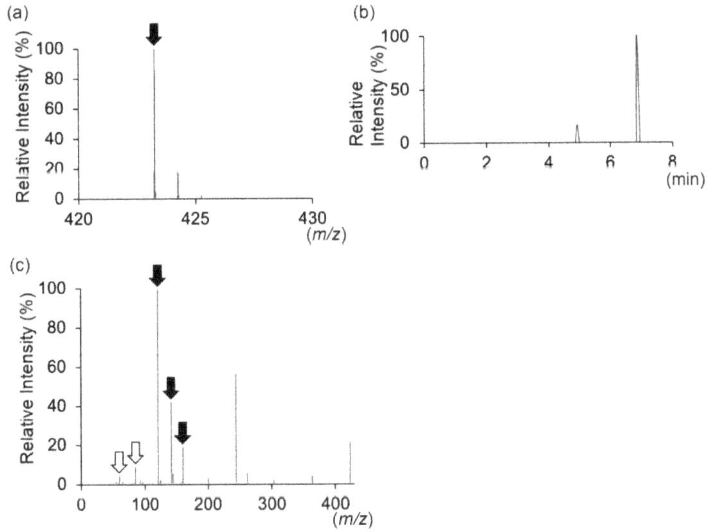

Figure 2. (**a**) MS spectra of m/z 420–430 in the urine during the period of oral semaglutide administration. (**b**) The chromatogram of m/z is 423.24. (**c**) MS/MS spectra of m/z 423.24, corresponding to the molecular weight of the SNAC–carnitine complex. The closed arrows indicate the product ions of SNAC, and the open arrows indicate those of carnitine.

3. Discussion

We report a case of carnitine deficiency in a patient with MADD treated with oral semaglutide. The patient, who had developed type 2 diabetes, was initially treated with semaglutide injections. Upon transitioning to oral semaglutide, the patient developed hypoglycemic symptoms accompanied by markedly reduced blood-free carnitine levels. Our LC-MS analysis detected a peak corresponding to the mass-to-charge ratio of the SNAC–carnitine complex exclusively in urine during oral semaglutide administration. The SNAC–carnitine complex eluted at 4.9 and 6.8 min, with both peaks exhibiting identical MS/MS spectra, suggesting they represent structural isomers. MS/MS analysis confirmed the presence of product ions corresponding to SNAC and carnitine. These findings suggested that carnitine deficiency resulted from SNAC and carnitine complex formation. Healthcare providers should monitor for SNAC-induced carnitine deficiency, particularly in vulnerable individuals.

Our study demonstrated that SNAC likely undergoes dehydration and condensation with free carnitine in the blood, forming a SNAC–carnitine complex, which leads to secondary carnitine deficiency. This finding is supported by the temporal association between the onset of the carnitine deficiency and the initiation of oral semaglutide treatment, as well as the recovery of blood carnitine levels after discontinuing oral semaglutide. No significant changes in the diet, body weight, medication adherence, or liver and kidney functions of the patient, ruling out alternative explanations for the carnitine deficiency. Oral semaglutide improves its absorption in the gastrointestinal tract by preventing enzymatic degradation in the stomach, a process facilitated by SNAC, which increases the local pH of the stomach [29]. SNAC, containing a fatty acid moiety, is rapidly absorbed in the stomach and excreted. In this study, the SNAC–carnitine complex was detected in urine, suggesting that stomach-absorbed SNAC undergoes conversion by acyl-CoA synthetase into a CoA ester of salcaprozate before excretion. The estimated quantity of SNAC entering systemic circulation aligned with the observed reduction in plasma carnitine levels, considering that the oral bioavailability of SNAC in monkeys has been reported to be approximately 15% [26]. However, the serum or plasma did not detect the SNAC–carnitine complex peak. Previous studies have shown the rapid absorption and elimination of semaglutide-contained SNAC, characterized by a short residence time after administration [29,33]. The absence of the SNAC–carnitine complex in the blood could be attributed to its high clearance rate, suggesting that it is excreted before blood collection. To further substantiate the relationship between SNAC and carnitine deficiency and clarify the clinical significance of the SNAC–carnitine complex from a different perspective, comprehensive data on carnitine profiles should be collected from larger and more diverse populations treated with SNAC-containing medications.

Healthcare providers should exercise vigilance regarding carnitine deficiency when administering SNAC-containing drugs to at-risk patients. Secondary carnitine deficiency caused by pivalate-containing medications occurs most commonly in 1-year-old infants because of their low biosynthesis, limited muscle volume, and unstable intake; however, older individuals with underlying medical conditions can also develop this condition [15,16]. For instance, individuals with type 2 diabetes, particularly those with diabetic complications, face an increased likelihood of carnitine deficiency and may develop additional risk factors, such as hemodialysis-requiring renal failure and nonalcoholic fatty liver disease, further exacerbating the condition [34–36]. Similarly, organic acidemias and FAODs may contribute to carnitine deficiency by lowering the renal threshold for carnitine and increasing renal carnitine excretion. As more individuals with inherited metabolic disorders survive into adulthood, the prevalence of adult-onset complications, such as diabetes, is expected to increase [37]. This trend suggests the potential emergence of previously unrecognized drug-related side effects, as observed in this case. In addition to SNAC, medium-chain fatty acids such as sodium caprylate and sodium caprate are utilized as permeation enhancers [38]. Given their structural similarity to SNAC, agents containing these substances could potentially also induce carnitine deficiency. Therefore, blood carnitine levels should be monitored following medication administration or changes in individuals at risk for carnitine deficiency, regardless of whether symptoms are present. No previous studies have linked oral semaglutide to carnitine deficiency induced by the use of oral semaglutide [30]. In this case, the diagnosis of carnitine deficiency was facilitated by the MADD condition of the patient and routine acylcarnitine profile measurements. In contrast, in individuals with diabetes experiencing hypoglycemia, medication side effects are often suspected first, potentially reducing opportunities to check blood carnitine levels. Quantitative analysis of carnitine and the SNAC–carnitine complex in the urine and blood of individuals administered SNAC-containing medications could provide insights into several aspects, such as the duration and dosage of SNAC-containing agents required to induce carnitine deficiency, as well as the characteristics of individuals susceptible to this condition.

This study has some limitations. First, this result was based on data from only a single case experienced at our institute. Second, we did not conduct a quantitative analysis of free carnitine or the SNAC–carnitine complex in the urine. Simultaneous and longitudinal quantification of these compounds would help to support the complex formation that induces carnitine deficiency. Specifically, such an analysis could clarify the extent of carnitine loss through renal excretion, thereby allowing for the assessment of the potential clinical significance and risk of carnitine deficiency in a practical and clinically relevant manner. In addition, conducting such studies in a more diverse population receiving oral semaglutide—with stratification by renal function and assessment of carnitine deficiency risk across different renal function levels—would further strengthen the evidence for the association between SNAC and carnitine deficiency.

4. Materials and Methods

Plasma, serum, and urine samples were collected during oral and subcutaneous semaglutide administration. Carnitine and SNAC were purchased from FUJIFILM Wako Pure Chemical Corporation (359-44361, Osaka, Japan) and Targetmol (T8926, Boston, MA, USA), respectively.

Samples (10 μL), carnitine (10 μL, 100 μmol/L), and SNAC (10 μL, 100 μmol/L) were added to the internal standard solution (10 μL) and acetonitrile (70 μL), and then centrifuged at 15,000× *g* for 5 min. The 90 μL of the supernatant was transferred to another tube, evaporated at 40 °C for 35 min, and dissolved in 20 μL of 50% methanol.

A hybrid quadrupole time-of-flight tandem mass spectrometer (TripleTOF 5600; SCIEX, Framingham, MA, USA) was used in conjunction with an ultrahigh-performance liquid chromatograph (Nexera; Shimadzu, Kyoto, Japan). A capcell pak ADME column (50 mm × 2.1 mm i.d., 2 μm) was used, with the temperature maintained at 40 °C. Measurements were conducted in the positive ion mode. Mobile phases consisted of (A) formic acid/water (0.1:100, *v*/*v*) and (B) formic acid/acetonitrile (0.1:100, *v*/*v*). The flow rate was 0.3 mL/min, with an injection volume of 5 μL.

Mean carnitine levels during semaglutide injection and oral semaglutide administration periods were compared using the Mann–Whitney U test. The data were analyzed with GraphPad Prism 10.0.2 (Dotmatics, Boston, MA, USA), and statistical significance was defined as $p < 0.05$.

5. Conclusions

In conclusion, this case highlights the potential for secondary carnitine deficiency caused by SNAC-containing agents in individuals at risk. A thorough investigation, considering the medical history of the individual and the excipients of the administered drug, explained the underlying causes of atypical or seemingly incomprehensible symptoms. Further research is required to establish the association and consequences between SNAC and carnitine deficiency by collecting and analyzing comprehensive data across diverse populations. Pharmaceutical excipients should be given attention for their potential to induce unexpected metabolic effects.

Supplementary Materials: The following supporting information can be downloaded at https://www.mdpi.com/article/10.3390/ijms26072962/s1.

Author Contributions: Conceptualization, methodology, validation, and investigation, Y.M.-S., M.M. and Y.W.; data acquisition, Y.M.-S., M.M., M.W., S.H., K.T., E.T., N.A.-I. and Y.W.; formal analysis and data curation, Y.M.-S., M.M., M.W. and Y.W.; original draft preparation, Y.M.-S.; visualization, Y.M.-S., M.M. and Y.W.; supervision, A.K., S.K., H.K. and Y.W.; project administration, Y.M.-S. and Y.W.; funding acquisition, Y.W. and M.M. All authors have read and agreed to the published version of the manuscript.

Funding: This research received no external funding.

Institutional Review Board Statement: The study was conducted in accordance with the Declaration of Helsinki and approved by the Institutional Review Board of the Faculty of Medicine, Tohoku University (approval number: 2022-1-1049 and 38222).

Informed Consent Statement: Written informed consent has been obtained from the patient to publish this paper.

Data Availability Statement: The data of this study are available within the article and in Supplemental Figures and Tables. The clinical data were de-identified.

Acknowledgments: We thank the patient for cooperating in this study.

Conflicts of Interest: The authors declare no conflicts of interest.

References

1. Vaz, F.M.; Wanders, R.J.A. Carnitine Biosynthesis in Mammals. *Biochem. J.* **2002**, *361*, 417–429.
2. Carter, A.L.; Abney, T.O.; Lapp, D.F. Biosynthesis and Metabolism of Carnitine. *J. Child Neurol.* **1995**, *10*, S3–S7. [PubMed]
3. McGarry, J.D.; Brown, N.F. The Mitochondrial Carnitine Palmitoyltransferase System. From Concept to Molecular Analysis. *Eur. J. Biochem.* **1997**, *244*, 1–14. [PubMed]
4. Ramsay, R.R.; Zammit, V.A. Carnitine Acyltransferases and Their Influence on CoA Pools in Health and Disease. *Mol. Asp. Med.* **2004**, *25*, 475–493.
5. Pons, R.; De Vivo, D.C. Primary and Secondary Carnitine Deficiency Syndromes. *J. Child Neurol.* **1995**, *10*, S8–S24.
6. Stanley, C.A. Carnitine Deficiency Disorders in Children. *Ann. N. Y. Acad. Sci.* **2004**, *1033*, 42–51.
7. Steiber, A.; Kerner, J.; Hoppel, C.L. Carnitine: A Nutritional, Biosynthetic, and Functional Perspective. *Mol. Asp. Med.* **2004**, *25*, 455–473.
8. Longo, N.; Frigeni, M.; Pasquali, M. Carnitine Transport and Fatty Acid Oxidation. *Biochim. Biophys. Acta* **2016**, *1863*, 2422–2435.
9. Bellinghieri, G.; Santoro, D.; Calvani, M.; Mallamace, A.; Savica, V. Carnitine and Hemodialysis. *Am. J. Kidney Dis.* **2003**, *41*, S116–S122.
10. Rebouche, C.J. Kinetics, Pharmacokinetics, and Regulation of L-Carnitine and Acetyl-L-Carnitine Metabolism. *Ann. N. Y. Acad. Sci.* **2004**, *1033*, 30–41.
11. Melegh, B.; Kerner, J.; Bieber, L.L. Pivampicillin-Promoted Excretion of Pivaloylcarnitine in Humans. *Biochem. Pharmacol.* **1987**, *36*, 3405–3409. [PubMed]
12. Ohtani, Y.; Endo, F.; Matsuda, I. Carnitine Deficiency and Hyperammonemia Associated with Valproic Acid Therapy. *J. Pediatr.* **1982**, *101*, 782–785. [PubMed]

13. Hockenberry, M.J.; Hooke, M.C.; Gregurich, M.; McCarthy, K. Carnitine Plasma Levels and Fatigue in Children/Adolescents Receiving Cisplatin, Ifosfamide, or Doxorubicin. *J. Pediatr. Hematol. Oncol.* **2009**, *31*, 664–669. [PubMed]
14. Brass, E.P.; Mayer, M.D.; Mulford, D.J.; Stickler, T.K.; Hoppel, C.L. Impact on Carnitine Homeostasis of Short-Term Treatment with the Pivalate Prodrug Cefditoren Pivoxil. *Clin. Pharmacol. Ther.* **2003**, *73*, 338–347.
15. Kobayashi, H.; Fukuda, S.; Yamada, K.; Hasegawa, Y.; Takahashi, T.; Purevsuren, J.; Yamaguchi, S. Clinical Features of Carnitine Deficiency Secondary to Pivalate-Conjugated Antibiotic Therapy. *J. Pediatr.* **2016**, *173*, 183–187.
16. Hanai, S.; Iwata, M.; Terasawa, T. Relapsing Hypoglycemia Associated with Hypocarnitinemia Following Treatment with Cefcapene Pivoxil in an Elderly Man. *Intern. Med.* **2019**, *58*, 2891–2894.
17. Nakajima, Y.; Ito, T.; Maeda, Y.; Ichiki, S.; Sugiyama, N.; Mizuno, M.; Makino, Y.; Sugiura, T.; Kurono, Y.; Togari, H. Detection of Pivaloylcarnitine in Pediatric Patients with Hypocarnitinemia after Long-Term Administration of Pivalate-Containing Antibiotics. *Tohoku J. Exp. Med.* **2010**, *221*, 309–313.
18. Abrahamsson, K.; Eriksson, B.O.; Holme, E.; Jodal, U.; Lindstedt, S.; Nordin, I. Impaired Ketogenesis in Carnitine Depletion Caused by Short-Term Administration of Pivalic Acid Prodrug. *Biochem. Med. Metab. Biol.* **1994**, *52*, 18–21.
19. Multiple Acyl-CoA Dehydrogenase Deficiency. Available online: https://www.ncbi.nlm.nih.gov/books/NBK558236/ (accessed on 23 January 2025).
20. Henriques, B.J.; Katrine Jentoft Olsen, R.; Gomes, C.M.; Bross, P. Electron Transfer Flavoprotein and Its Role in Mitochondrial Energy Metabolism in Health and Disease. *Gene* **2021**, *776*, 145407.
21. Zaric, B.L.; Macvanin, M.T.; Isenovic, E.R. Free Radicals: Relationship to Human Diseases and Potential Therapeutic Applications. *Int. J. Biochem. Cell Biol.* **2023**, *154*, 106346.
22. Pochini, L.; Scalise, M.; Galluccio, M.; Indiveri, C. OCTN Cation Transporters in Health and Disease: Role as Drug Targets and Assay Development: Role as Drug Targets and Assay Development. *J. Biomol. Screen.* **2013**, *18*, 851–867. [CrossRef] [PubMed]
23. Li, Q.; Yang, C.; Feng, L.; Zhao, Y.; Su, Y.; Liu, H.; Men, H.; Huang, Y.; Körner, H.; Wang, X. Glutaric Acidemia, Pathogenesis and Nutritional Therapy. *Front. Nutr.* **2021**, *8*, 704984. [CrossRef] [PubMed]
24. Twarog, C.; Fattah, S.; Heade, J.; Maher, S.; Fattal, E.; Brayden, D.J. Intestinal Permeation Enhancers for Oral Delivery of Macromolecules: A Comparison between Salcaprozate Sodium (SNAC) and Sodium Caprate (C10). *Pharmaceutics* **2019**, *11*, 78. [CrossRef] [PubMed]
25. Kommineni, N.; Sainaga Jyothi, V.G.S.; Butreddy, A.; Raju, S.; Shapira, T.; Khan, W.; Angsantikul, P.; Domb, A.J. SNAC for Enhanced Oral Bioavailability: An Updated Review. *Pharm. Res.* **2023**, *40*, 633–650. [CrossRef]
26. Leone-Bay, A.; Santiago, N.; Achan, D.; Chaudhary, K.; DeMorin, F.; Falzarano, L.; Haas, S.; Kalbag, S.; Kaplan, D.; Leipold, H. N-Acylated Alpha-Amino Acids as Novel Oral Delivery Agents for Proteins. *J. Med. Chem.* **1995**, *38*, 4263–4269. [CrossRef]
27. Castelli, M.C.; Friedman, K.; Sherry, J.; Brazzillo, K.; Genoble, L.; Bhargava, P.; Riley, M.G.I. Comparing the Efficacy and Tolerability of a New Daily Oral Vitamin B12 Formulation and Intermittent Intramuscular Vitamin B12 in Normalizing Low Cobalamin Levels: A Randomized, Open-Label, Parallel-Group Study. *Clin. Ther.* **2011**, *33*, 358–371.e2. [CrossRef]
28. Aroda, V.R.; Blonde, L.; Pratley, R.E. A New Era for Oral Peptides: SNAC and the Development of Oral Semaglutide for the Treatment of Type 2 Diabetes. *Rev. Endocr. Metab. Disord.* **2022**, *23*, 979–994.
29. Buckley, S.T.; Bækdal, T.A.; Vegge, A.; Maarbjerg, S.J.; Pyke, C.; Ahnfelt-Rønne, J.; Madsen, K.G.; Schéele, S.G.; Alanentalo, T.; Kirk, R.K.; et al. Transcellular Stomach Absorption of a Derivatized Glucagon-like Peptide-1 Receptor Agonist. *Sci. Transl. Med.* **2018**, *10*, eaar7047. [CrossRef]
30. Pratley, R.; Amod, A.; Hoff, S.T.; Kadowaki, T.; Lingvay, I.; Nauck, M.; Pedersen, K.B.; Saugstrup, T.; Meier, J.J.; PIONEER 4 investigators. Oral Semaglutide versus Subcutaneous Liraglutide and Placebo in Type 2 Diabetes (PIONEER 4): A Randomised, Double-Blind, Phase 3a Trial. *Lancet* **2019**, *394*, 39–50. [CrossRef]
31. Purevjav, E.; Kimura, M.; Takusa, Y.; Ohura, T.; Tsuchiya, M.; Hara, N.; Fukao, T.; Yamaguchi, S. Molecular Study of Electron Transfer Flavoprotein Alpha-Subunit Deficiency in Two Japanese Children with Different Phenotypes of Glutaric Acidemia Type II. *Eur. J. Clin. Investig.* **2002**, *32*, 707–712. [CrossRef]
32. Ikeda, N.; Wada, Y.; Izumi, T.; Munakata, Y.; Katagiri, H.; Kure, S. Stealthy Progression of Type 2 Diabetes Mellitus Due to Impaired Ketone Production in an Adult Patient with Multiple Acyl-CoA Dehydrogenase Deficiency. *Mol. Genet. Metab. Rep.* **2024**, *38*, 101061. [PubMed]
33. Granhall, C.; Donsmark, M.; Blicher, T.M.; Golor, G.; Søndergaard, F.L.; Thomsen, M.; Bækdal, T.A. Safety and Pharmacokinetics of Single and Multiple Ascending Doses of the Novel Oral Human GLP-1 Analogue, Oral Semaglutide, in Healthy Subjects and Subjects with Type 2 Diabetes. *Clin. Pharmacokinet.* **2019**, *58*, 781–791. [CrossRef] [PubMed]
34. Tamamoğullari, N.; Siliğ, Y.; Içağasioğlu, S.; Atalay, A. Carnitine Deficiency in Diabetes Mellitus Complications. *J. Diabetes Complicat.* **1999**, *13*, 251–253.
35. Poorabbas, A.; Fallah, F.; Bagdadchi, J.; Mahdavi, R.; Aliasgarzadeh, A.; Asadi, Y.; Koushavar, H.; Vahed Jabbari, M. Determination of Free L-Carnitine Levels in Type II Diabetic Women with and without Complications. *Eur. J. Clin. Nutr.* **2007**, *61*, 892–895.

36. Flanagan, J.L.; Simmons, P.A.; Vehige, J.; Willcox, M.D.; Garrett, Q. Role of Carnitine in Disease. *Nutr. Metab.* **2010**, *7*, 30. [CrossRef]
37. Sechi, A.; Urban, M.L.; Murphy, E.; Pession, A.; Scarpa, M.; SIMMESN Adult Metabolic Working Group and of MetabERN. Towards Needed Improvements in Inherited Metabolic Medicine in Adulthood: The SIMMESN Adult Metabolic Working Group and MetabERN Joint Position Statement. *Nutr. Metab. Cardiovasc. Dis.* **2024**, *34*, 2440–2445.
38. Kim, J.C.; Park, E.J.; Na, D.H. Gastrointestinal Permeation Enhancers for the Development of Oral Peptide Pharmaceuticals. *Pharmaceuticals* **2022**, *15*, 1585. [CrossRef]

International Journal of
Molecular Sciences

MDPI

Review

Exploring the Therapeutic Potential of N-Methyl-D-Aspartate Receptor Antagonists in Neuropathic Pain Management

Ciprian Pușcașu [1], Cornel Chiriță [1,*], Simona Negreș [1] and Nicoleta Mirela Blebea [2]

[1] Faculty of Pharmacy, "Carol Davila" University of Medicine and Pharmacy, Traian Vuia 6, 020956 Bucharest, Romania; ciprian.puscasu@umfcd.ro (C.P.); simona.negres@umfcd.ro (S.N.)
[2] Faculty of Pharmacy, "Ovidius" University of Constanța, Căpitan Aviator Al. Şerbănescu 6, 900470 Constanța, Romania; nicoleta.blebea@gmail.com
* Correspondence: cornel.chirita@umfcd.ro; Tel.: +40-740429281

Abstract: Neuropathic pain (NeP) is a complex and debilitating condition that impacts millions of people globally. Although various treatment options exist, their effectiveness is often limited, and they can be accompanied by significant side effects. In recent years, there has been increasing interest in targeting the N-methyl-D-aspartate receptor (NMDAR) as a potential therapeutic approach to alleviate different types of neuropathic pain. This narrative review aims to provide a comprehensive examination of NMDAR antagonists, specifically ketamine, memantine, methadone, amantadine, carbamazepine, valproic acid, phenytoin, dextromethorphan, riluzole, and levorphanol, in the management of NeP. By analyzing and summarizing current preclinical and clinical studies, this review seeks to evaluate the efficacy of these pharmacologic agents in providing adequate relief for NeP.

Keywords: neuropathic pain; NMDA receptor antagonist; ketamine; memantine; methadone

1. Introduction

Neuropathic pain (NeP) is "pain arising as a direct consequence of a lesion or disease of the somatosensory system" and is characterized by a paradoxical association with sensory loss [1]. NeP can be caused by traumatic nerve, spinal cord, or brain injury (including stroke) or may be linked to various conditions like diabetes, infection, multiple sclerosis [2], or cancer, as well as the harmful impact of chemotherapy drugs [3,4].

A systematic review of prevalence studies based on the general population determined that the prevalence of pain with neuropathic characteristics falls within the range of 7% to 10% [5]. Furthermore, recent research indicated that the age-standardized prevalence of chronic polyneuropathy is 3.3% for the European Union, 3.0% for the United States, and 2.3% for the global population and is projected to rise by approximately 25% over the next 20 years, taking into account anticipated age distribution [6].

Effective treatment of NeP presents significant challenges and is associated with significant decreases in quality of life, along with a substantial economic burden [7]. Patients suffering from chronic NeP typically experience poorer physical and mental health in comparison to those with other forms of chronic pain, even after adjusting the pain intensity [8–10]. The link to diminished physical and mental health implies that the quality of life is negatively impacted by the nature rather than just the intensity of NeP, highlighting the need for a comprehensive treatment approach [11]. Managing NeP can be challenging and often involves a trial-and-error process [12]. Conventional analgesics may not effectively relieve pain for NeP patients [13,14]. In a survey, NeP patients were more likely to use opioids and multiple pain medications but reported less pain relief [15]. However, it has been shown that adjunctive therapy with cannabidiol may have therapeutic potential in neuropathic pain [16]. First-line therapies for NeP, like tricyclic antidepressants, selective inhibitors of serotonin and norepinephrine reuptake, and gabapentinoids, require careful dosing due to potential side effects, particularly in elderly patients [17,18].

Nevertheless, in recent years there has been growing interest in antagonists targeting the presynaptic N-methyl-D-aspartate receptor (NMDAR) to alleviate pain from various types of NeP. The NMDAR is the most complex glutamatergic receptor, and its hyper/hypofunction leads to the development of various neurological disorders [19]. It is a ligand-dependent receptor, widely distributed in the brain and spinal cord, especially in the hippocampus and cerebral cortex, whose activation depends on the levels of glutamate and glycine. However, this ionic channel is the only one that allows the conduction of calcium (Ca^{2+}), sodium (Na^{+}), and potassium (K^{+}) [20]. Research indicates that the NMDAR within the dorsal horn is significant in both inflammation and central sensitization induced by nerve injury [21]. Activation of the NMDAR leads to disruptions in the sensory system, both peripheral and central, causing neuronal excitation and abnormal pain symptoms such as spontaneous pain, allodynia, and hyperalgesia [22,23]. While central NMDARs, particularly those in the spinal cord, remain a focus of study, growing evidence indicates that NMDARs in peripheral tissues and visceral pain pathways also contribute significantly to nociception. This suggests that, in chronic pain conditions, NMDAR activation occurs at various levels of the neural axis, making each level a potential target for therapeutic intervention [23].

Primary afferent neurons in the spinal dorsal horn release various neurotransmitters, including glutamate, IL-1β (interleukin 1β), ATP (adenosine triphosphate), TNF-α (tumor necrosis factor α), and NGF (nerve growth factor). Glutamate and IL-1β activate NMDAR on secondary neurons, causing Ca^{2+} influx and activating AMPARs (α-amino-3-hydroxy-5-methyl-4-isoxazolepropionic acid receptors), which reduces GABA (gamma-aminobutyric acid) release from inhibitory interneurons. ATP activates the P2Y4 receptor on secondary neurons, further stimulating NMDAR and leading to the activation of the JNK (c-Jun N-terminal kinase), P38, and MAPK (mitogen-activated protein kinase) pathways. These processes contribute to central sensitization and synaptic remodeling [24]. NeP resulting from nerve injury involves increased glutamate release from primary afferent terminals, activating AMPA and mGluR5 (metabotropic glutamate receptor 5) receptors in the spinal cord [25–27]. Glutamate interacts with postsynaptic receptors for excitatory neurotransmission, while NMDARs can enhance the release of neurotransmitters like substance P in the spinal dorsal horn [28]. Presynaptic NMDARs may boost glutamate release, and μ-opioid receptor stimulation initiates long-term potentiation in pain pathways [29]. Heightened activation of glutamate receptors contributes to the amplification of excitatory synaptic signaling in chronic pain's nociceptive pathways. Excessive activation of glutamate receptors in spinal dorsal horn neurons is primarily caused by increased glutamate release from primary afferent terminals within the spinal dorsal horn, enhanced numbers and functionality of glutamate receptors, and impaired clearance of glutamate [30–33]. In normal conditions, the NMDAR is blocked by Mg^{2+} and is only activated briefly [34]. However, in abnormal situations, the removal of Mg^{2+} leads to overactivation of the receptor, causing excessive Ca^{2+} influx and potentially triggering cell death processes like apoptosis or necrosis [35]. Moreover, in NeP, NMDARs are involved in a process called "wind-up", where repeated stimulation of pain signals leads to an increased response to pain. This phenomenon can result in chronic pain conditions where even non-painful stimuli are perceived as painful [36,37].

Considering these, the main objective of this narrative review is to comprehensively examine the use of the following NMDAR antagonists in managing NeP: ketamine, memantine, methadone, amantadine, carbamazepine, valproic acid, phenytoin, dextromethorphan, riluzole, and levorphanol (Table 1). This review aims to analyze and summarize the current literature, including preclinical and clinical studies, and to evaluate whether specific pharmacologic agents provide adequate relief for NeP, as there is no such review to date.

Table 1. NMDAR antagonists included in this review and their molecular formulas.

Drug	Molecular Formula
Amantadine	$C_{10}H_{17}N$
Carbamazepine	$C_{15}H_{12}N_2O$
Dextromethorphan	$C_{18}H_{25}NO$
Ketamine	$C_{13}H_{16}ClNO$
Levorphanol	$C_{17}H_{23}NO$
Memantine	$C_{12}H_{21}N$
Methadone	$C_{21}H_{27}NO$
Phenytoin	$C_{15}H_{12}N_2O_2$
Riluzole	$C_8H_{10}FN_3OS$
Valproic acid	$C_8H_{16}O_2$

2. Scope

This narrative review analyzes a combined total of 50 preclinical and 52 clinical studies. The findings from these studies are condensed and presented in the following tables in Section 3, arranged in chronological order. Among the NMDAR antagonists investigated in this study, ketamine emerges as the most studied for treating NeP, with 24 studies dedicated to its examination. Conversely, evidence for the efficacy of amantadine and levorphanol is notably scarce.

It should be noted that, in the preclinical studies, the most commonly assessed parameters were mechanical hypersensitivity and hypersensitivity to a hot stimulus. A variety of animal models were used to evaluate the benefits of NMDAR antagonists, including chronic constriction injury (CCI), streptozotocin (STZ)-induced diabetic neuropathy (DN), chemotherapy-induced polyneuropathy (CIPN), spinal cord injury (SCI), and spinal nerve ligation (SNL).

In clinical studies, pain severity was the primary outcome investigated in many of the studies, with the Numeric Rating Scale (NRS) and the Visual Analog Scale (VAS) being the most commonly used pain scales. However, other parameters were also investigated, such as the frequency of pain attacks, paresthesia, allodynia, walking ability, quality of life, depression, and anxiety. The results from clinical studies are based on various types of NeP conditions, including DN, post-herpetic neuralgia (PHN), trigeminal neuralgia (TN), cancer-related neuropathic pain (CRNP), radiculopathy, CIPN, and chronic idiopathic axonal polyneuropathy (CIAP).

3. Discussions

Since the late 1980s, it has been recognized that NMDAR antagonists can reduce neuronal hyperexcitability and alleviate pain. Various NMDAR antagonists have been studied in both preclinical and clinical pain research [38]. Despite numerous studies, there remains no consensus on the effectiveness of NMDAR antagonists for NeP, prompting the need for the current narrative review. This section explores the benefits of NMDAR antagonists for NeP treatment based on the analysis of preclinical and clinical studies included in this review.

3.1. Ketamine

Ketamine binds non-competitively to NMDARs at the phencyclidine site, affecting receptors through allosteric mechanisms [39]. It equally binds NMDARs subtypes 2A to 2D and surpasses the normal capacity of NMDAR magnesium-dependent gating [40]. Ketamine is rapidly absorbed, with high bioavailability (around 93%), but only 17% is absorbed after first-pass metabolism. It undergoes hepatic metabolism, with norketamine as a major metabolite [39]. Most of the administered dose is recovered in urine as metabolites.

Ketamine can also be eliminated through bile and feces [41]. Side effects include apnea, sedation, increased salivary secretions, hallucinations, dizziness, and drowsiness [39,42].

Ketamine, as a potent NMDAR antagonist, effectively blocks the receptor and reduces the neuronal hyperexcitability associated with NeP [43]. Moreover, research shows that ketamine levels in the brain are linked to pain relief in ischemic pain [44]. Additionally, studies suggest that ketamine decreases connectivity in brain regions involved in pain perception and emotional processing [45]. Ketamine is likely the most extensively studied NMDAR antagonist for treating NeP, resulting in the inclusion of the highest number of studies among all NMDAR antagonists in our review. We analyzed over 24 studies that examined the use of ketamine in treating NeP, including 13 preclinical studies and 11 clinical studies (Table 2).

Of the preclinical studies analyzed, 10 out of 13 indicated that ketamine has a positive effect in reducing pain sensitivity associated with NeP. In contrast, Salvat et al. showed that, in a CCI model of neuropathy in mice, ketamine at a dose of 15 mg/kg was effective as a treatment solely during the early postsurgical period [46]. In addition, the studies conducted by Kroin et al. [47] and Humo et al. [48] indicated that the effect of ketamine on NeP is not long-lasting. On the other hand, two studies showcased the anti-inflammatory potential of ketamine by effectively reducing the levels of pro-inflammatory cytokines like TNF-α, IL-6 (interleukin 6), and IL-1β [49,50]. Furthermore, Tai et al. [49] emphasized that enhancing the cage setup with additional space and varied objects for physical activity improved the effectiveness of ketamine in reducing pain sensitivity and promoted tissue integrity and locomotion by targeting the glutamatergic system.

Clinical studies support the results of preclinical research, with 8 out of 11 studies indicating the benefits of ketamine in reducing NeP. Over half of the studies (six studies) utilized intravenous (i.v.) administration, yielding contrasting results. Therefore, three studies documented that ketamine effectively relieved pain [51–53], whereas one study indicated that only half of the patients experienced pain reduction 1 month after treatment [54]. On the other hand, two studies reported no notable pain reduction after i.v. treatment with ketamine [55,56]. When given orally, ketamine showed clinically effective outcomes in two studies [57,58]. However, a study conducted by Fallon et al., which was the largest study involving 214 patients with CIPN, demonstrated poor results [59]. In other studies, topical administration [60] and subcutaneous (s.c.) infusion [61] of ketamine were also examined, showing positive results. Crucially, the effects of ketamine varied across different types of NeP, ranging from SCI to dentoalveolar NeP and CRNP. Rabi et al. [60] examined the impact of 10% topical ketamine on five patients with NeP due to SCI over a 2-week treatment period. They found that all patients experienced a decrease in their pain levels, as shown in the NPS. Moreno-Hay et al. [53] reported a case involving a patient with dentoalveolar NeP who underwent five regimens of ketamine infusion over 5 years. The patient's NeP symptoms were effectively managed, leading to the discontinuation of prior methadone treatment. Provido-Aljibe et al. [61] investigated 41 patients with CRNP, highlighting the positive effects of ketamine on alleviating neuropathic pain. Ketamine was administered at doses ranging from 75 to 475 mg via s.c. infusion, and the pain intensity was assessed using the NPS (Numerical Pain Score). Finally, Martin et al. [58] examined the advantages of incorporating either memantine or dextromethorphan into the treatment regimen of i.v. ketamine for 60 patients with refractory NeP. Dextromethorphan, but not memantine, was observed to extend the pain relief provided by ketamine by up to 1 month, as indicated by VAS and Neuropathic Pain Symptom Inventory (NPSI) ratings.

Table 2. Preclinical and clinical studies that evaluated the effect of ketamine on NeP.

Ketamine					
Preclinical Studies					
First Author/Reference	**Animals**	**Animal Model**	**Dosage**	**Route/Frequency**	**Results**
M'Dahoma et al. (2015) [62]	Male Sprague–Dawley rats	CCI	50 mg/kg bw	i.p. single dose	Alleviated mechanical hypersensitivity in von Frey test.
Mak et al. (2015) [63]	Male Wistar rats	STZ-induced DN	20 mg/kg bw	s.c. 5-day infusion	Demonstrated antinociceptive action in radiant heat plantar test and tail-flick test that lasted for 4 weeks.
Claudino et al. (2018) [64]	Male Wistar rats	CION	0.5–1 mg/kg bw	intranasal single dose	0.5 mg/kg effectively reversed heat-induced hypersensitivity in the radiant heat test, while 1 mg/kg was found to alleviate mechanical hypersensitivity in the von Frey test.
Doncheva et al. (2018) [65]	Male Wistar rats	CCI	50 mg/kg bw	i.p. single dose	Alleviated hypersensitivity in both hot-plate test and analgesia-meter test.
Pan et al. (2018) [66]	Male Sprague–Dawley rats	SNI	10 mg/kg bw	i.p. single dose	Reversed mechanical hypersensitivity in the von Frey test.
Salvat et al. (2018) [46]	Male 6J mice	CCI	15 mg/kg bw	i.p. 10 days	Provided analgesic effects only in the initial stages after surgery in the von Frey test.
Fang et al. (2019) [67]	Male Sprague–Dawley rats	SNI	10 mg/kg bw	i.p. single dose	Successfully alleviated the mechanical sensitivity in the von Frey test.
Kroin et al. (2019) [47]	Female D1 mice	SNI	10 mg/kg Bw	i.p. single dose	Did not produce long-lasting analgesia in von Frey test.
Humo et al. (2020) [48]	Male C75BL/6 mice	CCI	15 mg/kg bw	i.p. single dose	Provided temporary relief from increased sensitivity to mechanical stimuli in the von Frey test, with effects lasting less than 24 h.
Tai et al. (2021) [49]	Male Sprague–Dawley rats	SCI	30 mg/kg bw	i.m. for 10 days, starting from day 8 after SCI	In combination with environmental enrichment, improved the alleviation of pain in both von Frey test and plantar test, supported tissue health and mobility; reduced activation of MAPK family, NF-κB, and IL-1β signaling, while the levels of excitatory amino acid transporter 2 were restored.
Kim et al. (2022) [68]	Male Wistar rats	PSNL	5–10 mg/kg bw	i.p. 5 weeks, with 2 weeks pause after the first 4 weeks	The higher dose resulted in a significant increase in the mechanical withdrawal threshold during the von Frey test, which lasted for over 2 weeks.

Table 2. *Cont.*

Ketamine					
Preclinical Studies					
First Author/Reference	**Animals**	**Animal Model**	**Dosage**	**Route/Frequency**	**Results**
Seo et al. (2023) [69]	Male Sprague–Dawley rats	SNI	50 mg/kg bw	i.p. in the 15, 18, 21 day after SNI	Improved symptoms of NeP in the von Frey test and dry ice test, suppressed the presence of NMDA receptors and ATF-6 expression during ER stress.
Han et al. (2023) [50]	Male Sprague–Dawley rats	CCI	5–15 mg/kg bw	i.p. 14 days	Efficiently alleviated mechanical and thermal hyperalgesia in von Frey and radiant heat tests; decreased TNF-α, IL-6, IL-1β levels and p62 expression; upregulated C3II/LC3I and Beclin1 expression.
Clinical Studies					
First Author/Reference	**Population**	**Type of NeP**	**Dosage**	**Route/Frequency**	**Results**
Kim et al. (2015) [51]	N = 30	Severe NeP	Ketamine 1 mg/kg bw or Magnesium sulfate 30 mg/kg bw	i.v. for 1 h	Out of 15 patients, 10 recorded pain reduction according to VAS score.
Rabi et al. (2016) [60]	N = 5	SCI patients with NeP	10% cream	topical ×3 times/day 2 weeks	After the 2-week period, all five participants experienced a reduction in their pain levels as indicated in the NPS.
Rigo et al. (2017) [57]	N = 42	Refractory chronic NeP	Ketamine 30 mg or Methadone 3 mg or Methadone 3 mg + Ketamine 30 mg	orally 90 days	Only the group treated with ketamine alone demonstrated a noticeable pain reduction according to VAS and also an alleviation of allodynia.
Fallon et al. (2018) [59]	N = 214	CIPN	40–400 mg	orally 16 days	Showed no significant difference in pain reduction according to Sensory Component of the Short Form McGill Pain Questionnaire.
Czarnetzki et al. (2018) [55]	N = 160	NeP after back surgery	0.25 mg/kg bw preoperatively, 0.25 mg/kg bw intraoperatively, 0.1 mg/kg bw from 1 h before the end of surgery and continuing until the patient's discharge from the recovery room.	i.v.	The low-dose infusion administered during the perioperative period did not show any impact on the occurrence of neuropathic lower back pain 6 or 12 months after surgery, according to the DN4 questionnaire.

Table 2. *Cont.*

Clinical Studies					
First Author/Reference	Population	Type of NeP	Dosage	Route/Frequency	Results
Bosma et al. (2018) [54]	N = 30	Refractory NeP	0.5–2 mg/kg bw	i.v. 6 h/day 5 days	After 1 month post-treatment, about 50% of patients experienced pain reduction according to Brief Pain Inventory questionnaire.
Weber et al. (2018) [52]	N = 1	Bilateral neuropathic leg pain	7 μg/kg/min	i.v. 5 days	Demonstrated fast-acting pain-relieving effects, with 70% reduction of pain, according to rating scale of burning quality, that persisted for a duration of 5 months after the initial administration.
Moreno-Hay et al. (2018) [53]	N = 1	Dentoalveolar NeP	20–50 mg	i.v. 5 infusions over 4 years	The patient's NeP symptoms were efficiently treated, and methadone was eventually stopped.
Martin et al. (2019) [58]	N = 60	Refractory NeP	Ketamine 0.4–0.5 mg/kg bw, FOLLOWED BY Dextromethorphan 90 mg or Memantine 20 mg	Ketamine i.v. (infusion, 2 h) Dextromethorphan orally, 12 weeks Memantine orally, 12 weeks	Dextromethorphan, not memantine, was found to prolong the pain-relieving effects of ketamine for up to 1 month, according to VAS and NPSI.
Pickering et al. (2020) [56]	N = 20	Refractory chronic NeP	Ketamine 0.5 mg/kg bw or Ketamine 0.5 mg/kg bw + Magnesium sulfate 3 g	One infusion every 35 days for 3 times	At 35 days after infusion, ketamine did not provide pain relief according to four-item Neuropathic Pain Questionnaire; when combined with magnesium, the analgesic effects were not further enhanced.
Provido-Aljibe et al. (2022) [61]	N = 41	CRNP	75–475 mg	s.c. 5 days	Efficiently decreased the pain levels, according to NPS.

ATF-6—activating transcription factor-6; bw—body weight; CCI—chronic constriction injury; CION—constriction of intraorbital nerve; CIPN—chemotherapy-induced polyneuropathy; CRNP—cancer-related neuropathic pain; DN—diabetic neuropathy; DN4—Douleur Neuropathique en 4 Questions; ER—endoplasmic reticulum; i.p.—intraperitoneally; IL-1β—interleukin 1β; IL-6—interleukin 6; i.m.—intramuscular; i.v.—intravenous; MAPK—mitogen activated protein kinase; NF-κB—nuclear factor κB; NMDA—N-methyl-D-aspartate; NPS—numerical pain score; NPSI—Neuropathic Pain Symptom Inventory;; PSNL—partial sciatic nerve ligation; s.c.—subcutaneous; SCI—spinal cord injury; SNI—spared nerve injury; STZ—streptozotocin; TNF-α—tumor necrosis factor α; VAS—Visual Analog Scale.

3.2. Dextromethorphan

Dextromethorphan acts as a low-affinity uncompetitive NMDAR antagonist, a sigma-1 receptor agonist, and an antagonist of α3/β4 nicotinic receptors [70–72]. It has also been shown to decrease K^+-stimulated glutamate release, possibly through sigma receptor-related mechanisms [73,74]. Additional functions of dextromethorphan appear to involve mild inhibition of serotonin reuptake through suggested high-affinity binding to the serotonin transporter [75]. The bioavailability of dextromethorphan is both poor and inconsistent. This is due to its rapid first-pass metabolism and subsequent elimination [76]. Dextromethorphan is mainly metabolized into dextrorphan and has a half-life ranging from 3 to 30 h, and potential side effects may include confusion, agitation, memory loss, hallucinations, dysarthria, and ataxia [72,76–78].

The initial discovery of dextromethorphan's neuroprotective properties was made by Choi et al., who showed that the drug reduced glutamate-induced neurotoxicity in neocortical cell cultures [79]. The predominant mechanism underlying the neuroprotective potency is believed to be antagonism of the NMDA receptor/channel complex [80]. Moreover, in in vitro studies, both dextromethorphan and its primary metabolite, dextrorphan, have been shown to block the NMDAR in the central nervous system (CNS) and spinal regions. This action leads to the suppression of NMDA-induced convulsions and the reduction of hypoglycemic neuronal damage, with dextrorphan exhibiting a greater affinity for the NMDAR than dextromethorphan [81–83]. Among the four preclinical studies analyzed in this review (Table 3), the findings suggest a favorable outlook for the utilization of dextromethorphan in NeP. Thus, Zbârcea et al. [84] and Fahmi et al. [85] demonstrated the efficacy of dextromethorphan in reversing tactile allodynia when administered orally at a dose of 20 mg/kg and thermal hyperalgesia when administered intrathecally at a dose of 10 nmol. On the other hand, the research findings by Yang et al. [86] emphasized that the antiallodynic effect of dextromethorphan was enhanced when administered alongside oxycodone in mice with SNL. In addition, Shi et al. [87] evaluated the potency of dextromethorphan administered alone and in combination with gapabentin in two different models of NeP (photochemically-induced ischemic SCI and SNI (spared nerve injury)). The results indicated that dextromethorphan alone did not provide any pain relief. In contrast, the combination of dextromethorphan and gabapentin led to complete alleviation of allodynia.

Table 3. Preclinical and clinical studies that evaluated the effect of dextromethorphan on NeP.

			Dextromethorphan		
			Preclinical Studies		
First Author/Reference	**Animals**	**Animal Model**	**Dosage**	**Route/Frequency**	**Results**
Yang et al. (2015) [86]	Male C57BL/6J mice	SNL	Dextromethorphan 10, 20 mg/kg bw, or Oxycodone 1, 3, 5 mg/kg bw, or Dextromethorphan 10 mg/kg bw + Oxycodone 1, 3 mg/kg bw,	Dextromethorphan i.p. 14 days Oxycodone s.c. 14 days	Dextromethorphan alone did not demonstrate any notable long-term impacts. Administered in combination with oxycodone, dextromethorphan enhanced its antiallodynic effect in von Frey test.
Shi et al. (2018) [87]	Male and female Sprague–Dawley rats	Photochemically-induced ischemic SCI and SNI	Dextromethorphan 5–20 mg/kg bw or Gabapentin 7.5–30 mg/kg bw or Dextromethorphan 5–10 mg/kg bw + Gabapentin 7.5–30 mg/kg bw	i.p.	Dextromethorphan alone did not produce any pain relief in von Frey test and ethyl chloride cold test. In comparison, the dextromethorphan–gabapentin combination resulted in complete relief of allodynia, even in lower doses.
Zbârcea et al. (2018) [84]	Male Wistar rats	Vincristine-induced NeP	20 mg/kg bw	orally 7 days	Reversed tactile allodynia in Dynamic Plantar Aesthesiometer test.

Table 3. *Cont.*

Dextromethorphan					
Preclinical Studies					
First Author/Reference	**Animals**	**Animal Model**	**Dosage**	**Route/Frequency**	**Results**
Fahmi et al. (2021) [85]	Male mice	PSNL	10 nmol	intrathecally from day 8 to 14 after PSNL	Alleviated thermal hyperalgesia in stainless-steel heating plate test.
Clinical Studies					
First Author/Reference	**Population**	**Type of NeP**	**Dosage**	**Route/Frequency**	**Results**
Martin et al. (2019) [88]	N = 20	Freeze-injury-induced hyperalgesia model in healthy volunteers	30 mg	orally initially, at 5 h, 10 h, 14 h and once on day 1 after inducing the pain model	Demonstrated antihyperalgesic effects in humans, reversing sensitization in both peripheral and central neurons.

bw—body weight; i.p.—intraperitoneally; PSNL—partial sciatic nerve ligation; s.c.—subcutaneous; SCI—spinal cord injury; SNI—spared nerve injury; SNL—spinal nerve ligation.

The findings from clinical trials are limited, as only 1 study has explored the antihyperalgesic effects of dextromethorphan in humans (Table 2). Martin et al. [88] investigated the action of the NMDAR antagonist in a freeze-injury-induced hyperalgesia model in healthy volunteers and demonstrated that it reversed sensitization in both peripheral and central neurons.

3.3. Memantine

Memantine functions as an NMDAR antagonist with low to moderate affinity, pronounced voltage dependency, and fast blocking and unblocking kinetics [89–91]. Additionally, it also acts as an antagonist at the serotonergic type 3 receptors and nicotinic acetylcholine receptors [45]. Memantine is effectively absorbed after oral administration, reaching peak concentrations in 3–8 h [90]. It undergoes partial metabolism in the liver, with the hepatic CYP450 enzyme system playing a minor role. The primary route of excretion is through urine, with around 48% of the administered dose being excreted unchanged [92]. Agitation, constipation, urinary tract infections, diarrhea, headache, and confusion are the main side effects [93,94].

Animal studies indicate that memantine may serve as a promising alternative for treating NeP, as all five reviewed studies demonstrated positive outcomes (Table 4). A study conducted by Ciotu et al. investigated the effect of memantine in an animal model of paclitaxel induced NeP and showed that the NMDAR antagonist effectively reversed mechanical sensitivity in a dose-dependent manner [95]. Alomar et al. [96] administered a 10 mg/kg dose of memantine to mice with DN, showing that the drug not only effectively reduced thermal and mechanical hypersensitivity but also had the potential to decrease the release of proinflammatory cytokines TNF-α and IL-1β in the spinal cord. The findings of this study are consistent with those of the study carried out by Solmaz et al., which demonstrated the neuroprotective and antioxidant effects of memantine by significantly reducing TNF-α and MDA (malondialdehyde) levels in an animal model of NeP in rats [97]. On the other hand, 10 mg of oral memantine exhibited full neurobehavioral protection against the progression of neuropathy caused by cisplatin [98]. Additionally, preadministration of memantine intrathecally effectively prevented the onset of allodynia and decreased the overactivation of microglia in the spinal dorsal horn induced by SNI [99].

Table 4. Preclinical and clinical studies that evaluated the effect of memantine on NeP.

Memantine					
Preclinical Studies					
First Author/Reference	**Animals**	**Animal Model**	**Dosage**	**Route/Frequency**	**Results**
Solmaz et al. (2015) [97]	Male Sprague–Dawley rats	CIP	15–30 mg/kg bw	i.p. single administration	Significantly reduced TNF-α and MDA levels and CMAP distal latency.
Ciotu et al. (2016) [95]	Male Wistar rats	Paclitaxel-induced NeP	10–30 mg/kg bw	orally 24 days	The sensitivity thresholds returned to normal levels in Dynamic Plantar Aesthesiometer test.
Chen et al. (2019) [99]	Male C57BL/6J mice	SNI	10–30 nmol	intrathecally before SNI	Preadministration of the higher dose successfully blocked the development of allodynia in von Frey and paint brush test; 10 nmol of memantine exhibited a notable impact on reducing the excessive activation of microglia in the spinal dorsal horn caused by SNI.
Salih et al. (2020) [98]	Male BALB/c mice	Cisplatin-induced NeP	5–10 mg/kg bw	orally 30 days	The higher dose showed greater efficacy in protecting against neuropathy, demonstrating full neurobehavioral protection according to open field activity, negative geotaxis, hole-board, and swimming tests.
Alomar et al. (2021) [96]	Male Swiss albino mice	Alloxan-induced DN	10 mg/kg bw	orally . 5 weeks	Reduced pain symptoms in hot-plate and von Frey tests by inhibiting excessive activation of NMDAR1 receptors, lowering glutamate levels, and decreasing the release of TNF-α and IL-1β in the spinal cord.
Clinical Studies					
First Author/Reference	**Population**	**Type of NeP**	**Dosage**	**Route/Frequency**	**Results**
Ahmad-Sabry et al. (2015) [100]	N = 56	CRPS	5–10 mg increased every 4–7 days to a max. of 40–60 g	orally	A total of 13 individuals experienced full recovery, reporting a pain score of 0 on the VAS and the absence of allodynia for a minimum of 9 months; 18 patients displayed significant progress in reducing their VAS scores and managing allodynia symptoms.
Morel et al. (2016) [101]	N = 40	Mastectomy-associated NeP and CIPN	5–20 mg	orally 4 weeks, starting 2 weeks before mastectomy	At 3 months, patients exhibited a notable decrease in post-mastectomy pain intensity, as shown by the NRS. Moreover, in the group that received memantine, the symptoms of CIPN were greatly reduced.
Shaseb et al. (2023) [102]	N = 16	PHN	Memantine 5–10 mg + Gabapentin 300 mg	orally 8 weeks	The combination resulted in a decrease in the intensity of PHN symptoms, according to the DN4 questionnaire.
Jafarzadeh et al. (2023) [103]	N = 143	DN	Memantine 5 mg 1 week Followed by Memantine 10 mg + Gabapentin 300 mg or Gabapentin 300 mg 8 weeks	orally	The average DN4 questionnaire score in the memantine group was significantly lower, and the number of patients with DN in this group notably decreased by the end of the study.

bw—body weight; CIP—clinical illness polyneuropathy; CIPN—chemotherapy-induced polyneuropathy; CMAP—compound muscle action potentials; CRPS—complex regional pain syndrome; DN—diabetic neuropathy; DN4—Douleur Neuropathique 4 Questionnaire; i.p.—intraperitoneally; IL-1β—interleukin-1 beta; MDA—malondialdehyde; NMDAR1—N-methyl-D-aspartate type 1 receptor; NRS—numeric rating scale; PHN—post-herpetic neuralgia; SNI—spared nerve injury; TNF-α—tumor necrosis factor-α; VAS—Visual Analog Scale.

Clinical studies support the results of animal research, as all four studies analyzed in this review demonstrate the benefits of treating NeP with memantine, administered alone or in combination (Table 3). In a retrospective study conducted by Ahmad-Sabry et al. [100], the impact of memantine on 56 patients with CRPS (complex regional pain syndrome) was examined. The study revealed that 13 individuals achieved complete recovery, reporting a pain score of zero on the VAS scale and the absence of allodynia for at least 9 months. Additionally, 18 patients demonstrated significant improvement in reducing their VAS scores and managing symptoms of allodynia. Moreover, a randomized, blinded clinical trial indicates that administering memantine before surgery may help prevent the development of NeP 3 months post-mastectomy, as evidenced by NRS scores, impairment of cognition, and quality of life. Additionally, it hints at the potential for memantine to alleviate dysesthesia and paresthesia caused by chemotherapy [101]. Two studies examined the use of a combination of memantine and gabapentin in different types of NeP. In one study, 16 patients with PHN were treated with memantine 5–10 mg and gabapentin 300 mg, leading to a reduction in the intensity of pain [102]. Another study involved 143 patients with DN who initially received 5 mg of memantine for 1 week, followed by a combination of 10 mg memantine and 300 mg gabapentin for 8 weeks, resulting in a decrease in the severity of pain and a lowered number of patients with DN at the end of the study [103]. Both studies used the DN4 (Douleur Neuropathique 4) questionnaire to assess NeP symptoms.

3.4. *Amantadine*

Amantadine differs from other channel-blocking molecules by causing NMDAR channels to close faster [104]. Amantadine prompts NMDARs to adopt closed conformations after blocking open channels, increasing its affinity despite quick unbinding from open receptors. Therefore, amantadine primarily acts as a gating antagonist rather than a channel blocker, accelerating channel closure to stabilize closed states [104,105]. Furthermore, amantadine seems to induce the release of dopamine from brain cell nerve endings and inhibits M2 protein, which is found in viral membranes [106,107]. Amantadine is effectively absorbed through oral administration in the gastrointestinal tract; around 67% of it binds to plasma proteins, and its primary mode of excretion is through its unchanged form in the urine [108]. The most common side effects include confusion, hallucinations, tremors, seizures, nausea, and dizziness [104,109–111].

Although there is limited literature on the use of amantadine for NeP relief, three preclinical studies and only one clinical trial that were reviewed demonstrated the clinical benefits of amantadine (Table 5). In a rat model of NeP, amantadine reduced the hypersensitivity threshold and the frequency of hypersensitivity responses in a dose-dependent manner. Additionally, amantadine treatment decreased LP (lipid peroxidation) levels while increasing GSH (glutathione) levels in the injured tissue. The study's data indicated that the pain-relieving effects of amantadine treatment are influenced by the reduction of oxidative stress and excitotoxicity linked to the activation of NMDAR [112]. Furthermore, Dogan et al. showed that amantadine decreased TNF-α expression in inflammatory cells surrounding the blood vessels in the substantia grisea and alba, as well as MDA and MPO (myeloperoxidase) levels. Additionally, they observed negative Bax (Bcl-2-associated X protein) expression in neuron and glial cells and positive VEGF (vascular endothelial growth factor) expression in the vascular endothelium following amantadine treatment. These findings suggest that amantadine could improve SCI by promoting angiogenesis, influencing inflammation and apoptosis, reducing oxidative stress, and modulating signaling pathways [113]. Recently, Drummond et al. [114] explored the therapeutic potential of amantadine in a rat model of CIPN. The experimental groups received oral amantadine at doses of 2, 5, 12, 25, and 50 mg/kg daily for 14 days. Amantadine significantly reduced mechanical hyperalgesia in rats treated with 25 and 50 mg/kg, indicating a dose-dependent effect. Additionally, it demonstrated anti-inflammatory properties by activating anti-inflammatory cytokines and decreasing the expression of pro-inflammatory cytokines. Moreover, amantadine exhibited antioxidant effects by enhancing the expression of the antioxidant enzymes SOD (superoxide dismutase) and CAT (catalase) and by modulating apoptotic mediators.

Table 5. Preclinical and clinical studies that evaluated the effect of amantadine on NeP.

Amantadine					
Preclinical Studies					
First Author/Reference	**Animals**	**Animal Model**	**Dosage**	**Route/Frequency**	**Results**
Dogan et al. (2019) [113]	Male Sprague–Dawley rats	SCI	45 mg/kg bw	i.p. 7 days	Decreased MDA, MPO, and TNF-α levels; neuron and glial cell showed negative Bax expression, while vascular endothelium showed positive VEGF expression after treatment.
Mata-Bermudez et al. (2021) [112]	Female Wistar rats	SCI	6.25–50 mg/kg bw	i.p. 15 days	Effectively alleviated pain-related behavior in von Frey test; decreased LP and increased GSH levels in the damaged tissue.
Drummond et al. (2024) [114]	Male Wistar rats	CIPN	2, 5, 12, 25, and 50 mg/kg bw	orally 14 days	Higher doses efficiently reduced mechanical hyperalgesia in digital analgesia-meter test in a dose-dependent manner; decreased IL-6, TNF-α, MIP-1α, Perk, Bax, Casp 3, Casp 9, and CX3CR1 expression; increased Bcl-xl, CAT, SOD, and IL-10 expression.
First Author/Reference	**Population**	**Type of NeP**	**Dosage**	**Route/Frequency**	**Results**
Azimov et al. (2016) [115]	N = 64	Patients with neuropathy of the facial nerve	Amantadine 200 mg or Levodopa 125 mg or Amantadine 200 mg + Levodopa 125 mg	orally	There was a substantial increase in the enhancement of neurostatus dynamics when treated with the combination than with monotherapy, according to the scale of House–Brackmann.

Bax—Bcl-2-associated X protein; bw—body weight; GSH—glutathione; i.p.—intraperitoneally; LP—peroxidation levels; MDA—malondialdehyde; MPO—myeloperoxidase; SCI—spinal cord injury; TNF-α—tumor necrosis factor α; VEGF—vascular endothelial growth factor.

Azimov et al. [115] conducted a clinical trial examining the impact of amantadine, levodopa, and an amantadine–levodopa combination on 64 patients with facial nerve neuropathy. The findings indicated that both drugs had a comparable effect in reducing nerve dysfunction based on the House–Brackmann scale, with the combination showing an even greater effect.

3.5. Valproic Acid

Valproic acid's mechanisms of action are multifaceted, involving inhibition of excitatory responses triggered by NMDA both in vivo and in vitro, as well as NMDA-induced convulsions in vivo, and various other facets of brain glutamatergic activity [116–122]. Moreover, one study demonstrated that valproic acid decreases upregulated NMDA signaling involving arachidonic acid and its metabolites in the brain [123]. Valproic acid also affects the extracellular signal-regulated kinase (ERK) pathway, non-competitively inhibits myo-inositol-1-phosphate synthetase, directly inhibits histone deacetylase (HDAC), increases GABA synthesis, and reduces GABA degradation [124,125]. Both i.v. and oral forms of valproic acid are anticipated to have similar levels of exposure, peak concentration, and minimum concentration at steady state. The drug is primarily metabolized into glucuronide conjugates, with about 30–50% being eliminated through hepatic metabolism [126]. Dose-related side effects of valproic acid include weight gain, hair loss, nausea, vomiting, and rare idiosyncratic reactions such as hematological toxicity, hepatotoxicity, pancreatitis, and polycystic ovary syndrome [127,128]. In addition to these side effects, valproic acid is a known teratogen in humans, meaning it can cause birth defects. It is associated with an increased risk of spina bifida aperta, as well as heart deformities, cleft palate, and limb anomalies [128].

All preclinical studies reviewed suggest the potential advantages of using valproic acid in treating NeP (Table 6). Out of six studies, five showed that the NMDAR antagonist was able to decrease thermal sensitivity [129,130] and mechanical sensitivity [129–132] in various animal models of NeP. Four studies indicated that valproic acid could decrease the release of cytokines like TNF-α, IL-1β, and IL-6 [129,130,132,133]. Furthermore, Chen et al. [129] highlighted that the administration of 300 mg/kg of valproic acid demonstrated an anti-neuroinflammatory effect by reducing pNFκB (phosphorylated nuclear factor-κB)/iNOS (inducible nitric oxide synthase)/COX-2 (cyclooxygenase-2) activation and preventing pAKT (phosphorylated protein kinase B)/pGSK-3β (phosphorylated glycogen synthase kinase-3β)-mediated neuronal death resulting from peripheral nerve injury in rats with CCI (chronic constriction injury). On the other hand, another animal study illustrated that valproic acid shows promise as an anti-inflammatory agent for NeP therapy by regulating microglial activity and inhibiting spinal neuroinflammation through the STAT1 (signal transducer and activator of transcription 1)/NF-κB and JAK2 (Janus kinase 2)/STAT3 (signal transducer and activator of transcription 3) signaling pathways [132]. Furthermore, Wang et al. [133] showed that chitosan nanoparticles labeled with valproic acid facilitated tissue recovery and improved locomotor function. They also enhanced neural stem cell proliferation and the expression of neurotrophic factors (BDNF (brain-derived neurotrophic factor), NGF, and NTF-3 (neurotrophin-3)), while reducing the number of microglia. Additionally, there was an increase in Tuj 1 (class III beta-tubulin)-positive cells in the spinal cord of rats with SCI, indicating that valproic acid labeled chitosan nanoparticles could potentially enhance the differentiation of neural stem cells post-SCI.

The findings from clinical studies are limited, with only one study assessing the impact of valproic acid on NeP (Table 5). In a double-blind, randomized, placebo-controlled study involving 80 patients with radiculopathy, the NMDAR antagonist was assessed in combination with celecoxib and acetaminophen. The study findings indicated that low doses of Na^+ valproate (200 mg), particularly when combined with NSAIDs (nonsteroidal anti-inflammatory drugs), showed significant therapeutic effects in reducing or even eliminating chronic radicular pain. Pain levels were quantitatively assessed using VAS before the intervention and after 10 days [134].

Table 6. Preclinical and clinical studies that evaluated the effect of valproic acid on NeP.

Valproic Acid					
Preclinical Studies					
First Author/Reference	Animals	Animal Model	Dosage	Route/Frequency	Results
Chen et al. (2018) [129]	Male Sprague–Dawley rats	CCI	300 mg/kg bw	i.p. 14 days	Significantly reduced thermal sensitivity and mechanical sensitivity in plantar analgesia-meter and von Frey test; decreased pNFκB, iNOS, COX-2, pro-apoptotic protein, TNF-α, and IL-1β levels.
Elsherbiny et al. (2019) [130]	Male Swiss albino mice	Alloxan-induced DN	25–50 mg/kg bw	orally 5 days	Significantly alleviated thermal and mechanical sensitivity in hot-plate and von Frey test; decreased spinal histone deacetylases, TNF-α, and IL-1β levels.
Chu et al. (2020) [131]	Male Sprague–Dawley rats	SNI	200 mg/kg bw or 10, 20, 50 μg, in 0.5 μl	i.p. or into ventrolateral orbital cortex	Both i.p. injection and local administration demonstrated a significant analgesic effect in a dose-dependent manner in the paw withdrawal threshold test.

Table 6. *Cont.*

Valproic Acid					
Preclinical Studies					
First Author/Reference	**Animals**	**Animal Model**	**Dosage**	**Route/Frequency**	**Results**
Wang et al. (2020) [133]	Male Sprague–Dawley rats	SCI	80 mg/kg bw	i.v. 5 days	Greatly enhanced functional recovery and tissue repair; effectively suppressed reactive astrocytes post-SCI; decreased IL-1β, IL-6, and TNF-α levels.
Wang et al. (2021) [135]	Male Sprague–Dawley rats	SCI	80 mg/kg bw	i.v.	Facilitated the recovery of tissue and locomotor function in Basso–Beattie–Bresnahan test; decreased the number of microglia; increased neural stem cell growth, BDNF, NGF NTF-3, and Tuj-1 positive cells.
Guo et al. (2021) [132]	Male Sprague–Dawley rats	SNL	300 mg/kg bw	i.p. 3 days	I.p. administration effectively reduced mechanical allodynia in von Frey test; decreased TNF-α, IL-1β, and IL-6 levels, spinal cell apoptosis, NF-κB, JAK2, STAT3; increased STAT1.
Clinical Studies					
First Author/Reference	**Population**	**Type of NeP**	**Dosage**	**Route/Frequency**	**Results**
Ghasemian et al. (2020) [134]	N = 80	Radiculopathy	Na^+ valproate 200 mg + Celecoxib 100 mg + Acetaminophen 500 mg	orally 10 days	A low dosage of Na^+ valproate, particularly when combined with NSAIDs, showed promising effectiveness in reducing pain, according to VAS score.

BDNF—brain-derived neurotrophic factor; bw—body weight; CCI—chronic constriction injury; COX-2—cyclooxygenase-2; DN—diabetic neuropathy; i.p.—intraperitoneally; i.v.—intravenous; IL-1β—interleukin 1β; IL-6—interleukin 6; iNOS—inducible nitric oxide synthase; JAK2—Janus kinase 2; NGF—nerve growth factor; NSAIDs—nonsteroidal anti-inflammatory drugs; NF-κB—nuclear factor-κB; NTF-3—neurotrophin-3; SCI—spinal cord injury; SNI—spared nerve injury; SNL—spinal nerve ligation; STAT1—signal transducer and activator of transcription 1; STAT3—signal transducer and activator of transcription 3; TNF-α—tumor necrosis factor α; VAS—Visual Analog Score.

3.6. Carbamazepine

Carbamazepine inhibits NMDA-induced Ca^{2+} influx in cultured neurons and prevents NMDA-initiated convulsions in animals [136,137]. Furthermore, the drug protects against NMDA-mediated neurotoxicity and blocks NMDA-activated membrane currents in cultured spinal cord neurons [138,139]. Additionally, carbamazepine binds to voltage-dependent Na^+ channels, blocking action potentials that typically stimulate nerves, enhances dopamine turnover, and boosts GABA transmission [140–143]. Carbamazepine has a bioavailability of 75–85% when ingested. It undergoes significant metabolism in the liver, primarily by the CYP3A4 hepatic enzyme, which converts it to its active metabolite, carbamazepine-10,11-epoxid [144]. Carbamazepine is primarily excreted in urine as hydroxylated and conjugated metabolites, with minimal amounts of the unchanged drug [144,145]. Common side effects of carbamazepine include dizziness, ataxia, drowsiness, nausea, rash, and somnolence [146]. Agranulocytosis affects around six patients per 1 million annually, while aplastic anemia affects two patients per 1 million [146,147].

In this narrative review we investigated five preclinical studies (Table 7), with all of them demonstrating the potential of carbamazepine to reverse thermal sensitivity [148–151] and mechanical sensitivity [148,150,152], when administered alone [148,150–152] or in combination [149]. One study examined the impact of carbamazepine (20–40 mg/kg) given alone or with gabapentin (30–180 mg/kg) on NeP rats induced by STZ. The findings revealed that carbamazepine at 20 and 40 mg/kg did not show significant effects, but a combination of gabapentin at 90 mg/kg and carbamazepine at 20 mg/kg resulted in a notable increase in latency during the hot-plate test [149]. Dai et al. [150] investigated the effectiveness of incorporating carbamazepine into biodegradable microparticles for sustained perineural release as an analgesic for peripheral injuries. Animals treated with carbamazepine-loaded microparticles showed a two-fold increase in hindpaw withdrawal thresholds compared to controls for up to 14 days. This formulation significantly reduced systemic exposure to carbamazepine, offering substantial pain relief. In another study, the antiallodynic effects of s.c. carbamazepine, baclofen, morphine, and clomipramine were compared in an animal model of IoN-CCI (infraorbital nerve chronic constriction injury). The findings revealed that all drugs exhibited notable antiallodynic effects, with carbamazepine demonstrating the most potent effect [152].

Upon analyzing clinical studies, it was noted that, in recent years, carbamazepine has been assessed for a particular type of NeP (Table 7). Seven out of eight studies evaluated the drug's effectiveness in treating TN, resulting in quite positive outcomes. It is not surprising, as carbamazepine has been widely acknowledged as an effective treatment for this condition for several decades, with studies dating back to 1966 [153,154]. However, two studies compared the effectiveness of carbamazepine and oxcarbazepine in TN patients indicated that, while both drugs relieved pain, oxcarbazepine demonstrated a more significant impact [155,156]. Two additional studies showed that gabapentin was more effective than carbamazepine in TN patients [157,158]. Another study indicated that the combination of carbamazepine with baclofen was more efficient and effective in pain relief, with the carbamazepine–capsaicin combination also showing better results compared to carbamazepine alone, based on the VAS scores of 45 TN patients [159].

The only study that did not assess the effectiveness of carbamazepine on TN was conducted by Khan et al. [160]. In this clinical study, the use of 200 mg of carbamazepine for 8 weeks showed positive clinical outcomes, reducing pain by approximately 80% according to VAS in 50 patients with PHN, demonstrating a potency similar to amitriptyline.

Table 7. Preclinical and clinical studies that evaluated the effect of carbamazepine on NeP.

Carbamazepine					
Preclinical Studies					
First Author/Reference	**Animals**	**Animal Model**	**Dosage**	**Route/Frequency**	**Results**
Kohli et al. (2016) [148]	Sprague–Dawley rats	CCI	20 mg/kg bw	i.p. 14 days	Reversed thermal and mechanical hyperalgesia in hot-plate and pinprick tests.
AL-Mahmood et al. (2016) [149]	Female Sprague–Dawley rats	STZ- induced DN	Carbamazepine 20–40 mg/kg bw or Gabapentin 30–180 mg/kg bw or Carbamazepine 20–40 mg/kg bw + Gabapentin 30–180 mg/kg bw	orally 1 week	Carbamazepine at doses of 20 and 40 mg/kg did not result in a notable effect on hot-plate latency. Conversely, a combination of gabapentin at 90 mg/kg and carbamazepine at 20 mg/kg led to a significant increase in latency.
Deseure et al. (2017) [152]	Male Sprague–Dawley rats	IoN-CCI	Carbamazepine 30 mg or Baclofen 1.06 mg or Morphine 5 mg or Clomipramine 4.18 mg	s.c. 1 week	All medications exhibited significant antiallodynic effects; carbamazepine demonstrated the most potent effects in directed face grooming and von Frey testing.
Dai et al. (2018) [150]	Female Sprague–Dawley rats	CCI	Carbamazepine 100 μg/mL or Carbamazepine-loaded microparticles 10–20 mg in 150 μL saline	14 days local sustained perineural release	The administration of carbamazepine-loaded microparticles resulted in more notable pain relief in von Frey and thermal plantar tests.

Table 7. *Cont.*

Carbamazepine					
Preclinical Studies					
First Author/Reference	**Animals**	**Animal Model**	**Dosage**	**Route/Frequency**	**Results**
Bektas et al. (2019) [151]	Male Sprague –Dawley rats	Capsaicin-induced hyperalgesia	30 mg/kg bw	orally 45 min prior to capsaicin	There was a significant increase in thermal thresholds in plantar test.
Clinical Studies					
First Author/Reference	**Population**	**Type of NeP**	**Dosage**	**Route/Frequency**	**Results**
Shafiq et al. (2015) [155]	N = 202	TN	Carbamazepine 200 mg or Oxcarbazepine 200 mg	orally 8 months	Both medications alleviated pain as per VAS, with oxcarbazepine showing a more noticeable effect.
Syed et al. (2016) [161]	N = 9	TN	100–600 mg	orally 11 years	The pain perception significantly decreased according to FPS and NRS.
Puri et al. (2018) [159]	N = 45	TN	Carbamazepine 600–800 mg or Carbamazepine 600 mg + Baclofen 10–20 mg or Carbamazepine 600 mg orally + Capsaicin 0.25% cream	Carbamazepine/baclofen orally Capsaicin local application 1 month	The combination of carbamazepine with baclofen proves to be more efficient and effective in alleviating pain in patients with TN, with the carbamazepine-capsaicin combination following closely behind in comparison to carbamazepine alone according to VAS.
Kaur et al. (2018) [157]	N = 37	TN	Carbamazepine 400–1200 mg or Gabapentin 600–1800 mg	orally 3 months	Both medications demonstrated effectiveness in reducing pain, with gabapentin showing greater efficiency based on the frequency of the attacks.
Agarwal et al. (2020) [158]	N = 46	TN	Carbamazepine 400–1200 mg or Gabapentin 600–1800 mg	orally 3 months	Both drugs alleviated pain after 3 months of treatment according to VAS, with a more pronounced effect for gabapentin.
Tariq et al. (2021) [162]	N = 30	TN	100 mg	orally 28 days	The average VAS score decreased from 4.53 on day 7 to 3.27 on day 28 after treatment.
Iqbal et al. (2023) [156]	N = 56	TN	Carbamazepine 200 mg or Oxcarbazepine 200 mg	orally up to 7 months	Both medications demonstrated effectiveness based on the frequency of attacks, with oxcarbazepine showing a more pronounced effect.
Khan et al. (2023) [160]	N = 50	PHN	Carbamazepine 200 mg or Amitriptyline 25 mg	orally 8 weeks	Both drugs showed similar effectiveness, with carbamazepine reducing pain by 80% and amitriptyline by 86%, according to VAS.

bw—body weight; CCI—chronic constriction injury; DN—diabetic neuropathy; FPS—faces pain scale; i.p.—intraperitoneally; IoN-CCI—infraorbital nerve chronic constriction injury; NRS—numerical pain rating scale; PHN—postherpetic neuralgia; s.c.—subcutaneous; STZ—streptozotocin; TN—trigeminal neuralgia; VAS—Visual Analog Score.

3.7. Phenytoin

Studies have shown that phenytoin can effectively block NMDA responses, especially those induced by multiple applications of NMDA, as observed in research involving mouse neurons in culture [163]. Furthermore, phenytoin inhibits cortical NMDA-evoked [3H] norepinephrine efflux and NMDA-stimulated acetylcholine release from the striatum [164,165]. It has been shown that phenytoin does not affect NMDA-dependent LTP (long-term potentiation) or primed burst-induced LTP, both of which are NMDA receptor-mediated [166,167]. These findings suggest that the effects of phenytoin may be linked to its impact on voltage-dependent ion channels, while its influence on NMDA-mediated activity could be indirect or secondary. Phenytoin is fully absorbed, and approximately

90% is bound to proteins [168]. It undergoes extensive metabolism, initially converting into a reactive arene oxide intermediate. This reactive intermediate is believed to be accountable for numerous unwanted adverse effects of phenytoin [169]. The majority of phenytoin is eliminated as inactive metabolites through bile excretion [170]. Common adverse effects of phenytoin include rash, blood dyscrasias, hepatitis, nystagmus, ataxia, confusion, and memory loss [171].

This review analyzed two preclinical studies that assessed the effects of phenytoin in the CCI animal model of neuropathic pain (Table 8). Both studies showed that phenytoin effectively reversed thermal and mechanical sensitivity in rats [172,173]. Furthermore, Kocot-Kępska et al. [172] found that phenytoin reduced microglia/macrophage activation and/or infiltration at the spinal cord and dorsal root ganglion levels 7 days post-nerve injury. Significantly, the combination of phenytoin and morphine produced more effective antinociception compared to administering either drug individually.

In the clinical trials, all 12 studies demonstrated phenytoin's effectiveness in reducing NeP (Table 8). In two studies, i.v. phenytoin was administered to patients with TN, resulting in positive outcomes [174,175]. The remaining 10 studies utilized phenytoin cream in concentrations ranging from 5–10%, with the 10% concentration yielding superior results. These studies assessed phenytoin in various types of NeP, from DN [176,177] to small fiber neuropathy [178–180], symmetrical painful neuropathy [181], CIAP, and CIPN [182,183]. Most studies predominantly utilized the NRS as the pain screening tool.

Table 8. Preclinical and clinical studies that evaluated the effect of phenytoin on NeP.

Phenytoin					
Preclinical Studies					
First Author/Reference	**Animals**	**Animal Model**	**Dosage**	**Route/Frequency**	**Results**
Hesari et al. (2016) [173]	Male Wistar rats	CCI	50 mg/kg bw	i.p. 14 days	Significantly reversed thermal and mechanical sensitivity in von Frey, pinprick, acetone, and hot-plate tests.
Kocot-Kępska et al. (2023) [172]	Male Wistar rats	CCI	10–60 mg/kg bw	i.p. single dose day 7 after CCI	Administered in single and repeated doses, reduced thermal and mechanical sensitivity in von Frey and cold-plate tests; effectively decreased the activation and/or infiltration of microglia/macrophages in both the spinal cord and dorsal root ganglia; the phenytoin–morphine combination resulted in superior pain relief compared to administering each drug separately.
			30 mg/kg bw	i.p. 16 h and 1 h before CCI	
			Phenytoin 30 mg/kg bw Day 8 after CCI FOLLOWED BY Morphine 10 mg/kg bw	i.p.	
Clinical Studies					
First Author/Reference	**Population**	**Type of NeP**	**Dosage**	**Route/Frequency**	**Results**
Kopsky et al. (2017) [182]	N = 1	60 years old male with peripheral NeP	5–10% cream	2 times daily 3 months	The 5% cream quickly reduced allodynia on the NRS. With the 10% cream, the person experienced complete relief from allodynia for the entire night.
	N = 1	71 years old with CIAP+CINP	5% cream	3 times daily 2 months	After the application, the patient scored 0 on the NRS.
	N = 1	54 years old with CIPN	5–10% cream	2–3 times daily 1 months	Both concentrations of the cream resulted in a reduction of pain levels on the NRS, with the 10% cream showing a more pronounced effect.
Kopsky et al. (2018) [184]	N = 70	Different types of NeP	5–10% cream	Up to 41 weeks	Resulted in a significant reduction in NeP, with more pronounced effects for the 10% concentration, according to the NRS.
Kopsky et al. (2018) [185]	N = 21	Localized NeP	10% cream	-	After 30 min, the average decrease in pain as recorded by the NRS within the region treated was 3.3.

Table 8. *Cont.*

Phenytoin					
Preclinical Studies					
First Author/Reference	Animals	Animal Model	Dosage	Route/Frequency	Results
Hesselink et al. (2017) [178]	N = 5	SFN	10% cream	-	In every instance, the time it took for the pain relief to become noticeable was less than 20 min, with four out of five cases experiencing relief within just 10 min.
Kopsky et al. (2020) [181]	N = 12	Symmetrical painful polyneuropathy	10–20% cream	6 weeks	Half of the patients exhibited positive responses to treatment on the NRS.
Hesselink et al. (2017) [179]	N = 1	SFN	10% cream	several weeks	The application of 10% cream resulted in a significant (50%) reduction in pain. The pain-relieving effects of 10% cream typically begin to take effect within approximately 5 min of application, providing relief for up to 20 h in this particular instance. The pain screening tool used was the NRS.
Hesselink et al. (2018) [183]	N = 1	CIAP	10% cream	-	Within 20 min of applying the cream, the pain in the right foot stayed constant, but the pain in the left foot decreased from a score of 7 to 2 on the NRS.
Hesselink et al. (2024) [180]	N = 3	SFN	5% cream	-	The pain experienced by two patients was significantly reduced (by more than 50%), while one patient reported complete disappearance of the pain. The pain screening tool used was the NRS.
Hesselink et al. (2016) [176]	N = 1	DN	5% cream	-	The outcome led to a significant decrease (of 50%) in neuropathic pain, according to the DN4.
Hesselink et al. (2018) [177]	N = 1	DN	10–30% cream	-	Phenytoin cream, applied in a single-blind manner, decreased pain levels on the NRS within just 5 min of application.
Schnell et al. (2020) [174]	N = 39	TN	10–20 mg/kg	i.v.	Nearly 90% of individuals experienced instant relief from pain in TN crisis.
Vargas et al. (2015) [175]	N = 1	TN	15 mg/kg	i.v.	After the infusion, the patient reported that his pain level as 2 out of 10; he was able to communicate clearly and effortlessly.

bw—body weight; CCI—chronic constriction injury; CIAP—chronic idiopathic axonal polyneuropathy; CIPN—chemotherapy-induced polyneuropathy; DN4—Douleur Neuropathique 4 Questionnaire; DN—diabetic neuropathy; i.p.—intraperitoneally; NRS—numeric rating scale; SFN—small fiber neuropathy; TN—trigeminal neuralgia.

3.8. Riluzole

Riluzole stabilizes voltage-dependent Na^+ channels in their inactivated state and activates a G-protein-dependent process, leading to reduced glutamate release and inhibition of postsynaptic events mediated by NMDARs [186]. These synergistic mechanisms block excitotoxicity, providing powerful neuroprotection with minimal side effects compared to excitatory amino acid receptor antagonists. By directly and non-competitively inhibiting NMDAR activity, riluzole reduces the excitotoxic effects of excessive glutamate, thereby preventing neuronal damage and death [187–190]. Riluzole has an estimated oral bioavailability of 60%. After absorption, it is metabolized in the liver by cytochrome P450 into N-hydroxyl riluzole, which is then glucuronidated. The drug's metabolites are mainly excreted through the kidneys, with less than 1% of the absorbed dose eliminated in urine, and about 10% of the metabolized riluzole is excreted in the feces [191]. The most frequently observed adverse effects include weakness, nausea, dizziness, cough, and abdominal discomfort [192].

Out of the 13 preclinical studies reviewed (Table 9), 7 showed that riluzole effectively reduced thermal and mechanical sensitivity in various types of NeP [193–199]. The only study that found riluzole to have no effect on mechanical sensitivity in an animal model of NeP in rats was conducted by Thompson et al. [200], using doses ranging from 2 to 8 mg/kg. Additionally, two animal studies highlighted that riluzole-treated rats exhibited improved motor function recovery [201,202]. The mechanisms by which riluzole provides neuroprotection are complex. They include preventing the downregulation of GLT-1 (glutamate transporter-1), the increase of glutamate concentration, and the activation of NMDAR [195], downregulating P2X7R (P2X purinoceptor 7) expression, inhibiting microglial activation [194], activating the SK (small-conductance Ca^{2+}-activated K^{+}) channel in the amygdala [200], inhibiting TRPM8 (transient receptor potential melastatin 8) overexpression in the dorsal root ganglions [197], and activating the GSK-3β (glycogen synthase kinase-3 beta)/CRMP-2 signaling pathway (collapsin response mediator protein-2) [203]. Furthermore, riluzole demonstrated an anti-inflammatory effect by reducing the levels of proinflammatory cytokines in rats with SCI [202]. Martins et al. [201] demonstrated that, in rats with SCI, the combination of riluzole and dantrolene provided enhanced neuroprotection by more effectively reducing apoptotic cell death compared to when each drug was administered individually. On the other hand, Ghayour et al. found that both acute and chronic administration of riluzole slowed the regeneration process and delayed the recovery of motor function rats with SNI [204].

Although animal studies suggested some advantages of riluzole in NeP, none of the three clinical trials analyzed yielded positive outcomes for riluzole therapy (Table 9). Consequently, riluzole treatment did not lead to notable improvement in patients with NeP linked to secondary progressive multiple sclerosis and cervical spine injury [205,206]. Moreover, in patients experiencing oxaliplatin-induced neuropathy, riluzole exacerbated neuropathic symptoms [207].

Table 9. Preclinical and clinical studies that evaluated the effect of riluzole on NeP.

Riluzole					
Preclinical Studies					
First Author/Reference	**Animals**	**Animal Model**	**Dosage**	**Route/Frequency**	**Results**
Karadimas et al. (2015) [193]	Female Sprague–Dawley rats	CSM	8 mg/kg bw	i.p. 2 weeks	Attenuated pain sensitivity in von Frey and tail flick tests.
Jiang et al. (2016) [194]	Male Sprague–Dawley rats	CCI	4 mg/kg bw	i.p. 5 days	Reduced thermal hyperalgesia and mechanical allodynia in plantar analgesia meter and von Frey tests; decreased the expression of P2X7R; suppressed microglial activation in the spinal cord dorsal horn.
Ghayour et al. 2017 [204]	Male Wistar rats	SNI	6–8 mg/kg bw and 4–6 mg/kg bw	i.p. single dose and i.p. 8 weeks	Acute and chronic treatment slowed the regeneration process and delayed the recovery of motor function.
Yamamoto et al. (2017) [195]	Male Sprague–Dawley rats	Oxaliplatin-induced neuropathy	12 mg/kg bw	orally 27 days	Alleviated mechanical allodynia in the von Frey test, suppressed the rise in glutamate concentration, and prevented the reduction of GLT-1 expression.

Table 9. *Cont.*

Riluzole					
Preclinical Studies					
First Author/Reference	**Animals**	**Animal Model**	**Dosage**	**Route/Frequency**	**Results**
Thompson et al. (2018) [200]	Male Sprague–Dawley rats	SNL	2–8 mg/kg bw	i.p. 14 days	Inhibited vocalizations and depression-like behaviors in FST; did not affect withdrawal thresholds in the von Frey test; enhanced the mAHP mediated by SK channels in amygdala neurons.
Poupon et al. (2018) [196]	57Bl/6JRj mice	Oxaliplatin-induced neuropathy	60 µg/mL	in drinking water 28 days	Prevented cold and mechanical hypersensitivities in various tests (tail immersion, acetone von Frey, and tail brush), dexterity impairment (beam walk and adhesive removal tests), and depression-like symptoms chemotherapy (FST test); significantly prevented the decrease of NCV.
Yamamoto et al. (2018) [197]	Male Sprague–Dawley rats	Oxaliplatin-induced neuropathy	12 mg/kg bw	orally 4 days	Reduced cold allodynia in acetone test via inhibition of TRPM8 overexpression in the dorsal root ganglions.
Martins et al. (2018) [201]	Male Wistar rats	SCI	Riluzole 4 mg/kg bw or Dantrolene 10 mg/kg bw or Riluzole 4 mg/kg bw + Dantrolene 10 mg/kg bw	i.p. 15 min and 1 h before SCI	The combination synergistically enhanced neuroprotection by reducing apoptotic cell death; significantly improved motor recovery as measured by the BBB locomotor rating scale.
Zhang et al. (2018) [198]	Male Sprague–Dawley rats	SNL	12 mg/kg bw	i.p. single dose at 5 days post SNL surgery	Decreased mechanical sensitivity in von Frey test for at least 14 days; prompted LTD of spinal nociceptive signaling by acting on postsynaptic GluR2 receptors.
Wu et al. (2020) [202]	Female Wistar rats	SCI	4 mg/kg bw	i.p. 7 days	Significant increased locomotor scores (BBB score, inclined plane test); reduced spinal cavity size, increased levels of MPB and neurofilament 200; decreased levels of proinflammatory cytokines (IL-13, IL-1β, IL-6, TNF-α, TGF-β1); induced polarization of M2 microglia/macrophages.
Taiji et al. (2021) [199]	Male Sprague–Dawley rats	SNI	4 mg/kg bw	i.p. single dose at 7 days after surgery	Reduced mechanical allodynia in von Frey test.

Table 9. *Cont.*

Clinical Studies					
First Author/Reference	**Population**	**Type of NeP**	**Dosage**	**Route/Frequency**	**Results**
Wu et al. (2022) [208]	Female Wistar rats	SCI	6 mg/kg bw	i.p. single dose	Decreased IL-1β mRNA, protected neurons from damage, and reduced the activation of microglia/macrophage M1 expression; increased the levels of IL-33 and its receptor ST2 in microglia/macrophages in the spinal cord.
Xu et al. (2022) [203]	Female Wistar rats	SCI	4 mg/kg bw	i.p. 7 days	Promotes neurological functional restoration by activating the GSK-3β/CRMP-2 signaling pathway.
Trinh et al. 2021 [207]	N = 52	Oxaliplatin-induced neuropathy	50 mg	orally prior to the second oxaliplatin dose, continuing to the end of treatment	According to TNS and FACT-GOG NTX scores, riluzole worsens neuropathy symptoms, neurotoxicity, and quality of life associated with oxaliplatin treatment.
Foley et al. (2022) [205]	N = 445	NeP associated with secondary progressive multiple sclerosis	50 mg	orally 1/day for 4 weeks, then 2/day until week 96	Riluzole showed no positive effect on any NeP outcome measure (NPS and Brief Pain Inventory).
Kumarasam et al. (2022) [206]	N = 52	Cervical spine injury	100 mg FOLLOWED BY 50 mg	orally, 3 days FOLLOWED BY orally 13 days	Riluzole therapy did not result in a significant improvement in the severity of NeP, as measured by the NRS.

BBB—Basso–Beattie–Bresnahan; bw—body weight; CCI—chronic constriction injury; CRMP-2—collapsin response mediator protein-2; CSM—cervical spondylotic myelopathy; FACT-GOG NTX—Functional Assessment of Cancer Therapy/Gynecologic Oncology Group–Neurotoxicity; FST—forced swim test; GLT-1—glutamate transporter 1; GluR2—glutamate receptor 2; GSK-3β—glycogen synthase kinase-3 beta; i.p.—intraperitoneally, IL-13—interleukin 13; IL-1β—interleukin 1β; IL-33—interleukin 33; IL-6—interleukin 6; LTD—long-term depression; mAHP—medium after hyperpolarization; NCV—nerve conduction velocity; NPS—neuropathic pain score; NRS—numerical rating scale for neuropathic pain; P2X7R—P2X purinoceptor 7; SCI—spinal cord injury; SK—small-conductance Ca^{2+}-activated K+; SNI—spared nerve injury; SNL—spinal nerve ligation; TGF-β1—transforming growth factor beta 1; TNF-α—tumor necrosis factor α; TNS—total neuropathy score-reduced; TRPM8—transient receptor potential melastatin 8.

3.9. Levorphanol

Levorphanol exhibits unique pharmacological properties by acting as a mu, delta, kappa1, and kappa3 receptor agonist, while also inhibiting the reuptake of norepinephrine and serotonin [209]. It was also found to selectively block NMDAR-mediated neurotoxicity in cortical neurons in mice and inhibit the excitatory response of rat spinal neurons to NMDA [210,211]. Its affinity for the NMDAR is stronger than methadone and comparable to ketamine. Due to its potent NMDA antagonism and specific inhibitory action on norepinephrine uptake, levorphanol is considered a strong candidate for treating NeP [212–214]. Levorphanol is well absorbed when taken orally and is metabolized to the inactive levorphanol-3-glucuronide. The glucuronide metabolite is excreted through the kidney [209]. Side effects include nausea, vomiting, mood and mental changes, itching, flushing, urinary difficulties, constipation, and biliary spasm [215,216].

In this narrative review, only one study, comprising two case reports (Table 10), was analyzed, and it showed promising results [217]. Pain intensity was assessed in both cases using the Edmonton Symptom Assessment System (ESAS) pain score. The first case involved a patient with osteosarcoma who was later diagnosed with phantom limb pain. Despite receiving a combination of medications consisting of hydromorphone extended-release 16 mg once daily, hydromorphone 4 mg six times a day, and gabapentin 300 mg three times a day, the pain level persists at a notable intensity, with a score of 7 out of 10 on the ESAS pain scale. However, after adding levorphanol 2 mg to the existing regimen of hydromorphone, the pain intensity decreased to 0–1 on the ESAS pain scale. The second case involved a breast cancer patient diagnosed with Brown–Sequard syndrome. Despite being on a regimen of 1200 mg of gabapentin three times daily, venlafaxine extended-release 75 mg once daily, and hydrocodone/acetaminophen 10/325 mg every 6 h, the patient continued to report uncontrolled pain. A low dose of levorphanol at 1 mg every 8 h was added alongside the existing hydrocodone/acetaminophen. After 1 month, the patient noted significant improvement in pain, reducing to a 2 out of 10 on the ESAS pain scale. This improvement continued over several months, with decreased reliance on hydrocodone until its discontinuation.

Table 10. Clinical studies that evaluated the effect of levorphanol on NeP.

Levorphanol					
Clinical Studies					
First Author/Reference	**Population**	**Type of NeP**	**Dosage**	**Route/Frequency**	**Results**
Reddy et al. (2018) [217]	N = 1	Phantom limb pain	Levorphanol 2 mg + Hydromorphone 4 mg every	Levorphanol every 8 h Hydromorphone every 4 h as necessary for breakthrough pain several months	One week later, pain had nearly disappeared, with a pain intensity rating of 0–1 out of 10 on the ESAS pain scale.
	N = 1	Brown–Sequard syndrome	Levorphanol 1 mg + Hydrocodone 10 mg and Acetaminophen 325 mg	Levorphanol every 8 h Hydrocodone and Acetaminophen taken as needed several months	After 1 month, pain significantly improved, scoring 2 out of 10 on the ESAS pain scale.

ESAS—Edmonton Symptom Assessment System.

3.10. Methadone

Methadone is a synthetic opioid analgesic that functions as a full agonist of the μ-opioid receptor [209]. Notably, methadone also antagonizes the NMDAR and strongly inhibits the uptake of serotonin and norepinephrine, which likely enhances its pain-relief properties [209,218]. Methadone is a highly lipid-soluble opioid that is well absorbed from the gastrointestinal tract [218]. It binds extensively to plasma proteins and undergoes significant first-pass metabolism. Its elimination involves extensive biotransformation, followed by excretion through the kidneys and feces [219]. Side effects include constipation, sedation, nausea, vomiting, and respiratory depression [220].

Evidence supporting the benefits of methadone in treating NeP is derived from clinical studies, with all 10 studies included in this narrative review reporting positive outcomes (Table 11). Notably, seven of these studies focused on the effect of methadone on CRNP, with all demonstrating that the NMDAR antagonist significantly reduced pain severity [221–227]. Additionally, case reports by Bach et al. [228] demonstrated the safe and effective use of low-dose methadone as an adjuvant treatment in frail elderly patients with various types of NeP who could not tolerate higher doses of conventional opioids and adjuvant pain medications. Methadone helped reduce the required dosage of hydromorphone in these patients. In contrast, Madden et al. [223] demonstrated that, in children with CRNP, adding a very low dose of methadone (0.03–0.04 mg/kg) to their existing gabapentin treatment regimen effectively managed NeP syndrome. Another study compared the effectiveness of oral methadone to fentanyl patches in patients with CRNP and found that the reduction in NRS scores was significantly greater with methadone than with fentanyl [221]. Moreover, Adumala et al. [227] reported that methadone provided superior analgesic effects and good overall tolerability compared to morphine for managing CRNP. The pain reduction was measured using the NRS and DN4 questionnaires. Combining duloxetine (40–60 mg) with methadone (15–30 mg) effectively reduced CRNP and alleviated emotional symptoms in patients more than either medication used alone, as shown by ESAS scores [224].

Table 11. Clinical studies that evaluated the effect of methadone on NeP.

Methadone					
Clinical Studies					
First Author/Reference	**Population**	**Type of NeP**	**Dosage**	**Route/Frequency**	**Results**
Rasmussen et al. (2015) [229]	N = 1	Vincristine-induced neuropathy	32.7 mg/kg bw	i.v. 182 days	Decreased pain by as much as 4 points on the NRS scale.
	N = 1		24 mg/kg bw	i.v. 180 days	Decreased pain by as much as 5 points on the NRS scale.
Haumann et al. (2016) [221]	N = 52	CRNP	Methadone 2 mg or Fentanyl 12 μg/h	Methadone orally Fentanyl patch 5 weeks	The decrease in NRS scores was notably superior when methadone was utilized in comparison to fentanyl.
Sugiyama et al. (2016) [222]	N = 28	Severe CRNP	7.5–150 mg	orally 14 days	In this study involving patients who switched from other strong opioids like oxycodone and fentanyl to methadone, 22 patients experienced a significant reduction in their mean FPS score.
Bach et al. (2016) [228]	N = 1	A 94-year-old patient with intractable back pain secondary to spinal stenosis and disc protrusion	0.5 mg	orally every 12 h	The co-administration of methadone relieved chronic nonmalignant NeP and reduced the dosage of hydromorphone in elderly patients.
		An 88-year-old patient with phantom limb pain in right leg and NeP in the left	1–2 mg	orally every 12 h	
		A 94-year-old patient with end-stage renal disease and a C5 injury experiencing burning pain that extends from the neck down to both arms	0.5–2.5 mg	orally	

Table 11. *Cont.*

Methadone					
Clinical Studies					
First Author/Reference	Population	Type of NeP	Dosage	Route/Frequency	Results
Madden et al. (2017) [223]	N = 2	Refractory CRNP in children	Methadone 0.03–0.04 mg/kg bw + Gabapentin 45 mg/kg	orally 1 year	Refractory NeP syndrome effectively managed by adding very low dose of methadone to their existing gabapentin treatment regimen.
Lynch et al. (2019) [230]	N = 9	Moderate to severe chronic NeP	5–60 mg	orally 11 weeks	All individuals demonstrated a decrease in average pain intensity based on the NPRS.
Curry et al. (2021) [224]	N = 43	CRNP	Methadone 33.75 mg (dose range) or Duloxetine 60 mg (dose range) FOLLOWED BY Methadone 15–30 mg + Duloxetine 40–60 mg	orally 2–8 weeks	After patients transitioned from monotherapy to combination therapy, there was a reduction in both the total ESAS scores and subscores. Additionally, 28% of patients on combination therapy reported a minimum two-point decrease in pain scores.
Matsuda et al. (2022) [225]	N = 3	NeP due to NBP	15–60 mg	orally 5–57 days	Pain scores decreased according to NRS in all three cases.
Fawoubo et al. (2023) [226]	N = 48	CRNP	21–60 mg	orally 28 days	By day 28, the pain intensity was notably reduced, with 53% of patients reporting a lower VAS score. Additionally, the NPSI score decreased in 50% of patients.
Adumala et al. (2023) [227]	N = 74	CRNP	Methadone 2.5–20 mg or Morphine 30–360 mg	orally 12 weeks	All participants exhibited a decrease in the average values for NRS and DN4, with a superior analgesic effect for methadone compared to morphine.

bw—body weight; CRNP—cancer-related neuropathic pain; DN4—Douleur Neuropathique 4; ESAS—Edmonton Symptom Assessment System; FPS—faces pain scale; NBP—neoplastic brachial plexopathy; NPRS—Numeric Rating Scale for Pain Intensity; NPSI—Neuropathic Pain Symptom Inventory; NRS—numerical rating scores.

4. Materials and Methods

A literature survey was conducted using PubMed to explore the most relevant articles containing preclinical and clinical research findings on the impact of various NMDAR antagonists on NeP. We restricted the search to articles that were published in English from 2015 to 2024. We used the following keywords and MeSH terms: "ketamine" OR "memantine" OR "methadone" OR "amantadine" OR "carbamazepine" OR "valproic acid" OR "phenytoin" OR "dextromethorphan" OR "riluzole" OR "levorphanol" AND "neuropathic pain" OR "neuropathy". After careful analysis and cross-checking, we chose the most suitable studies.

5. Summary

Ketamine, the most extensively studied NMDAR antagonist for NeP, effectively reduced the pain sensitivity associated with NeP and exhibited anti-inflammatory potential by reducing pro-inflammatory cytokines in preclinical studies [49,50]. However, its effects may not be long-lasting [47]. Clinical studies on ketamine for NeP revealed mixed results across different administration methods. Out of 11 studies, 8 supported ketamine's benefits. Among these, six studies used i.v. administration, with varying outcomes: three reported effective pain relief [51–53], one exhibited partial success [54], and two showed no significant reduction in pain [55,56]. Orally administered ketamine showed effectiveness in two studies [57,58], but a large study on CIPN yielded poor results. Topical and subcutaneous administration both showed positive outcomes in one study [60,61].

In preclinical studies, dextromethorphan effectively reversed tactile allodynia and thermal hyperalgesia [84,85]. However, combining dextromethorphan with gabapentin or oxycodone enhanced its antiallodynic effect [86,87]. Clinical studies are limited, with one study showing dextromethorphan's antihyperalgesic effects in a freeze-injury-induced hyperalgesia model [88].

Memantine showed efficacy in preclinical studies by reversing mechanical sensitivity and reducing proinflammatory cytokine levels [96,99]. Clinical trials support its benefits in treating NeP, particularly CRPS and post-mastectomy NeP, when administered alone [100,101]. Amantadine reduced hypersensitivity and oxidative stress in preclinical studies [112,113], with positive outcomes reported in clinical trials for facial nerve neuropathy [115].

Valproic acid demonstrates potential in preclinical studies by reducing sensitivity and cytokine release [129,130]. Limited clinical evidence suggests its effectiveness when combined with NSAIDs for chronic radicular pain [134]. Carbamazepine shows efficacy in preclinical studies for thermal and mechanical sensitivity reversal, particularly in TN [148,152]. Clinical studies mainly focus on TN, with varying effectiveness compared to other medications like oxcarbazepine or gabapentin [155,157].

In preclinical studies, phenytoin effectively reversed thermal and mechanical sensitivity in rats and reduced microglia/macrophage activation post-nerve injury [172]. Combining phenytoin with morphine provided more effective pain relief than either drug alone [172]. Clinical trials demonstrated phenytoin's effectiveness in reducing NeP, with i.v. administration benefiting TN patients [174,175] and topical phenytoin cream (5–10%) showing superior results in various types of NeP, including DN [176], small fiber neuropathy [180], and CIPN [182]. Riluzole, despite promising preclinical results [193–195], showed no positive outcomes in clinical trials for NeP treatment [205,207].

Levorphanol, as seen in case reports, holds promise in pain management for conditions like osteosarcoma and Brown–Sequard syndrome [217]. Clinical studies indicated that methadone is highly effective in treating NeP, particularly in cancer-related cases [224,227]. It was safe and effective at low doses for frail elderly patients, helping to reduce hydromorphone usage [228], and effectively managed pain in children with CRNP when combined with gabapentin [223]. Methadone also showed superior pain relief compared to fentanyl patches and morphine [221,227], and combining it with duloxetine further alleviated CRNP and emotional symptoms [224].

This narrative review acknowledges several limitations that warrant discussion. While preclinical studies offer promising insights into the mechanisms and potential benefits of the various NMDAR antagonists, their direct translation to clinical applications is hindered by various factors. Preclinical trials frequently use animal models that may not accurately represent human physiology and pathology. Moreover, variations in dosages, administration methods, and experimental designs across studies lead to a lack of standardized protocols.

Moreover, supplementary limitations exist within the current body of research. Many studies included in the review have small sample sizes, posing challenges in drawing definitive conclusions applicable to broader patient populations. Furthermore, there is significant variability in dosing among clinicians, leading to inconsistent treatment outcomes. Additionally, neuropathic pain encompasses a wide spectrum of conditions, meaning a pharmacologic agent that is effective for one condition may not be suitable for another. Clinicians need to recognize this variability when devising treatment plans for patients with specific types of neuropathic pain. Lastly, the considerable variation in follow-up periods among analyzed studies, ranging from weeks to months, may impact the assessment of treatment efficacy and requires careful consideration by clinicians.

6. Conclusions

After our thorough examination of the 10 NMDAR antagonists, it is evident that clinicians have multiple options for treating NeP. While some agents have stronger evidence supporting their efficacy, it is essential for physicians to recognize alternative choices in cases where pharmacologic drugs fail to provide sufficient relief or are limited by side effects. However, additional research is necessary to enhance understanding of the mechanisms and clinical utility of these 10 agents.

Author Contributions: Conceptualization, S.N.; data curation, C.P.; writing—original draft preparation, C.P.; writing—review and editing, N.M.B. and C.C. All authors have read and agreed to the published version of the manuscript.

Funding: This research received no external funding.

Institutional Review Board Statement: Not applicable.

Informed Consent Statement: Not applicable.

Data Availability Statement: All data generated or analyzed during this study are included in this published article.

Acknowledgments: The publication of this paper was supported by the University of Medicine and Pharmacy "Carol Davila", through the institutional program Publish not Perish.

Conflicts of Interest: The authors declare no conflicts of interest.

References

1. Terminology | International Association for the Study of Pain. Available online: https://www.iasp-pain.org/resources/terminology/ (accessed on 17 May 2024).
2. Mitsikostas, D.-D.; Moka, E.; Orrillo, E.; Aurilio, C.; Vadalouca, A.; Paladini, A.; Varrassi, G. Neuropathic Pain in Neurologic Disorders: A Narrative Review. *Cureus* **2022**, *14*, e22419. [CrossRef] [PubMed]
3. Was, H.; Borkowska, A.; Bagues, A.; Tu, L.; Liu, J.Y.H.; Lu, Z.; Rudd, J.A.; Nurgali, K.; Abalo, R. Mechanisms of Chemotherapy-Induced Neurotoxicity. *Front. Pharmacol.* **2022**, *13*, 750507. [CrossRef] [PubMed]
4. Shkodra, M.; Caraceni, A. Treatment of Neuropathic Pain Directly Due to Cancer: An Update. *Cancers* **2022**, *14*, 1992. [CrossRef]
5. Smith, B.H.; Hébert, H.L.; Veluchamy, A. Neuropathic Pain in the Community: Prevalence, Impact, and Risk Factors. *Pain* **2020**, *161*, S127–S137. [CrossRef]
6. Pușcașu, C.; Zanfirescu, A.; Negreș, S. Recent Progress in Gels for Neuropathic Pain. *Gels* **2023**, *9*, 417. [CrossRef] [PubMed]
7. Langley, P.C.; Van Litsenburg, C.; Cappelleri, J.C.; Carroll, D. The Burden Associated with Neuropathic Pain in Western Europe. *J. Med. Econ.* **2013**, *16*, 85–95. [CrossRef]
8. Freynhagen, R.; Baron, R.; Gockel, U.; Tölle, T.R. PainDETECT: A New Screening Questionnaire to Identify Neuropathic Components in Patients with Back Pain. *Curr. Med. Res. Opin.* **2006**, *22*, 1911–1920. [CrossRef] [PubMed]
9. Smith, B.H.; Torrance, N.; Bennett, M.I.; Lee, A.J. Health and Quality of Life Associated with Chronic Pain of Predominantly Neuropathic Origin in the Community. *Clin. J. Pain* **2007**, *23*, 143–149. [CrossRef]
10. Jensen, M.P.; Chodroff, M.J.; Dworkin, R.H. The Impact of Neuropathic Pain on Health-Related Quality of Life: Review and Implications. *Neurology* **2007**, *68*, 1178–1182. [CrossRef]
11. Freynhagen, R.; Bennett, M.I. Diagnosis and Management of Neuropathic Pain. *BMJ* **2009**, *339*, 391–395. [CrossRef]
12. van Velzen, M.; Dahan, A.; Niesters, M. Neuropathic Pain: Challenges and Opportunities. *Front. Pain Res.* **2020**, *1*, 1. [CrossRef] [PubMed]
13. O'Connor, A.B. Neuropathic Pain. *PharmacoEconomics* **2012**, *27*, 95–112. [CrossRef] [PubMed]

14. Hans, G.; Masquelier, E.; De Cock, P. The Diagnosis and Management of Neuropathic Pain in Daily Practice in Belgium: An Observational Study. *BMC Public Health* **2007**, *7*, 170. [CrossRef] [PubMed]
15. Torrance, N.; Smith, B.H.; Watson, M.C.; Bennett, M.I. Medication and Treatment Use in Primary Care Patients with Chronic Pain of Predominantly Neuropathic Origin. *Fam. Pract.* **2007**, *24*, 481–485. [CrossRef]
16. Blebea, N.M.; Mihai, D.P.; Andrei, C.; Stoica, D.M.; Chiriță, C.; Negreș, S. Evaluation of Therapeutic Potential of Cannabidiol-Based Products in Animal Models of Epileptic Seizures, Neuropathic Pain and Chronic Inflammation. *Farmacia* **2022**, *70*, 1185–1193. [CrossRef]
17. Pușcașu, C.; Ungurianu, A.; Șeremet, O.C.; Andrei, C.; Mihai, D.P.; Negreș, S. The Influence of Sildenafil–Metformin Combination on Hyperalgesia and Biochemical Markers in Diabetic Neuropathy in Mice. *Medicina* **2023**, *59*, 1375. [CrossRef]
18. Bates, D.; Carsten Schultheis, B.; Hanes, M.C.; Jolly, S.M.; Chakravarthy, K.V.; Deer, T.R.; Levy, R.M.; Hunter, C.W. A Comprehensive Algorithm for Management of Neuropathic Pain. *Pain Med.* **2019**, *20*, S2–S12. [CrossRef]
19. Jang, K.; Garraway, S.M. A Review of Dorsal Root Ganglia and Primary Sensory Neuron Plasticity Mediating Inflammatory and Chronic Neuropathic Pain. *Neurobiol. Pain* **2024**, *15*, 100151. [CrossRef]
20. Cohen, S.P.; Mao, J. Neuropathic Pain: Mechanisms and Their Clinical Implications. *BMJ* **2014**, *348*, f7656. [CrossRef]
21. Bleakman, D.; Alt, A.; Nisenbaum, E.S. Glutamate Receptors and Pain. *Semin. Cell Dev. Biol.* **2006**, *17*, 592–604. [CrossRef]
22. Bennett, G.J. Update on the Neurophysiology of Pain Transmission and Modulation. *J. Pain Symptom Manag.* **2000**, *19*, 2–6. [CrossRef] [PubMed]
23. Petrenko, A.B.; Yamakura, T.; Baba, H.; Shimoji, K. The Role of N-Methyl-d-Aspartate (NMDA) Receptors in Pain: A Review. *Anesth. Analg.* **2003**, *97*, 1108–1116. [CrossRef] [PubMed]
24. Ma, Y.; Kang, Z.; Shi, Y.; Ji, W.; Zhou, W.; Nan, W. The Complexity of Neuropathic Pain and Central Sensitization: Exploring Mechanisms and Therapeutic Prospects. *J. Integr. Neurosci.* **2024**, *23*, 89. [CrossRef]
25. Zhou, H.-Y.; Chen, S.-R.; Chen, H.; Pan, H.-L. Functional Plasticity of Group II Metabotropic Glutamate Receptors in Regulating Spinal Excitatory and Inhibitory Synaptic Input in Neuropathic Pain. *J. Pharmacol. Exp. Ther.* **2011**, *336*, 254–264. [CrossRef] [PubMed]
26. Zhang, H.-M.; Chen, S.-R.; Pan, H.-L. Effects of Activation of Group III Metabotropic Glutamate Receptors on Spinal Synaptic Transmission in a Rat Model of Neuropathic Pain. *Neuroscience* **2009**, *158*, 875–884. [CrossRef]
27. Wang, X.L.; Zhang, H.M.; Chen, S.R.; Pan, H.L. Altered Synaptic Input and GABAB Receptor Function in Spinal Superficial Dorsal Horn Neurons in Rats with Diabetic Neuropathy. *J. Physiol.* **2007**, *579*, 849–861. [CrossRef] [PubMed]
28. Liu, H.; Mantyh, P.W.; Basbaum, A.I. NMDA-Receptor Regulation of Substance P Release from Primary Afferent Nociceptors. *Nature* **1997**, *386*, 721–724. [CrossRef]
29. Zhou, H.-Y.; Chen, S.-R.; Chen, H.; Pan, H.-L. Opioid-Induced Long-Term Potentiation in the Spinal Cord Is a Presynaptic Event. *J. Neurosci.* **2010**, *30*, 4460–4466. [CrossRef]
30. Chen, Y.; Balasubramanyan, S.; Lai, A.Y.; Todd, K.G.; Smith, P.A. Effects of Sciatic Nerve Axotomy on Excitatory Synaptic Transmission in Rat Substantia Gelatinosa. *J. Neurophysiol.* **2009**, *102*, 3203–3215. [CrossRef]
31. Yang, K.; Takeuchi, K.; Wei, F.; Dubner, R.; Ren, K. Activation of Group I MGlu Receptors Contributes to Facilitation of NMDA Receptor Membrane Current in Spinal Dorsal Horn Neurons after Hind Paw Inflammation in Rats. *Eur. J. Pharmacol.* **2011**, *670*, 509–518. [CrossRef]
32. Doolen, S.; Blake, C.B.; Smith, B.N.; Taylor, B.K. Peripheral Nerve Injury Increases Glutamate-Evoked Calcium Mobilization in Adult Spinal Cord Neurons. *Mol. Pain* **2012**, *8*, 1744-8069-8-56. [CrossRef] [PubMed]
33. Sung, B.; Lim, G.; Mao, J. Altered Expression and Uptake Activity of Spinal Glutamate Transporters after Nerve Injury Contribute to the Pathogenesis of Neuropathic Pain in Rats. *J. Neurosci.* **2003**, *23*, 2899–2910. [CrossRef] [PubMed]
34. Zeevalk, G.D.; Nicklas, W.J. Evidence That the Loss of the Voltage-Dependent Mg 2+ Block at the N-Methyl-D-Aspartate Receptor Underlies Receptor Activation During Inhibition of Neuronal Metabolism. *J. Neurochem.* **1992**, *59*, 1211–1220. [CrossRef] [PubMed]
35. Zipfel, G.J.; Babcock, D.J.; Lee, J.-M.; Choi, D.W. Neuronal Apoptosis after CNS Injury: The Roles of Glutamate and Calcium. *J. Neurotrauma* **2000**, *17*, 857–869. [CrossRef]
36. Malmberg, A.B.; Chen, C.; Tonegawa, S.; Basbaum, A.I. Preserved Acute Pain and Reduced Neuropathic Pain in Mice Lacking PKCγ. *Science* **1997**, *278*, 279–283. [CrossRef]
37. Mao, J.; Price, D.D.; Hayes, R.L.; Lu, J.; Mayer, D.J. Differential Roles of NMDA and Non-NMDA Receptor Activation in Induction and Maintenance of Thermal Hyperalgesia in Rats with Painful Peripheral Mononeuropathy. *Brain Res.* **1992**, *598*, 271–278. [CrossRef] [PubMed]
38. Childers, W.E.; Baudy, R.B. N-Methyl-d-Aspartate Antagonists and Neuropathic Pain: The Search for Relief. *J. Med. Chem.* **2007**, *50*, 2557–2562. [CrossRef]
39. Karunarathna, I.; Kusumarathna, K.; Jayathilaka, P.; Kusumarathna, K.; Gunarathna, I.; Jayathilaka, P.; Senarathna, R.; Wijayawardana, K.; Priyalath, N.; Karunananda, S.; et al. Ketamine: Mechanisms, Applications, and Future Directions. Uva Clinical Lab.: Charlottesville, VA, USA, 2024.
40. Harden, R.N.; Oaklander, A.L.; Burton, A.W.; Perez, R.S.G.M.; Richardson, K.; Swan, M.; Barthel, J.; Costa, B.; Graciosa, J.R.; Bruehl, S. Complex Regional Pain Syndrome: Practical Diagnostic and Treatment Guidelines, 4th Edition. *Pain Med.* **2013**, *14*, 180–229. [CrossRef]

41. Pypendop, B.H.; Ilkiw, J.E. Pharmacokinetics of Ketamine and Its Metabolite, Norketamine, after Intravenous Administration of a Bolus of Ketamine to Isoflurane-Anesthetized Dogs. *Am. J. Vet. Res.* **2005**, *66*, 2034–2038. [CrossRef]
42. Finnerup, N.B.; Attal, N.; Haroutounian, S.; McNicol, E.; Baron, R.; Dworkin, R.H.; Gilron, I.; Haanpää, M.; Hansson, P.; Jensen, T.S.; et al. Pharmacotherapy for Neuropathic Pain in Adults: A Systematic Review and Meta-Analysis. *Lancet Neurol.* **2015**, *14*, 162–173. [CrossRef]
43. Mothet, J.-P.; Parent, A.T.; Wolosker, H.; Brady, R.O.; Linden, D.J.; Ferris, C.D.; Rogawski, M.A.; Snyder, S.H. D -Serine Is an Endogenous Ligand for the Glycine Site of the *N*.-Methyl-D-Aspartate Receptor. *Proc. Natl. Acad. Sci. USA* **2000**, *97*, 4926–4931. [CrossRef] [PubMed]
44. Eldufani, J.; Nekoui, A.; Blaise, G. Nonanesthetic Effects of Ketamine: A Review Article. *Am. J. Med.* **2018**, *131*, 1418–1424. [CrossRef] [PubMed]
45. Rogawski, M.A.; Wenk, G.L. The Neuropharmacological Basis for the Use of Memantine in the Treatment of Alzheimer's Disease. *CNS Drug Rev.* **2003**, *9*, 275–308. [CrossRef] [PubMed]
46. Salvat, E.; Yalcin, I.; Muller, A.; Barrot, M. A Comparison of Early and Late Treatments on Allodynia and Its Chronification in Experimental Neuropathic Pain. *Mol. Pain* **2018**, *14*, 174480691774968. [CrossRef]
47. Kroin, J.S.; Das, V.; Moric, M.; Buvanendran, A. Efficacy of the Ketamine Metabolite (2R,6R)-Hydroxynorketamine in Mice Models of Pain. *Reg. Anesth. Pain Med.* **2019**, *44*, 111–117. [CrossRef]
48. Humo, M.; Ayazgök, B.; Becker, L.J.; Waltisperger, E.; Rantamäki, T.; Yalcin, I. Ketamine Induces Rapid and Sustained Antidepressant-like Effects in Chronic Pain Induced Depression: Role of MAPK Signaling Pathway. *Prog. Neuro-Psychopharmacol. Biol. Psychiatry* **2020**, *100*, 109898. [CrossRef]
49. Tai, W.L.; Sun, L.; Li, H.; Gu, P.; Joosten, E.A.; Cheung, C.W. Additive Effects of Environmental Enrichment and Ketamine on Neuropathic Pain Relief by Reducing Glutamatergic Activation in Spinal Cord Injury in Rats. *Front. Neurosci.* **2021**, *15*, 635187. [CrossRef]
50. Han, J.; Zhang, X.; Xia, L.; Liao, O.; Li, Q. S-Ketamine Promotes Autophagy and Alleviates Neuropathic Pain by Inhibiting PI3K/Akt/MTOR Signaling Pathway. *Mol. Cell. Toxicol.* **2023**, *19*, 81–88. [CrossRef]
51. Kim, Y.H.; Lee, P.B.; Oh, T.K. Is Magnesium Sulfate Effective for Pain in Chronic Postherpetic Neuralgia Patients Comparing with Ketamine Infusion Therapy? *J. Clin. Anesth.* **2015**, *27*, 296–300. [CrossRef]
52. Weber, G.; Yao, J.; Binns, S.; Namkoong, S. Case Report of Subanesthetic Intravenous Ketamine Infusion for the Treatment of Neuropathic Pain and Depression with Suicidal Features in a Pediatric Patient. *Case Rep. Anesthesiol.* **2018**, *2018*, 9375910. [CrossRef]
53. Moreno-Hay, I.; Lindroth, J. Continuous Dentoalveolar Neuropathic Pain Response to Repeated Intravenous Ketamine Infusions: A Case Report. *J. Oral Facial Pain Headache* **2018**, *32*, e22–e27. [CrossRef] [PubMed]
54. Bosma, R.L.; Cheng, J.C.; Rogachov, A.; Kim, J.A.; Hemington, K.S.; Osborne, N.R.; Venkat Raghavan, L.; Bhatia, A.; Davis, K.D. Brain Dynamics and Temporal Summation of Pain Predicts Neuropathic Pain Relief from Ketamine Infusion. *Anesthesiology* **2018**, *129*, 1015–1024. [CrossRef] [PubMed]
55. Czarnetzki, C.; Desmeules, J.; Tessitore, E.; Faundez, A.; Chabert, J.; Daali, Y.; Fournier, R.; Dupuis-Lozeron, E.; Cedraschi, C.; Richard Tramèr, M. Perioperative Intravenous Low-dose Ketamine for Neuropathic Pain after Major Lower Back Surgery: A Randomized, Placebo-controlled Study. *Eur. J. Pain* **2020**, *24*, 555–567. [CrossRef]
56. Pickering, G.; Pereira, B.; Morel, V.; Corriger, A.; Giron, F.; Marcaillou, F.; Bidar-Beauvallot, A.; Chandeze, E.; Lambert, C.; Bernard, L.; et al. Ketamine and Magnesium for Refractory Neuropathic Pain. *Anesthesiology* **2020**, *133*, 154–164. [CrossRef] [PubMed]
57. Rigo, F.K.; Trevisan, G.; Godoy, M.C.; Rossato, M.F.; Dalmolin, G.D.; Silva, M.A.; Menezes, M.S.; Caumo, W.; Ferreira, J. Management of Neuropathic Chronic Pain with Methadone Combined with Ketamine: A Randomized, Double Blind, Active-Controlled Clinical Trial. *Pain Physician* **2017**, *20*, 207–215.
58. Martin, E.; Sorel, M.; Morel, V.; Marcaillou, F.; Picard, P.; Delage, N.; Tiberghien, F.; Crosmary, M.-C.; Najjar, M.; Colamarino, R.; et al. Dextromethorphan and Memantine after Ketamine Analgesia: A Randomized Control Trial. *Drug Des. Devel. Ther.* **2019**, *13*, 2677–2688. [CrossRef]
59. Fallon, M.T.; Wilcock, A.; Kelly, C.A.; Paul, J.; Lewsley, L.-A.; Norrie, J.; Laird, B.J.A. Oral Ketamine vs. Placebo in Patients with Cancer-Related Neuropathic Pain. *JAMA Oncol.* **2018**, *4*, 870–872. [CrossRef]
60. Rabi, J.; Minori, J.; Abad, H.; Lee, R.; Gittler, M. Topical Ketamine 10% for Neuropathic Pain in Spinal Cord Injury Patients: An Open-Label Trial. *Int. J. Pharm. Compd.* **2016**, *20*, 517–520.
61. Provido-Aljibe, M.T.; Yee, C.M.; Low, Z.J.C.; Hum, A. The Impact of a Standardised Ketamine Step Protocol for Cancer Neuropathic Pain. *Prog. Palliat. Care* **2022**, *30*, 4–10. [CrossRef]
62. M'Dahoma, S.; Barthélemy, S.; Tromilin, C.; Jeanson, T.; Viguier, F.; Michot, B.; Pezet, S.; Hamon, M.; Bourgoin, S. Respective Pharmacological Features of Neuropathic-like Pain Evoked by Intrathecal BDNF versus Sciatic Nerve Ligation in Rats. *Eur. Neuropsychopharmacol.* **2015**, *25*, 2118–2130. [CrossRef]
63. Mak, P.; Broadbear, J.H.; Kolosov, A.; Goodchild, C.S. Long-Term Antihyperalgesic and Opioid-Sparing Effects of 5-Day Ketamine and Morphine Infusion ("Burst Ketamine") in Diabetic Neuropathic Rats. *Pain Med.* **2015**, *16*, 1781–1793. [CrossRef] [PubMed]
64. Claudino, R.; Nones, C.; Araya, E.; Chichorro, J. Analgesic Effects of Intranasal Ketamine in Rat Models of Facial Pain. *J. Oral Facial Pain Headache* **2018**, *32*, 238–246. [CrossRef]

65. Doncheva, N.; Vasileva, L.; Saracheva, K.; Dimitrova, D.; Getova, D. Study of Antinociceptive Effect of Ketamine in Acute and Neuropathic Pain Models in Rats. *Adv. Clin. Exp. Med.* **2018**, *28*, 573–579. [CrossRef]
66. Pan, W.; Zhang, G.-F.; Li, H.-H.; Ji, M.-H.; Zhou, Z.-Q.; Li, K.-Y.; Yang, J.-J. Ketamine Differentially Restores Diverse Alterations of Neuroligins in Brain Regions in a Rat Model of Neuropathic Pain-Induced Depression. *Neuroreport* **2018**, *29*, 863–869. [CrossRef] [PubMed]
67. Fang, X.; Zhan, G.; Zhang, J.; Xu, H.; Zhu, B.; Hu, Y.; Yang, C.; Luo, A. Abnormalities in Inflammatory Cytokines Confer Susceptible to Chronic Neuropathic Pain-Related Anhedonia in a Rat Model of Spared Nerve Injury. *Clin. Psychopharmacol. Neurosci.* **2019**, *17*, 189–199. [CrossRef] [PubMed]
68. Kim, N.; Kim, B.; Lee, J.; Chung, H.; Kwon, H.; Kim, Y.; Choi, J.; Song, J. Response after Repeated Ketamine Injections in a Rat Model of Neuropathic Pain. *Physiol. Res.* **2022**, *71*, 297–303. [CrossRef]
69. Seo, E.-H.; Piao, L.; Cho, E.-H.; Hong, S.-W.; Kim, S.-H. The Effect of Ketamine on Endoplasmic Reticulum Stress in Rats with Neuropathic Pain. *Int. J. Mol. Sci.* **2023**, *24*, 5336. [CrossRef]
70. Chen, M.-H.; Tsai, S.-J. Maintenance of Antidepressant Effect by Dextromethorphan in Patients with Treatment-Resistant Depression Who Respond to Ketamine Intervention. *Med. Hypotheses* **2024**, *182*, 111242. [CrossRef]
71. Damaj, M.I.; Flood, P.; Ho, K.K.; May, E.L.; Martin, B.R. Effect of Dextrometorphan and Dextrorphan on Nicotine and Neuronal Nicotinic Receptors: In Vitro and in Vivo Selectivity. *J. Pharmacol. Exp. Ther.* **2005**, *312*, 780–785. [CrossRef]
72. Werling, L.L.; Lauterbach, E.C.; Calef, U. Dextromethorphan as a Potential Neuroprotective Agent with Unique Mechanisms of Action. *Neurologist* **2007**, *13*, 272–293. [CrossRef]
73. Annels, S.J.; Ellis, Y.; Davies, J.A. Non-Opioid Antitussives Inhibit Endogenous Glutamate Release from Rabbit Hippocampal Slices. *Brain Res.* **1991**, *564*, 341–343. [CrossRef] [PubMed]
74. Maurice, T.; Lockhart, B.P. Neuroprotective and Anti-Amnesic Potentials of Sigma (σ) Receptor Ligands. *Prog. Neuro-Psychopharmacol. Biol. Psychiatry* **1997**, *21*, 69–102. [CrossRef] [PubMed]
75. Maji, S.; Mohapatra, D.; Jena, M.; Srinivasan, A.; Maiti, R. Repurposing of Dextromethorphan as an Adjunct Therapy in Patients with Major Depressive Disorder: A Randomised, Group Sequential Adaptive Design, Controlled Clinical Trial Protocol. *BMJ Open* **2024**, *14*, e080500. [CrossRef]
76. Majhi, P.K.; Sayyad, S.; Gaur, M.; Kedar, G.; Rathod, S.; Sahu, R.; Pradhan, P.K.; Tripathy, S.; Ghosh, G.; Subudhi, B.B. Tinospora Cordifolia Extract Enhances Dextromethorphan Bioavailability: Implications for Alzheimer's Disease. *ACS Omega* **2024**, *9*, 23634–23648. [CrossRef]
77. Zhou, S.-F.; Zhou, Z.-W.; Yang, L.-P.; Cai, J.-P. Substrates, Inducers, Inhibitors and Structure-Activity Relationships of Human Cytochrome P450 2C9 and Implications in Drug Development. *Curr. Med. Chem.* **2009**, *16*, 3480–3675. [CrossRef]
78. Journey, J.D.; Agrawal, S.; Stern, E. Dextromethorphan Toxicity. In *StatPearls*; StatPearls Publishing: Treasure Island, FL, USA, 2023.
79. Choi, D.W. Dextrorphan and Dextromethorphan Attenuate Glutamate Neurotoxicity. *Brain Res.* **1987**, *403*, 333–336. [CrossRef] [PubMed]
80. Trube, G.; Netzer, R. Dextromethorphan: Cellular Effects Reducing Neuronal Hyperactivity. *Epilepsia* **1994**, *35*, S62–S67. [CrossRef]
81. Hollander, D.; Pradas, J.; Kaplan, R.; McLeod, H.L.; Evans, W.E.; Munsat, T.L. High-dose Dextromethorphan in Amyotrophic Lateral Sclerosis: Phase I Safety and Pharmacokinetic Studies. *Ann. Neurol.* **1994**, *36*, 920–924. [CrossRef]
82. Fleming, P.M. Dependence on Dextromethorphan Hydrobromide. *BMJ* **1986**, *293*, 597. [CrossRef]
83. Capon, D.A.; Bochner, F.; Kerry, N.; Mikus, G.; Danz, C.; Somogyi, A.A. The Influence of CYP2D6 Polymorphism and Quinidine on the Disposition and Antitussive Effect of Dextromethorphan in Humans*. *Clin. Pharmacol. Ther.* **1996**, *60*, 295–307. [CrossRef]
84. Zbârcea, C.E.; Chiriţă, C.; Ştefănescu, E.; Şeremet, O.C.; Velescu, B.Ş.; Marineci, C.D.; Muşat, O.; Giuglea, C.; Negreş, Ş. Therapeutic potential of tramadol and dextromethorphan on vincristine induced peripheral neuropathy in rats. *Farmacia* **2018**, *66*, 1037–1043. [CrossRef]
85. Fahmi, A.; Aji, Y.K.; Aprianto, D.R.; Wido, A.; Asadullah, A.; Roufi, N.; Indiastuti, D.N.; Subianto, H.; Turchan, A. The Effect of Intrathecal Injection of Dextromethorphan on the Experimental Neuropathic Pain Model. *Anesthesiol. Pain Med.* **2021**, *11*, e114318. [CrossRef] [PubMed]
86. Yang, P.-P.; Yeh, G.-C.; Huang, E.Y.-K.; Law, P.-Y.; Loh, H.H.; Tao, P.-L. Effects of Dextromethorphan and Oxycodone on Treatment of Neuropathic Pain in Mice. *J. Biomed. Sci.* **2015**, *22*, 81. [CrossRef] [PubMed]
87. Shi, T.; Hao, J.-X.; Wiesenfeld-Hallin, Z.; Xu, X.-J. Gabapentin and NMDA Receptor Antagonists Interacts Synergistically to Alleviate Allodynia in Two Rat Models of Neuropathic Pain. *Scand. J. Pain* **2018**, *18*, 687–693. [CrossRef]
88. Martin, E.; Narjoz, C.; Decleves, X.; Labat, L.; Lambert, C.; Loriot, M.-A.; Ducheix, G.; Dualé, C.; Pereira, B.; Pickering, G. Dextromethorphan Analgesia in a Human Experimental Model of Hyperalgesia. *Anesthesiology* **2019**, *131*, 356–368. [CrossRef]
89. Witt, A.; Macdonald, N.; Kirkpatrick, P. Memantine Hydrochloride. *Nat. Rev. Drug Discov.* **2004**, *3*, 109–110. [CrossRef]
90. Makino, K.M.; Porsteinsson, A.P. Memantine: A Treatment for Alzheimer's Disease with a New Formulation. *Aging Health* **2011**, *7*, 349–362. [CrossRef]
91. Karimi Tari, P.; Parsons, C.G.; Collingridge, G.L.; Rammes, G. Memantine: Updating a Rare Success Story in pro-Cognitive Therapeutics. *Neuropharmacology* **2024**, *244*, 109737. [CrossRef]
92. Kreutzwiser, D.; Tawfic, Q.A. Expanding Role of NMDA Receptor Antagonists in the Management of Pain. *CNS Drugs* **2019**, *33*, 347–374. [CrossRef]

93. Nakamura, Z.M.; Deal, A.M.; Park, E.M.; Stanton, K.E.; Lopez, Y.E.; Quillen, L.J.; O'Hare Kelly, E.; Heiling, H.M.; Nyrop, K.A.; Ray, E.M.; et al. A Phase II Single-arm Trial of Memantine for Prevention of Cognitive Decline during Chemotherapy in Patients with Early Breast Cancer: Feasibility, Tolerability, Acceptability, and Preliminary Effects. *Cancer Med.* **2023**, *12*, 8172–8183. [CrossRef]
94. Ballard, C.; Thomas, A.; Gerry, S.; Yu, L.-M.; Aarsland, D.; Merritt, C.; Corbett, A.; Davison, C.; Sharma, N.; Khan, Z.; et al. A Double-Blind Randomized Placebo-Controlled withdrawal Trial Comparing Memantine and Antipsychotics for the Long-Term Treatment of Function and Neuropsychiatric Symptoms in People with Alzheimer's Disease (MAIN-AD). *J. Am. Med. Dir. Assoc.* **2015**, *16*, 316–322. [CrossRef] [PubMed]
95. Ciotu, I.C.; Lupuliasa, D.; Chiriță, C.; Zbârcea, C.E.; Negreș, S. The Antihyperalgic Effect of Memantine in a Rat Model of Paclitaxel Induced Neuropathic Pain. *Farmacia* **2016**, *64*, 809–812.
96. Alomar, S.Y.; El Gheit, R.E.A.; Enan, E.T.; El-Bayoumi, K.S.; Shoaeir, M.Z.; Elkazaz, A.Y.; Al Thagfan, S.S.; Zaitone, S.A.; El-Sayed, R.M. Novel Mechanism for Memantine in Attenuating Diabetic Neuropathic Pain in Mice via Downregulating the Spinal HMGB1/TRL4/NF-KB Inflammatory Axis. *Pharmaceuticals* **2021**, *14*, 307. [CrossRef] [PubMed]
97. Solmaz, V.; Cinar, B.; Uyanikgil, Y.; Cavusoglu, T.; Peker, G.; Erbas, O. The Presentation of Therapeutic Effect Of Memantine in Sepsis Induced Critical Illness Polyneuropathy. *Yeni Symp.* **2015**, *53*, 23. [CrossRef]
98. Salih, N.A.; Al-Baggou, B.K. Effect of Memantine Hydrochloride on Cisplatin-Induced Neurobehavioral Toxicity in Mice. *Acta Neurol. Belg.* **2020**, *120*, 71–82. [CrossRef] [PubMed]
99. Chen, Y.; Shi, Y.; Wang, G.; Li, Y.; Cheng, L.; Wang, Y. Memantine Selectively Prevented the Induction of Dynamic Allodynia by Blocking Kir2.1 Channel and Inhibiting the Activation of Microglia in Spinal Dorsal Horn of Mice in Spared Nerve Injury Model. *Mol. Pain* **2019**, *15*, 174480691983894. [CrossRef]
100. Ahmad-Sabry, M.-H.; Shareghi, G. Effects of memantine on pain in patients with complex regional pain syndrome—A retrospective study. *Middle East J. Anaesthesiol.* **2015**, *23*, 51–54.
101. Morel, V.; Joly, D.; Villatte, C.; Dubray, C.; Durando, X.; Daulhac, L.; Coudert, C.; Roux, D.; Pereira, B.; Pickering, G. Memantine before Mastectomy Prevents Post-Surgery Pain: A Randomized, Blinded Clinical Trial in Surgical Patients. *PLoS ONE* **2016**, *11*, e0152741. [CrossRef]
102. Shaseb, E.; Farashi, E.; Ghadim, H.H.; Asdaghi, A.; Sarbakhsh, P.; Ghaffary, S. Memantine as a Potential Therapy in Subacute Herpetic Neuralgia: A Randomized Clinical Trial. *J. Pharm. Care* **2023**, *10*, 205–210. [CrossRef]
103. Jafarzadeh, E.; Beheshtirouy, S.; Aghamohammadzadeh, N.; Ghaffary, S.; Sarbakhsh, P.; Shaseb, E. Management of Diabetic Neuropathy with Memantine: A Randomized Clinical Trial. *Diabetes Vasc. Dis. Res.* **2023**, *20*, 14791641231191093. [CrossRef]
104. Blanpied, T.A.; Clarke, R.J.; Johnson, J.W. Amantadine Inhibits NMDA Receptors by Accelerating Channel Closure during Channel Block. *J. Neurosci.* **2005**, *25*, 3312–3322. [CrossRef]
105. Guedes, P.E.B.; Pinto, T.M.; Corrêa, J.M.X.; Niella, R.V.; dos Anjos, C.M.; de Oliveira, J.N.S.; Marques, C.S.d.C.; de Souza, S.S.; da Silva, E.B.; de Lavor, M.S.L. Efficacy of Preemptive Analgesia with Amantadine for Controlling Postoperative Pain in Cats Undergoing Ovariohysterectomy. *Animals* **2024**, *14*, 643. [CrossRef]
106. Saini, N.; Singh, N.; Kaur, N.; Garg, S.; Kaur, M.; Kumar, A.; Verma, M.; Singh, K.; Sohal, H.S. Motor and Non-Motor Symptoms, Drugs, and Their Mode of Action in Parkinson's Disease (PD): A Review. *Med. Chem. Res.* **2024**, *33*, 580–599. [CrossRef]
107. Kumar, G.; Sakharam, K.A. Tackling Influenza A Virus by M2 Ion Channel Blockers: Latest Progress and Limitations. *Eur. J. Med. Chem.* **2024**, *267*, 116172. [CrossRef]
108. Shuklinova, O.; Neuhoff, S.; Polak, S. How PBPK Can Help to Understand Old Drugs and Inform Their Dosing in Elderly: Amantadine Case Study. *Clin. Pharmacol. Ther.* **2024**, *116*, 225–234. [CrossRef]
109. Stiver, G. The Treatment of Influenza with Antiviral Drugs. *Can. Med. Assoc. J. CMAJ* **2003**, *168*, 49–56.
110. Rissardo, J.P.; Fornari Caprara, A.L. Myoclonus Secondary to Amantadine: Case Report and Literature Review. *Clin. Pract.* **2023**, *13*, 830–837. [CrossRef]
111. Harandi, A.A.; Pakdaman, H.; Medghalchi, A.; Kimia, N.; Kazemian, A.; Siavoshi, F.; Barough, S.S.; Esfandani, A.; Hosseini, M.H.; Sobhanian, S.A. A Randomized Open-Label Clinical Trial on the Effect of Amantadine on Post Covid 19 Fatigue. *Sci. Rep.* **2024**, *14*, 1343. [CrossRef]
112. Mata-Bermudez, A.; Ríos, C.; Burelo, M.; Pérez-González, C.; García-Martínez, B.A.; Jardon-Guadarrama, G.; Calderón-Estrella, F.; Manning-Balpuesta, N.; Diaz-Ruiz, A. Amantadine Prevented Hypersensitivity and Decreased Oxidative Stress by NMDA Receptor Antagonism after Spinal Cord Injury in Rats. *Eur. J. Pain* **2021**, *25*, 1839–1851. [CrossRef]
113. Dogan, G.; Onur, K. N-Methyl-d-Aspartate Receptor Antagonists May Ameliorate Spinal Cord Injury by Inhibiting Oxidative Stress: An Experimental Study in Rats. *Turk. Neurosurg.* **2019**, *30*, 60–68. [CrossRef]
114. Drummond, I.S.A.; de Oliveira, J.N.S.; Niella, R.V.; Silva, Á.J.C.; de Oliveira, I.S.; de Souza, S.S.; da Costa Marques, C.S.; Corrêa, J.M.X.; Silva, J.F.; de Lavor, M.S.L. Evaluation of the Therapeutic Potential of Amantadine in a Vincristine-Induced Peripheral Neuropathy Model in Rats. *Animals* **2024**, *14*, 1941. [CrossRef]
115. Azimov, A.; Sadykov, R.; Dadajonov, S.; Azizova, O.; Ismoilov, R. Dopaminergic Medicines Are Drugs of Choice for Medicament-Resistant Facial Nerve Neuropathy. *Park. Relat. Disord.* **2016**, *22*, e139–e140. [CrossRef]
116. Gean, P.-W.; Huang, C.-C.; Hung, C.-R.; Tsai, J.-J. Valproic Acid Suppresses the Synaptic Response Mediated by the NMDA Receptors in Rat Amygdalar Slices. *Brain Res. Bull.* **1994**, *33*, 333–336. [CrossRef]
117. Gobbi, G.; Janiri, L. Sodium- and Magnesium-Valproate In Vivo Modulate Glutamatergic and GABAergic Synapses in the Medial Prefrontal Cortex. *Psychopharmacology* **2006**, *185*, 255–262. [CrossRef]

118. Yi-Ping Lee Ko, G.; Brown-Croyts, L.M.; Teyler, T.J. The Effects of Anticonvulsant Drugs on NMDA-EPSP, AMPA-EPSP, and GABA-IPSP in the Rat Hippocampus. *Brain Res. Bull.* **1997**, *42*, 297–302. [CrossRef]
119. Turski, L.; Niemann, W.; Stephens, D.N. Differential Effects of Antiepileptic Drugs and β-Carbolines on Seizures Induced by Excitatory Amino Acids. *Neuroscience* **1990**, *39*, 799–807. [CrossRef]
120. Zeise, M.L.; Kasparow, S.; Zieglgänsberger, W. Valproate Suppresses N-Methyl-d-Aspartate-Evoked, Transient Depolarizations in the Rat Neocortex in Vitro. *Brain Res.* **1991**, *544*, 345–348. [CrossRef]
121. Lan, M.J.; McLoughlin, G.A.; Griffin, J.L.; Tsang, T.M.; Huang, J.T.J.; Yuan, P.; Manji, H.; Holmes, E.; Bahn, S. Metabonomic Analysis Identifies Molecular Changes Associated with the Pathophysiology and Drug Treatment of Bipolar Disorder. *Mol. Psychiatry* **2009**, *14*, 269–279. [CrossRef]
122. Shao, L.; Young, L.T.; Wang, J.-F. Chronic Treatment with Mood Stabilizers Lithium and Valproate Prevents Excitotoxicity by Inhibiting Oxidative Stress in Rat Cerebral Cortical Cells. *Biol. Psychiatry* **2005**, *58*, 879–884. [CrossRef]
123. Basselin, M.; Chang, L.; Chen, M.; Bell, J.M.; Rapoport, S.I. Chronic Administration of Valproic Acid Reduces Brain NMDA Signaling via Arachidonic Acid in Unanesthetized Rats. *Neurochem. Res.* **2008**, *33*, 2229–2240. [CrossRef]
124. Tursunov, A.N.; Vasilyev, D.S.; Nalivaeva, N.N. Molecular Mechanisms of Valproic Acid Action on Signalling Systems and Brain Functions. *J. Evol. Biochem. Physiol.* **2023**, *59*, 1740–1755. [CrossRef]
125. Wisłowska-Stanek, A.; Turzyńska, D.; Sobolewska, A.; Kołosowska, K.; Szyndler, J.; Skórzewska, A.; Maciejak, P. The Effect of Valproate on the Amino Acids, Monoamines, and Kynurenic Acid Concentrations in Brain Structures Involved in Epileptogenesis in the Pentylenetetrazol-Kindled Rats. *Pharmacol. Rep.* **2024**, *76*, 348–367. [CrossRef]
126. Zaccara, G.; Messori, A.; Moroni, F. Clinical Pharmacokinetics of Valproic Acid—1988. *Clin. Pharmacokinet.* **1988**, *15*, 367–389. [CrossRef]
127. Forrest, L.F.; Smith, M.; Quevedo, J.; Frey, B.N. Bipolar Disorder in Women: Menstrual Cycle, Perinatal Period, and Menopause Transition. In *Women's Mental Health*; Springer International Publishing: Cham, Switzerland, 2020; pp. 59–71.
128. Safdar, A.; Ismail, F. A Comprehensive Review on Pharmacological Applications and Drug-Induced Toxicity of Valproic Acid. *Saudi Pharm. J.* **2023**, *31*, 265–278. [CrossRef]
129. Chen, J.-Y.; Chu, L.-W.; Cheng, K.-I.; Hsieh, S.-L.; Juan, Y.-S.; Wu, B.-N. Valproate Reduces Neuroinflammation and Neuronal Death in a Rat Chronic Constriction Injury Model. *Sci. Rep.* **2018**, *8*, 16457. [CrossRef]
130. Elsherbiny, N.M.; Ahmed, E.; Kader, G.A.; Abdel-mottaleb, Y.; ElSayed, M.H.; Youssef, A.M.; Zaitone, S.A. Inhibitory Effect of Valproate Sodium on Pain Behavior in Diabetic Mice Involves Suppression of Spinal Histone Deacetylase 1 and Inflammatory Mediators. *Int. Immunopharmacol.* **2019**, *70*, 16–27. [CrossRef]
131. Chu, Z.; Liu, P.; Li, X.; Liu, Y.; Liu, F.; Lei, G.; Yang, L.; Deng, L.; Dang, Y. Microinjection of Valproic Acid into the Ventrolateral Orbital Cortex Exerts an Antinociceptive Effect in a Rat of Neuropathic Pain. *Psychopharmacology* **2020**, *237*, 2509–2516. [CrossRef]
132. Guo, A.; Li, J.; Luo, L.; Chen, C.; Lu, Q.; Ke, J.; Feng, X. Valproic Acid Mitigates Spinal Nerve Ligation-Induced Neuropathic Pain in Rats by Modulating Microglial Function and Inhibiting Neuroinflammatory Response. *Int. Immunopharmacol.* **2021**, *92*, 107332. [CrossRef]
133. Wang, D.; Wang, K.; Liu, Z.; Wang, Z.; Wu, H. Valproic Acid-Labeled Chitosan Nanoparticles Promote Recovery of Neuronal Injury after Spinal Cord Injury. *Aging* **2020**, *12*, 8953–8967. [CrossRef]
134. Ghasemian, M.; Owlia, M.B.; Mosaddegh, M.H.; Nakhaie nejad, M.; Sohrevardi, S.M. Evaluation of Sodium Valproate Low Dose Efficacy in Radicular Pain Management and It's Relation with Pharmacokinetics Parameters. *BioMedicine* **2020**, *10*, 33–39. [CrossRef]
135. Wang, D.; Wang, K.; Liu, Z.; Wang, Z.; Wu, H. Valproic Acid Labeled Chitosan Nanoparticles Promote the Proliferation and Differentiation of Neural Stem Cells after Spinal Cord Injury. *Neurotox. Res.* **2021**, *39*, 456–466. [CrossRef]
136. Sofia, R.D.; Gordon, R.; Gels, M.; Diamantis, W. Comparative Effects of Felbamate and Other Compounds on N-Methyl-D-Aspartic Acid-Induced Convulsions and Lethality in Mice. *Pharmacol. Res.* **1994**, *29*, 139–144. [CrossRef]
137. Ambrósio, A.F.; Silva, A.P.; Malva, J.O.; Soares-da-Silva, P.; Carvalho, A.P.; Carvalho, C.M. Carbamazepine Inhibits L-Type Ca2+ Channels in Cultured Rat Hippocampal Neurons Stimulated with Glutamate Receptor Agonists. *Neuropharmacology* **1999**, *38*, 1349–1359. [CrossRef]
138. Farber, N.B.; Jiang, X.-P.; Heinkel, C.; Nemmers, B. Antiepileptic Drugs and Agents That Inhibit Voltage-Gated Sodium Channels Prevent NMDA Antagonist Neurotoxicity. *Mol. Psychiatry* **2002**, *7*, 726–733. [CrossRef]
139. Lampe, H.; Bigalke, H. Carbamazepine Blocks NMDA-Activated Currents in Cultured Spinal Cord Neurons. *NeuroReport* **1990**, *1*, 26–28. [CrossRef]
140. Wang, L.; Wang, Y.; Zhang, R.-Y.; Wang, Y.; Liang, W.; Li, T.-G. Management of Acute Carbamazepine Poisoning: A Narrative Review. *World J. Psychiatry* **2023**, *13*, 816–830. [CrossRef]
141. Kuo, C.-C.; Chen, R.-S.; Lu, L.; Chen, R.-C. Carbamazepine Inhibition of Neuronal Na + Currents: Quantitative Distinction from Phenytoin and Possible Therapeutic Implications. *Mol. Pharmacol.* **1997**, *51*, 1077–1083. [CrossRef]
142. Maan, J.S.; Duong, T.v.H.; Saadabadi, A. Carbamazepine. In *StatPearls*; StatPearls Publishing: Treasure Island, FL, USA, 2024. Available online: http://www.ncbi.nlm.nih.gov/books/NBK482455 (accessed on 15 July 2024).
143. Ayano, G. Bipolar Disorders and Carbamazepine: Pharmacokinetics, Pharmacodynamics, Therapeutic Effects and Indications of Carbamazepine: Review of Articles. *J. Neuropsychopharmacol. Ment. Health* **2016**, *1*, 2. [CrossRef]

144. Chen, Y.; Ke, M.; Fang, W.; Jiang, Y.; Lin, R.; Wu, W.; Huang, P.; Lin, C. Physiologically Based Pharmacokinetic Modeling to Predict Maternal Pharmacokinetics and Fetal Carbamazepine Exposure during Pregnancy. *Eur. J. Pharm. Sci.* **2024**, *194*, 106707. [CrossRef]
145. Fuhr, L.M.; Marok, F.Z.; Hanke, N.; Selzer, D.; Lehr, T. Pharmacokinetics of the CYP3A4 and CYP2B6 Inducer Carbamazepine and Its Drug–Drug Interaction Potential: A Physiologically Based Pharmacokinetic Modeling Approach. *Pharmaceutics* **2021**, *13*, 270. [CrossRef]
146. Guo, M.; Shen, W.; Zhou, M.; Song, Y.; Liu, J.; Xiong, W.; Gao, Y. Safety and Efficacy of Carbamazepine in the Treatment of Trigeminal Neuralgia: A Metanalysis in Biomedicine. *Math. Biosci. Eng.* **2024**, *21*, 5335–5359. [CrossRef]
147. Delcker, A.; Wilhelm, H.; Timmann, D.; Diener, H. Side Effects from Increased Doses of Carbamazepine on Neuropsychological and Posturographic Parameters of Humans. *Eur. Neuropsychopharmacol.* **1997**, *7*, 213–218. [CrossRef]
148. Kohli, S.; Sharma, T.; Kalra, J.; Dhasmana, D. A Comparative Study of the Anti-Nociceptive Potential of Duloxetine and Carbamazepine in an Animal Model of Neuropathic Pain. *Int. J. Basic Clin. Pharmacol.* **2016**, *5*, 1–5. [CrossRef]
149. AL-Mahmood, S.M.A.; Abdullah, S.T.B.C.; Ahmad, N.N.F.N.; Bin Mohamed, A.H.; Razak, T.A. Analgesic Synergism of Gabapentin and Carbamazepine in Rat Model of Diabetic Neuropathic Pain. *Trop. J. Pharm. Res.* **2016**, *15*, 1191. [CrossRef]
150. Dai, H.; Tilley, D.M.; Mercedes, G.; Doherty, C.; Gulati, A.; Mehta, N.; Khalil, A.; Holzhaus, K.; Reynolds, F.M. Opiate-Free Pain Therapy Using Carbamazepine-Loaded Microparticles Provides Up to 2 Weeks of Pain Relief in a Neuropathic Pain Model. *Pain Pract.* **2018**, *18*, 1024–1035. [CrossRef]
151. Bektas, N.; Tekes, F.A.; Eken, H.; Arslan, R. The Antihyperalgesic Effects of Gabapentinoids, Carbamazepine and Its Keto Analogue, Oxcarbazepine, in Capsaicin-Induced Thermal Hyperalgesia. *World J. Pharm. Sci.* **2019**, *7*, 1–6.
152. Deseure, K.; Hans, G. Differential Drug Effects on Spontaneous and Evoked Pain Behavior in a Model of Trigeminal Neuropathic Pain. *J. Pain Res.* **2017**, *10*, 279–286. [CrossRef]
153. Graham, J.G.; Zilkha, K.J. Treatment of Trigeminal Neuralgia with Carbamazepine: A Follow-up Study. *BMJ* **1966**, *1*, 210–211. [CrossRef]
154. Cruccu, G.; Truini, A. A Review of Neuropathic Pain: From Guidelines to Clinical Practice. *Pain Ther.* **2017**, *6*, 35–42. [CrossRef]
155. Shafiq, H.; Badar, M.A.; Qayyum, Z.; Saeed, M. Effectiveness of carbamazepine versus oxcarbazepine in the management of trigeminal neuralgia. *J. Khyber Coll. Dent.* **2015**, *6*, 32–35. [CrossRef]
156. Iqbal, S.; Rashid, W.; Ul Ain, Q.; Atiq, T.; Noor, R.; Irfan, F. Efficacy of Carbamazepine and Oxcarbazepine for Treating Trigeminal Neuralgia. *BMC J. Med. Sci.* **2022**, *3*, 55–59.
157. Kaur, B.; Dhir, P. Evaluation of the Efficacy of Carbamazepine and Gabapentin in the Management of Trigeminal Neuralgia: A Clinical Study. *J. Indian Acad. Oral Med. Radiol.* **2018**, *30*, 253. [CrossRef]
158. Agarwal, R.; Kumari, S. Comparison of Carbamazepine and Gabapentin in Management of Trigeminal Neuralgia-A Clinical Study. *J. Adv. Med. Dent. Sci. Res.* **2020**, *8*, 38–41. [CrossRef]
159. Puri, N.; Rathore, A.; Dharmdeep, G.; Vairagare, S.; Prasad, B.; Priyadarshini, R.; Singh, H. A Clinical Study on Comparative Evaluation of the Effectiveness of Carbamazepine and Combination of Carbamazepine with Baclofen or Capsaicin in the Management of Trigeminal Neuralgia. *Niger. J. Surg.* **2018**, *24*, 95–99. [CrossRef] [PubMed]
160. Mehran, K.; Naimat, U.; Irfan, U.; Dawood, K.; Hurriya, K. Comparison of Amitriptyline and Carbamazepine Effectiveness in the Treatment of Postherpetic Neuralgia. *J. Pak. Assoc. Dermatol.* **2023**, *33*, 190–195.
161. Syed, S.O.; Mazlam, N.A.; Kallarakkal, T.G. Evaluation of Carbamazepine Pharmacotherapy In Patients With Trigeminal Neuralgia. *Ann. Dent.* **2016**, *23*, 19–26. [CrossRef]
162. Tariq, R.; Janjua, O.S.; Mehmood, S.; Khalid, M.U.; Zafar, K.J.; Hameed, S. Comparison of effectiveness of carbamazepine versus topiramate for the management of trigeminal neuralgia. *Pak. Armed Forces Med. J. PAFMJ* **2021**, *71*, 1360–1363. [CrossRef]
163. Wamil, A.W.; McLean, M.J. Phenytoin Blocks N-Methyl-D-Aspartate Responses of Mouse Central Neurons. *J. Pharmacol. Exp. Ther.* **1993**, *267*, 218–227.
164. Brown, L.M.; Lee, Y.-P.; Teyler, T.J. Antiepileptics Inhibit Cortical N-Methyl-D-Aspartate-Evoked [3H]Norepinephrine Efflux. *Eur. J. Pharmacol.* **1994**, *254*, 307–309. [CrossRef]
165. Sethy, V.H.; Sage, G.P. Modulation of Release of Acetylcholine from the Striatum by a Proposed Excitatory Amino Acid Antagonist U-54494A: Comparison with Known Antagonists, Diazepam and Phenytoin. *Neuropharmacology* **1992**, *31*, 111–114. [CrossRef]
166. Lee, G.Y.-P.; Brown, L.M.; Teyler, T.J. The Effects of Anticonvulsant Drugs on Long-Term Potentiation (LTP) in the Rat Hippocampus. *Brain Res. Bull.* **1996**, *39*, 39–42. [CrossRef] [PubMed]
167. Leung, L.S.; Shen, B. Long-Term Potentiation in Hippocampal CA1: Effects of Afterdischarges, NMDA Antagonists, and Anticonvulsants. *Exp. Neurol.* **1993**, *119*, 205–214. [CrossRef] [PubMed]
168. Luke, M. Therapeutic Drug Monitoring of Classical and Newer Anticonvulsants. In *Therapeutic Drug Monitoring*; Elsevier: Amsterdam, The Netherlands, 2024; pp. 133–161.
169. Mardhiani, R.; Harahap, Y.; Wiratman, W. Association between CYP2C9 and CYP2C19 Polymorphism, Metabolism, and Neurotoxicity after Administration of Phenytoin: A Systematic Review. *Keluwih J. Kesehat. Dan Kedokt.* **2023**, *5*, 33–34. [CrossRef]
170. Jones, G.L.; Wimbish, G.H.; McIntosh, W.E. Phenytoin: Basic and Clinical Pharmacology. *Med. Res. Rev.* **1983**, *3*, 383–434. [CrossRef]
171. Ujefa, S.; Lakshmi, E.S.; Supriya, G.; Bellapu, D.; Padmalatha, K. Phenytoin toxicity and it's management—A systematic review. *World J. Pharm. Res.* **2023**, *12*, 70–79. [CrossRef]

172. Kocot-Kępska, M.; Pawlik, K.; Ciapała, K.; Makuch, W.; Zajączkowska, R.; Dobrogowski, J.; Przeklasa-Muszyńska, A.; Mika, J. Phenytoin Decreases Pain-like Behaviors and Improves Opioid Analgesia in a Rat Model of Neuropathic Pain. *Brain Sci.* **2023**, *13*, 858. [CrossRef]
173. Hesari, F.; Fereidoni, M.; Moghimi, A.; Abdolmaleki, A.; Ghayour, M.B. Phenytoin Intraperitoneal Administration Effects on Neuropathic Pain Induced by Chronic Constriction Injury in Male Rats. In Proceedings of the 5th Basic and Clinical Neuroscience Congress 2016, Tehran, Iran, 7–9 December 2016.
174. Schnell, S.; Marrodan, M.; Acosta, J.N.; Bonamico, L.; Goicochea, M.T. Trigeminal Neuralgia Crisis—Intravenous Phenytoin as Acute Rescue Treatment. *Headache J. Head Face Pain* **2020**, *60*, 2247–2253. [CrossRef]
175. Vargas, A.; Thomas, K. Intravenous Fosphenytoin for Acute Exacerbation of Trigeminal Neuralgia: Case Report and Literature Review. *Ther. Adv. Neurol. Disord.* **2015**, *8*, 187–188. [CrossRef]
176. Hesselink, J.M.K. Topical Phenytoin in Painful Diabetic Neuropathy: Rationale to Select a Non-Selective Sodium Channel Blocker. *Clin. Res. Neurol.* **2016**, *3*, 1–5. [CrossRef]
177. Keppel Hesselink, J.M.; Kopsky, D.J. The Individualized N-of-1 Trial: Dose-Response in a Single-Blind Cross-Over Response Test of Phenytoin 10% and 30% Cream in Neuropathic Pain. *EC Anaesth.* **2018**, *4*, 245–249.
178. Hesselink, K.; Kopsky, J.M. Topical Phenytoin Cream in Small Fiber Neuropathic Pain: Fast Onset of Perceptible Pain Relief. *J. Pain Relief* **2017**, *1*, 15–019.
179. Keppel Hesselink, J.M.; Kopsky, D.J. Topical Phenytoin Cream Reduces Burning Pain Due to Small Fiber Neuropathy in Sarcoidosis Case Report. *J. Anesth. Pain Med.* **2017**, *2*, 1–3.
180. Keppel Hesselink, J.M.; Kopsky, D.J. Burning Pain in Small Fibre Neuropathy Treated with Topical Phenytoin: Rationale and Case Presentations. *J. Clin. Anesth. Pain Med.* **2017**, *4*, 2.
181. Kopsky, D.J.; Vrancken, A.F.; Keppel Hesselink, J.M.; van Eijk, R.P.; Notermans, N.C. Usefulness of a Double-Blind Placebo-Controlled Response Test to Demonstrate Rapid Onset Analgesia with Phenytoin 10% Cream in Polyneuropathy. *J. Pain Res.* **2020**, *13*, 877–882. [CrossRef]
182. Kopsky, D.; Keppel Hesselink, J. Topical Phenytoin for the Treatment of Neuropathic Pain. *J. Pain Res.* **2017**, *10*, 469–473. [CrossRef] [PubMed]
183. Keppel Hesselink, J.; Kopsky, D. Neuropathic Pain Due to Chronic Idiopathic Axonal Neuropathy: Fast Pain Reduction after Topical Phenytoin Cream Application. *Open J. Pain Med.* **2018**, *2*, 007–008. [CrossRef]
184. Kopsky, D.; Keppel Hesselink, J. Phenytoin Cream for the Treatment for Neuropathic Pain: Case Series. *Pharmaceuticals* **2018**, *11*, 53. [CrossRef]
185. Kopsky, D.; Keppel Hesselink, J. Single-Blind Placebo-Controlled Response Test with Phenytoin 10% Cream in Neuropathic Pain Patients. *Pharmaceuticals* **2018**, *11*, 122. [CrossRef]
186. Debono, M.-W.; Le Guern, J.; Canton, T.; Doble, A.; Pradier, L. Inhibition by Riluzole of Electrophysiological Responses Mediated by Rat Kainate and NMDA Receptors Expressed in Xenopus Oocytes. *Eur. J. Pharmacol.* **1993**, *235*, 283–289. [CrossRef]
187. Moon, E.S.; Karadimas, S.K.; Yu, W.-R.; Austin, J.W.; Fehlings, M.G. Riluzole Attenuates Neuropathic Pain and Enhances Functional Recovery in a Rodent Model of Cervical Spondylotic Myelopathy. *Neurobiol. Dis.* **2014**, *62*, 394–406. [CrossRef]
188. Cheah, B.C.; Vucic, S.; Krishnan, A.; Kiernan, M.C. Riluzole, Neuroprotection and Amyotrophic Lateral Sclerosis. *Curr. Med. Chem.* **2010**, *17*, 1942–1959. [CrossRef]
189. Gurney, M.E.; Cutting, F.B.; Zhai, P.; Doble, A.; Taylor, C.P.; Andrus, P.K.; Hall, E.D. Benefit of Vitamin E, Riluzole, and Gababapentin in a Transgenic Model of Familial Amyotrophic Lateral Sclerosis. *Ann. Neurol.* **1996**, *39*, 147–157. [CrossRef]
190. Doble, A. The Pharmacology and Mechanism of Action of Riluzole. *Neurology* **1996**, *47*, S233–S241. [CrossRef]
191. Golmohammadi, M.; Mahmoudian, M.; Hasan, E.K.; Alshahrani, S.H.; Romero-Parra, R.M.; Malviya, J.; Hjazi, A.; Najm, M.A.A.; Almulla, A.F.; Zamanian, M.Y.; et al. Neuroprotective Effects of Riluzole in Alzheimer's Disease: A Comprehensive Review. *Fundam. Clin. Pharmacol.* **2024**, *38*, 225–237. [CrossRef]
192. Kawano, C.; Isozaki, Y.; Nakagawa, A.; Hirayama, T.; Nishiyama, K.; Kuroyama, M. Liver Injury Risk Factors in Amyotrophic Lateral Sclerosis Patients Treated with Riluzole. *YAKUGAKU ZASSHI* **2020**, *140*, 923–928. [CrossRef]
193. Karadimas, S.K.; Laliberte, A.M.; Tetreault, L.; Chung, Y.S.; Arnold, P.; Foltz, W.D.; Fehlings, M.G. Riluzole Blocks Perioperative Ischemia-Reperfusion Injury and Enhances Postdecompression Outcomes in Cervical Spondylotic Myelopathy. *Sci. Transl. Med.* **2015**, *7*, 316ra194. [CrossRef]
194. Jiang, K.; Zhuang, Y.; Yan, M.; Chen, H.; Ge, A.Q.; Sun, L.; Miao, B. Effects of Riluzole on P2 × 7R Expression in the Spinal Cord in Rat Model of Neuropathic Pain. *Neurosci. Lett.* **2016**, *618*, 127–133. [CrossRef]
195. Yamamoto, S.; Ushio, S.; Egashira, N.; Kawashiri, T.; Mitsuyasu, S.; Higuchi, H.; Ozawa, N.; Masuguchi, K.; Ono, Y.; Masuda, S. Excessive Spinal Glutamate Transmission Is Involved in Oxaliplatin-Induced Mechanical Allodynia: A Possibility for Riluzole as a Prophylactic Drug. *Sci. Rep.* **2017**, *7*, 9661. [CrossRef]
196. Poupon, L.; Lamoine, S.; Pereira, V.; Barriere, D.A.; Lolignier, S.; Giraudet, F.; Aissouni, Y.; Meleine, M.; Prival, L.; Richard, D.; et al. Targeting the TREK-1 Potassium Channel via Riluzole to Eliminate the Neuropathic and Depressive-like Effects of Oxaliplatin. *Neuropharmacology* **2018**, *140*, 43–61. [CrossRef] [PubMed]
197. Yamamoto, S.; Egashira, N.; Tsuda, M.; Masuda, S. Riluzole Prevents Oxaliplatin-Induced Cold Allodynia via Inhibition of Overexpression of Transient Receptor Potential Melastatin 8 in Rats. *J. Pharmacol. Sci.* **2018**, *138*, 214–217. [CrossRef] [PubMed]

198. Zhang, X.; Gao, Y.; Wang, Q.; Du, S.; He, X.; Gu, N.; Lu, Y. Riluzole Induces LTD of Spinal Nociceptive Signaling via Postsynaptic GluR2 Receptors. *J. Pain Res.* **2018**, *11*, 2577–2586. [CrossRef]
199. Taiji, R.; Yamanaka, M.; Taniguchi, W.; Nishio, N.; Tsutsui, S.; Nakatsuka, T.; Yamada, H. Anti-Allodynic and Promotive Effect on Inhibitory Synaptic Transmission of Riluzole in Rat Spinal Dorsal Horn. *Biochem. Biophys. Rep.* **2021**, *28*, 101130. [CrossRef]
200. Thompson, J.M.; Yakhnitsa, V.; Ji, G.; Neugebauer, V. Small Conductance Calcium Activated Potassium (SK) Channel Dependent and Independent Effects of Riluzole on Neuropathic Pain-Related Amygdala Activity and Behaviors in Rats. *Neuropharmacology* **2018**, *138*, 219–231. [CrossRef]
201. Martins, B.D.C.; Torres, B.B.J.; de Oliveira, K.M.; Lavor, M.S.; Osório, C.M.; Fukushima, F.B.; Rosado, I.R.; de Melo, E.G. Association of Riluzole and Dantrolene Improves Significant Recovery after Acute Spinal Cord Injury in Rats. *Spine J.* **2018**, *18*, 532–539. [CrossRef]
202. Wu, Q.; Zhang, Y.; Zhang, Y.; Zhang, W.; Zhang, W.; Liu, Y.; Xu, S.; Guan, Y.; Chen, X. Riluzole Improves Functional Recovery after Acute Spinal Cord Injury in Rats and May Be Associated with Changes in Spinal Microglia/Macrophages Polarization. *Neurosci. Lett.* **2020**, *723*, 134829. [CrossRef]
203. Xu, S.; Wu, Q.; Zhang, W.; Liu, T.; Zhang, Y.; Zhang, W.; Zhang, Y.; Chen, X. Riluzole Promotes Neurite Growth in Rats after Spinal Cord Injury through the GSK-3β/CRMP-2 Pathway. *Biol. Pharm. Bull.* **2022**, *45*, 569–575. [CrossRef]
204. Ghayour, M.B.; Abdolmaleki, A.; Behnam-Rassouli, M. The Effect of Riluzole on Functional Recovery of Locomotion in the Rat Sciatic Nerve Crush Model. *Eur. J. Trauma Emerg. Surg.* **2017**, *43*, 691–699. [CrossRef]
205. Foley, P.; Parker, R.A.; de Angelis, F.; Connick, P.; Chandran, S.; Young, C.; Weir, C.J.; Chataway, J. Efficacy of Fluoxetine, Riluzole and Amiloride in Treating Neuropathic Pain Associated with Secondary Progressive Multiple Sclerosis. Pre-Specified Analysis of the MS-SMART Double-Blind Randomised Placebo-Controlled Trial. *Mult. Scler. Relat. Disord.* **2022**, *63*, 103925. [CrossRef]
206. Kumarasamy, D.; Viswanathan, V.K.; Shetty, A.P.; Pratheep, G.K.; Kanna, R.M.; Rajasekaran, S. The Role of Riluzole in Acute Traumatic Cervical Spinal Cord Injury with Incomplete Neurological Deficit: A Prospective, Randomised Controlled Study. *Indian J. Orthop.* **2022**, *56*, 2160–2168. [CrossRef]
207. Trinh, T.; Park, S.B.; Murray, J.; Pickering, H.; Lin, C.S.-Y.; Martin, A.; Friedlander, M.; Kiernan, M.C.; Goldstein, D.; Krishnan, A.V. Neu-Horizons: Neuroprotection and Therapeutic Use of Riluzole for the Prevention of Oxaliplatin-Induced Neuropathy—A Randomised Controlled Trial. *Support. Care Cancer* **2021**, *29*, 1103–1110. [CrossRef] [PubMed]
208. Wu, Q.; Zhang, W.; Yuan, S.; Zhang, Y.; Zhang, W.; Zhang, Y.; Chen, X.; Zang, L. A Single Administration of Riluzole Applied Acutely after Spinal Cord Injury Attenuates Pro-Inflammatory Activity and Improves Long-Term Functional Recovery in Rats. *J. Mol. Neurosci.* **2022**, *72*, 730–740. [CrossRef]
209. Haider, A.; Reddy, A. Levorphanol versus Methadone Use: Safety Considerations. *Ann. Cardiothorac. Surg.* **2020**, *9*, 579–585. [CrossRef]
210. Choi, D.W.; Peters, S.; Viseskul, V. Printedin USA. Dextrorphan and Levorphanol Selectively Block N-Methyl-D.-Aspartate Receptor-Mediated Neurotoxicity on Cortical Neurons. *J. Pharmacol. Exp. Ther.* **1987**, *242*, 713–720.
211. Church, J.; Lodge, D.; Berry, S.C. Differential Effects of Dextrorphan and Levorphanol on the Excitation of Rat Spinal Neurons by Amino Acids. *Eur. J. Pharmacol.* **1985**, *111*, 185–190. [CrossRef]
212. Pham, T.C.; Fudin, J.; Raffa, R.B. Is Levorphanol a Better Option than Methadone? *Pain Med.* **2015**, *16*, 1673–1679. [CrossRef]
213. Rowbotham, M.C.; Twilling, L.; Davies, P.S.; Reisner, L.; Taylor, K.; Mohr, D. Oral Opioid Therapy for Chronic Peripheral and Central Neuropathic Pain. *N. Engl. J. Med.* **2003**, *348*, 1223–1232. [CrossRef]
214. Codd, E.E.; Shank, R.P.; Schupsky, J.J.; Raffa, R.B. Serotonin and Norepinephrine Uptake Inhibiting Activity of Centrally Acting Analgesics: Structural Determinants and Role in Antinociception. *J. Pharmacol. Exp. Ther.* **1995**, *274*, 1263–1270.
215. Gudin, J.; Fudin, J.; Nalamachu, S. Levorphanol Use: Past, Present and Future. *Postgrad. Med.* **2016**, *128*, 46–53. [CrossRef]
216. Reddy, A.; Haider, A.; Arthur, J.; Hui, D.; Dalal, S.; Dev, R.; Tanco, K.; Amaram-Davila, J.; Hernandez, F.; Chavez, P.; et al. Levorphanol as a Second Line Opioid in Cancer Patients Presenting to an Outpatient Supportive Care Center: An Open-Label Study. *J. Pain Symptom Manag.* **2023**, *65*, e683–e690. [CrossRef] [PubMed]
217. Reddy, A.; Ng, A.; Mallipeddi, T.; Bruera, E. Levorphanol for Treatment of Intractable Neuropathic Pain in Cancer Patients. *J. Palliat. Med.* **2018**, *21*, 399–402. [CrossRef] [PubMed]
218. Eap, C.B.; Buclin, T.; Baumann, P. Interindividual Variability of the Clinical Pharmacokinetics of Methadone. *Clin. Pharmacokinet.* **2002**, *41*, 1153–1193. [CrossRef]
219. Volpe, D.A.; Xu, Y.; Sahajwalla, C.G.; Younis, I.R.; Patel, V. Methadone Metabolism and Drug-Drug Interactions: In Vitro and In Vivo Literature Review. *J. Pharm. Sci.* **2018**, *107*, 2983–2991. [CrossRef]
220. Boisvert-Plante, V.; Poulin-Harnois, C.; Ingelmo, P.; Einhorn, L.M. What We Know and What We Don't Know about the Perioperative Use of Methadone in Children and Adolescents. *Pediatr. Anesth.* **2023**, *33*, 185–192. [CrossRef]
221. Haumann, J.; Geurts, J.W.; van Kuijk, S.M.J.; Kremer, B.; Joosten, E.A.; van den Beuken-van Everdingen, M.H.J. Methadone Is Superior to Fentanyl in Treating Neuropathic Pain in Patients with Head-and-Neck Cancer. *Eur. J. Cancer* **2016**, *65*, 121–129. [CrossRef]
222. Sugiyama, Y.; Sakamoto, N.; Ohsawa, M.; Onizuka, M.; Ishida, K.; Murata, Y.; Iio, A.; Sugano, K.; Maeno, K.; Takeyama, H.; et al. A Retrospective Study on the Effectiveness of Switching to Oral Methadone for Relieving Severe Cancer-Related Neuropathic Pain and Limiting Adjuvant Analgesic Use in Japan. *J. Palliat. Med.* **2016**, *19*, 1051–1059. [CrossRef]

223. Madden, K.; Bruera, E. Very-Low-Dose Methadone To Treat Refractory Neuropathic Pain in Children with Cancer. *J. Palliat. Med.* **2017**, *20*, 1280–1283. [CrossRef]
224. Curry, Z.A.; Dang, M.C.; Sima, A.P.; Abdullaziz, S.; Del Fabbro, E.G. Combination Therapy with Methadone and Duloxetine for Cancer-Related Pain: A Retrospective Study. *Ann. Palliat. Med.* **2021**, *10*, 2505–2511. [CrossRef]
225. Matsuda, Y.; Okayama, S. Oral Methadone for Patients with Neuropathic Pain Due to Neoplastic Brachial Plexopathy. *J. Palliat. Care* **2022**, *37*, 77–82. [CrossRef]
226. Fawoubo, A.; Perceau-Chambard, É.; Ruer, M.; Filbet, M.; Tricou, C.; Economos, G. Methadone and Neuropathic Cancer Pain Subcomponents: A Prospective Cohort Pilot Study. *BMJ Support. Palliat. Care* **2023**, *13*, e273–e277. [CrossRef]
227. Adumala, A.; Palat, G.; Vajjala, A.; Brun, E.; Segerlantz, M. Oral Methadone versus Morphine IR for Patients with Cervical Cancer and Neuropathic Pain: A Prospective Randomised Controlled Trial. *Indian J. Palliat. Care* **2023**, *29*, 200–206. [CrossRef] [PubMed]
228. Bach, T.V.; Pan, J.; Kirstein, A.; Grief, C.J.; Grossman, D. Use of Methadone as an Adjuvant Medication to Low-Dose Opioids for Neuropathic Pain in the Frail Elderly: A Case Series. *J. Palliat. Med.* **2016**, *19*, 1351–1355. [CrossRef] [PubMed]
229. Rasmussen, V.F.; Lundberg, V.; Jespersen, T.W.; Hasle, H. Extreme Doses of Intravenous Methadone for Severe Pain in Two Children with Cancer. *Pediatr. Blood Cancer* **2015**, *62*, 1087–1090. [CrossRef] [PubMed]
230. Lynch, M.; Moulin, D.; Perez, J. Methadone vs. Morphine SR for Treatment of Neuropathic Pain: A Randomized Controlled Trial and the Challenges in Recruitment. *Can. J. Pain* **2019**, *3*, 180–189. [CrossRef] [PubMed]

International Journal of
Molecular Sciences

Review

Molecular Therapeutics for Diabetic Kidney Disease: An Update

Man Guo, Fangfang He * and Chun Zhang *

Department of Nephrology, Union Hospital, Tongji Medical College, Huazhong University of Science and Technology, Wuhan 430022, China
* Correspondence: hefang_105@hust.edu.cn (F.H.); drzhangchun@hust.edu.cn (C.Z.)

Abstract: Diabetic kidney disease (DKD) is a common microvascular complication of diabetes mellitus (DM). With the increasing prevalence of DM worldwide, the incidence of DKD remains high. If DKD is not well controlled, it can develop into chronic kidney disease or end-stage renal disease (ESRD), which places considerable economic pressure on society. Traditional therapies, including glycemic control, blood pressure control, blood lipid control, the use of renin–angiotensin system blockers and novel drugs, such as sodium–glucose cotransporter 2 inhibitors, mineralocorticoid receptor inhibitors and glucagon-like peptide-1 receptor agonists, have been used in DKD patients. Although the above treatment strategies can delay the progression of DKD, most DKD patients still ultimately progress to ESRD. Therefore, new and multimodal treatment methods need to be explored. In recent years, researchers have continuously developed new treatment methods and targets to delay the progression of DKD, including miRNA therapy, stem cell therapy, gene therapy, gut microbiota-targeted therapy and lifestyle intervention. These new molecular therapy methods constitute opportunities to better understand and treat DKD. In this review, we summarize the progress of molecular therapeutics for DKD, leading to new treatment strategies.

Keywords: diabetic kidney disease; therapeutics; miRNA; stem cell; gene therapy; gut microbiota

1. Introduction

Diabetic kidney disease (DKD) is a chronic disease caused by diabetes mellitus (DM). The early manifestation of DKD is microalbuminuria, which gradually progresses to massive proteinuria and progressive renal dysfunction. If DKD is not well controlled, it eventually progresses to end-stage renal disease (ESRD) [1], which is the main cause of increased morbidity and mortality in diabetic patients. Because DKD is a common complication of DM, and the prevalence of DM is increasing worldwide, the prevalence of DKD has also increased. Approximately 40% of patients with type 2 DM (T2DM) and 30% of patients with type 1 DM (T1DM) progress to DKD [2–4]. More than 700 million people worldwide are estimated to be currently affected by DKD, and this number will continue to increase in the coming decades [5]. The high medical costs and poor prognosis of people in end-stage DKD have placed a heavy burden on society and families [6,7].

The pathogenesis of DKD is complex, and the therapeutic targets are not yet clear. Traditionally, glomerular hypertension, altered renal hemodynamics, inflammation, oxidative stress stimulation, and activation of the renin–aldosterone system are believed to contribute to the progression and poor prognosis of DKD [8]. At present, the treatment methods for DKD mainly include classic methods, such as glycemic control, blood lipid control, blood pressure control, and other symptomatic supportive treatment methods. In addition, various drugs, such as metformin, mineralocorticoid receptor antagonists (MRAs), sodium–glucose cotransporter 2 (SGLT2) inhibitors, and glucagon-like peptide-1 receptor (GLP-1R) agonists, have also entered clinical use. However, the benefits of these treatments for DKD remain limited and they cannot prevent DKD from progressing into to ESRD. We need to search for new targets and strategies for treating DKD.

In recent years, treatment methods for DKD have also been continuously advancing. As research on new treatment methods, such as gene therapy, stem cell therapy, miRNA therapy, gut microbiota therapy and lifestyle therapy has progressed, new options have been provided for the treatment of DKD. This study focuses on the therapeutic potential of the above therapies for DKD to provide a new theoretical basis for the clinical treatment of DKD.

2. Conventional Treatment and Newer Medications for DKD

In the past 20 years, renin–angiotensin–aldosterone system (RAAS) blockers, angiotensin-converting enzyme inhibitors (ACEIs) and angiotensin receptor blockers (ARBs) have been the only drug choices for the treatment of DKD [9]. The above treatment has limited efficacy in preventing the progression of DKD. With the development of therapeutic targets for DKD, new therapeutic drugs for DKD are being clinically applied.

2.1. Renin–Angiotensin–Aldosterone System Blockades

RAAS inhibitors are the oldest and most advanced drugs used for the treatment of DKD. In many clinical trials, RAAS inhibitors have been proven to effectively treat DKD. ACEIs and ARBs can reduce glomerular hypertension, prevent structural damage to glomeruli, and reduce proteinuria levels by blocking the formation and function of angiotensin II [10]. These drugs have been shown to have renal and cardiovascular protective functions [11] and can prevent the progression of renal dysfunction in patients with diabetic or nondiabetic kidney disease [12].

2.2. Mineralocorticoid Receptor Antagonists

The activation of mineralocorticoid receptors can lead to renal inflammation and fibrosis reactions [13], whereas MRAs inhibit aldosterone binding and its subsequent biological effects by inducing different conformational changes in the mineralocorticoid receptor [14]. Clinical trial results suggested that finerenone, a novel nonsteroidal MRA, reduced the risk of renal damage and cardiovascular events in patients with DKD. Guidelines recommend the use of nonsteroidal MRAs as an adjunct therapy in patients with normal potassium levels <4.8 mmol/L, who are already taking SGLT2 inhibitors and RAAS blockers, as well as in patients at high risk of DKD progression with proteinuria (>30 mcg/mg) [13,15].

2.3. Glucagon-like Peptide-1 Receptor Agonist

Glucagon-like peptide-1 (GLP-1) is an intestinal insulinotropic hormone secreted by lower intestinal L cells that regulates glycemic levels by activating GLP-1R in the pancreas [4]. Research has shown that GLP-1R agonists exert renal protection in T2DM patients by reducing proteinuria and improving the impaired estimated glomerular filtration rate (eGFR) [16]. Recent experiments have further demonstrated the ability of GLP-1R agonists to reduce the occurrence of cardiovascular complications in DKD patients, making this drug a potential early option for use in DKD patients [9]. In addition, these drugs have the additional benefit of inducing weight loss [9], and KDIGO recommends adding GLP-1R agonists for patients who are already using metformin and SGLT2 inhibitors, but have not achieved their glycemic control goals [17].

2.4. Sodium–Glucose Cotransporter 2 Inhibitors

SGLT2 inhibitors are novel oral hypoglycemic drugs that activate the tubuloglomerular feedback mechanism to reduce glomerular hyperfiltration, especially when used in combination with ACEIs or ARBs [18]. SGLT2 inhibitors can inhibit the reabsorption of glucose in the proximal tubules and promote the excretion of urinary glucose [19]. In addition, SGLT2 inhibitors reportedly reduce the production of circulating inflammatory factors, such as interleukins (IL) -6, TNF receptor-1, and TNF receptor-2, and decrease the production of ketones [20]. Many clinical trials have shown that SGLT2 inhibitors can reduce the risk of DKD progression and renal failure, as well as heart failure, atherosclerotic cardiovascular disease and death [21–25]. SGLT2 inhibitors are now recommended as first-line treatments for DKD patients.

3. Gene Therapy

The pathological process of DKD is complex, and the core mechanism involves inflammation and oxidative stress induced by high glucose in the blood, resulting in damage to tubules and glomeruli which ultimately leads to kidney injury and fibrosis [8]. Therefore, the treatment of inflammation and oxidative stress is very valuable in DKD, which may lead to a potential new strategy for DKD treatment.

3.1. *Inflammation-Based Gene Therapy*

Increasing evidence shows that inflammation is a key factor in the pathogenesis of DKD [26,27]. In recent years, regulating the inflammatory response has become a new method to prevent the development of DKD. At present, therapeutic targets based on inflammatory molecules, including chemokines, adhesion factors, cytokines, and intracellular signaling pathways involved in inflammatory responses have been surveyed [28].

3.1.1. Adhesion and Chemokine Molecules Inhibition

Adhesive molecules are proteins on the surface of cells that participate in binding cells together or attaching them to the extracellular matrix [27]. Research has shown that various adhesion molecules, such as ICAM1, VCAM-1, and VAP-1, are involved in the occurrence and development of DKD, mainly by increasing the excretion of urinary proteins [29–31]. At present, treatments for adhesion factors remain limited. A Phase II clinical trial revealed that the oral VAP-1 inhibitor ASP8232 could delay the progression of kidney injury [32].

Chemokines and their receptors are soluble small-molecule proteins that play crucial roles in mediating the migration of immune cells to renal tissue and subsequently triggering inflammatory responses [33]. In DKD, the chemokine C-C motif-ligand 2 (CCL2), drives the recruitment of T cells and macrophages to the kidneys, and promotes the occurrence and development of DKD by binding to chemokine receptor 2 (CCR2) [34]. A Phase II clinical trial revealed that emapticap pegol could improve proteinuria and exert renal protection by blocking the CCL2-CCR2 signaling pathway [35]. In addition, CXCL12, a chemokine produced by podocytes, has also been reported to cause proteinuria and glomerulosclerosis in diabetic mice. NOX-A12, a CXCL12-specific inhibitor, can improve proteinuria and glomerulosclerosis in db/db mice [36]. However, relevant clinical research needs to be carried out.

3.1.2. Inhibiting Inflammatory Cytokines

Cytokines, including interleukins and interferons, are immune modulators that play a vital role in the occurrence of DM-related complications. Currently, many cytokines are involved in DKD research, including IL-1β, IL-18, and TNF-α [37], which trigger an inflammatory response in DKD. Specific compounds that target these inflammatory factors have been reported. Research has shown that alprostadil and thrombomodulin lectin domains alleviate inflammation in DKD by inhibiting the expression of IL-18 [38,39]. The inhibition of TNF-α with infliximab could reduce the excretion of urinary albumin in a streptozotocin (STZ)-induced diabetic rat model [40]. Neutralizing IL-1β antibodies improved the level of blood glucose and reduced inflammation in T2DM patients by restoring insulin production [41]. However, a clinical trial suggested that canakinumab, an inhibitor of IL-1β, did not provide significant clinical benefits or cause substantial harm in patients with chronic kidney disease (CKD) [42]. Anti-inflammatory therapy targeting specific molecules may be a promising treatment strategy for DKD.

3.1.3. JAK/STAT Inhibitors

Numerous studies have shown that the Janus kinase (JAK)/signal transducer and activator of transcription (STAT) signaling pathway is associated with the progression of DKD. JAK/STAT mediates the occurrence and development of DKD by inducing the transcription of genes that encode inflammatory factors, adhesion molecules, growth factors, extracellular matrix proteins, and pro-oxidant enzymes [37]. At present, many selective compounds, such as AG-490 which targets JAK2, fludarabine which targets STAT1, and nifuroxazide which selectively acts on STAT3, have been reported to alleviate kidney damage by inhibiting hyperglycemia-induced damage to the kidney [43–45]. A Phase II clinical trial revealed that balitinib, a selective inhibitor of JAK1 and JAK2, reduced proteinuria in DKD patients [46]. These findings suggest that targeting JAK/STAT has broad prospects in the treatment of DKD.

3.1.4. Inhibition of NF-κB

NF-κB is a key transcription factor in the pathogenesis of the inflammatory response in DKD, and activation of NF-κB can induce the expression of various inflammatory mediators, which play an important role in the pathogenesis of DKD [47]. Celastrol, an inhibitor of NF-κB, can improve insulin resistance and play an anti-inflammatory role, making it a new useful method for treating DKD [48]. In addition, the NF-κB inhibitor, BAY 11-7082 and a small molecule analog of resveratrol, have been reported to reduce renal inflammation in mouse models of DKD by inhibiting the activation of NF-κB, exerting renoprotective effects. However, owing to the complexity of NF-κB signaling [49,50], a clinical study revealed that PTX indirectly inhibited the activity of downstream NF-κB signaling by suppressing phosphodiesterase activity and activating protein kinase A/cAMP response element binding protein, reducing the progression of proteinuria in DKD [26,51]. At present, specific drugs that target NF-κB still need further research.

3.2. *Antioxidative Gene Therapy*

The oxidative stress process during DKD is caused by increased reactive oxygen species (ROS) induced by peroxidase in a high-glucose environment, accompanied by the depletion of antioxidants, which results in an imbalance of the oxidative/antioxidant system and causes cell and tissue damage [52]. The regulation of antioxidant genes is highly important for controlling the occurrence and development of DKD.

3.2.1. NOX Inhibitors

Nicotinamide adenine dinucleotide phosphate (NADPH) oxidase (NOX) is the most important source of ROS in diabetic patients [53]. Inhibiting NOX expression can reduce the production of ROS and is considered a new option for the treatment of DKD. Animal experiments revealed that the deletion of the NOX4 gene in DKD mice significantly reduced the production of ROS in the renal cortex and protected the kidney from damage [54]. GKT137831 is a small molecule inhibitor of NOX1 and NOX4. GKT137831 provided a similar degree of renal protection as NOX4 knockout in diabetic mice [55]. However, in a clinical trial, it failed to effectively reduce proteinuria (NCT02010242). Recently, a new pan-nitrogen oxide inhibitor, APX-115, was shown to reduce proteinuria, glomerular hypertrophy, renal tubular injury, podocyte injury, fibrosis, inflammation and oxidative stress to prevent renal injury in diabetic mice [56]. However, further clinical trials are needed to evaluate the clinical benefits of APX-115 for DKD patients.

3.2.2. AGE Inhibitors

Under high-glucose conditions, the body produces a large amount of advanced glycation end products (AGEs), which subsequently bind to AGE receptors (RAGE) to active downstream target molecules, such as NOX, and produce a large amount of ROS [52]. AGE inhibitors, such as aminoguanidine, pyridoxamine, and alagebrium, have been shown to protect the kidneys. However, these therapies have not yet been translated into clinical practice. Another method for targeting the AGE/RAGE axis is focused on inhibiting the biological activity of AGEs by interfering with RAGE. RAGE inhibitors, such as azeliragon, have been developed for the treatment of Alzheimer's disease [57]. In 2018, a Phase II randomized controlled trial was launched on the effects of green tea extract on soluble RAGE and kidney disease in patients with T2DM (NCT03622762), but the status of this study remains unclear.

3.2.3. Activation of Nrf2 Pathways

Nrf2 is a key regulatory factor in the cellular antioxidant response. When cells are stimulated by ROS, Nrf2 enters the nucleus and can activate the expression of various downstream target proteins, such as heme oxygenase-1 and superoxide dismutase. These activated target proteins can regulate the redox balance in the body, to restore the body from an oxidative state to a normal level [58]. In vitro experiments have shown that Nrf2 knockout induces ROS production by upregulating the expression of NOX2 [59]. In diabetic mouse models, dietary supplementation with Nrf2-targeted compounds, such as sulforaphane or cinnamic aldehyde, can alleviate the progression of DKD by reducing the oxidative stress response [60]. Bardoxolone methyl is an activator of Nrf2, whereas a Phase 3 study (AYAME study) in patients with DKD was terminated because the early instances of heart failure in patients participating in the study were surprisingly increased [61]. Further in-depth studies of novel Nrf2 regulators are highly valuable.

3.3. *Antifibrotic Therapy*

The inflammatory response and oxidative stress are also involved in activating the excessive production and deposition of extracellular matrix proteins in renal tissue, leading to renal fibrosis, which is the main outcome of almost all CKD cases and causes irreversible kidney damage. The TGF-β1 signaling pathway is the core molecule of renal fibrosis in DKD and can regulate fibrosis by promoting the overexpression of the extracellular matrix, which enhances the dedifferentiation of tubular epithelial cells and glomerular endothelial cells, and promotes cross-linking between collagen and elastin fibers [4,62]. The renoprotective effects of conventional therapies for DKD, such as RAAS inhibitors (including ACEIs, ARBs, and MRAs) are, at least partially, attributed to a reduction in TGF-β signaling. In animal experiments, TGF-β neutralizing monoclonal antibodies and drugs such as pirfenidone effectively alleviated renal fibrosis by reducing the activity of the TGF-β1 gene promoter and the release of the TGF-β1 protein in DKD [63–65]. Due to the complexity of TGF-β molecules, targeting them may inhibit important physiological functions, such as systemic anti-inflammatory activity. Therefore, current efforts have shifted toward targeting the intrinsic downstream components and molecules that can affect the profibrotic activity of the TGF-β pathway, such as SMAD proteins and related miRNAs [66]. A new strategy for reducing pathologically high levels of TGF-β activity to physiological levels has potential value for developing more effective and safe treatment methods to specifically delay the progression of DKD [66].

Overall, gene therapy can modify specific genes, which is expected to completely cure the disease, and an increasing number of clinical trials that target related genes are being conducted (Table 1). However, gene therapy still has limitations, theoretical research on gene therapy is still relatively weak, and gene technology is difficult, unstable, and has significant differences in effectiveness. Developing other novel treatment strategies to provide more possibilities for the prevention and treatment of DKD is necessary.

Table 1. Summary of selected clinical trials evaluating molecular therapies with anti-inflammatories and antioxidants in DKD.

Drug	Proposed Effect	Kidney Outcome	Reference
ASP8232	Inhibit the activity of VAP-1, improve oxidative stress damage and cellular toxicity	Reduced urinary ACR	[32]
Emapticap pegol	Inhibit the pro-inflammatory chemokine C-C motif-ligand 2	Improved the ACR and HbA1c	[35]

Table 1. *Cont.*

Drug	Proposed Effect	Kidney Outcome	Reference
Baricitinib	Inhibit the inflammation signal JAK1/JAK2	Decreased albuminuria	[46]
Gevokizumab	Anti-IL-1β antibody	Improved glycemia, decreased inflammation in diabetic patients	[41]
Bardoxolone methyl	An activator of Nrf2	Improved renal function, but the study was terminated because of increased early heart failure events	[61]
Pirfenidone	Inhibit the activity of TGF-β1,2,3	Increased eGFR	[65]
Pentoxifylline	Downregulate NF-κB signaling	Reduced proteinuria	[51]
GKT137831	Inhibitor of NOX 1/4	Failed to reduce albuminuria	NCT02010242
Canakinumab	Inhibitor of IL-1β	Neither clinical benefit nor substantial harm	[42]
Green tea	Interference with AGE receptor	No status of the study	NCT03622762

ACR: albumin-to-creatinine ratio, HbA1c: glycosylated hemoglobin type A1C, VAP-1: vascular adhesion protein-1, Nrf2: nuclear factor erythroid 2-related factor 2, JAK: Janus kinase, NOX: Nicotinamide adenine dinucleotide phosphate oxidase, eGFR: estimated glomerular filtration rate, AGE: advanced glycation end product.

4. MiRNA Therapy

MiRNAs are single-stranded small-molecule RNAs that are highly conserved, tissue specific and tissue targeted [67]. MiRNAs in the human genome regulate more than 60% of human protein-coding genes [68] and regulate gene expression at the post-transcriptional level by forming RNA-induced silencing complexes that recognize the 3′ UTRs of target mRNAs. They degrade target mRNAs when fully paired and inhibit translation when incompletely paired, resulting in negative regulation of target genes [69]. MiRNAs have been shown to regulate various biological processes and are closely related to the occurrence and development of DKD [70]. Currently, various RNA family members, such as the miR-192, miR-21, and miR-29 families, are reportedly specifically overexpressed in DKD renal tissue and play key roles in regulating glomerular basement membrane injury, mesangial damage, podocyte damage, and renal interstitial fibrosis [71]. Although clinical studies targeting miRNAs in DKD are limited, laboratory findings provide new possibilities for small molecule therapy in DKD. Thus, miRNA inhibitors, agonists, and mimics that control the expression levels of miRNAs can be developed for use in DKD patients in the future [72].

4.1. MiRNA Inducers

MiRNA inducers are small molecules that can upregulate the expression of target miRNAs. At present, many studies have focused on the role of miRNA inducers in protecting renal function in DKD. Triptolide is a monomeric compound, that is isolated from traditional Chinese medicine. Han et al. reported that triptolide could induce the expression of miRNA-137, reduce the level of proteinuria in DKD rats by inhibiting the activation of the Notch1 pathway, and improve glomerulosclerosis [73]. Linagliptin, a dipeptidyl peptidase-4 inhibitor, can delay the progression of renal fibrosis by increasing the level of miR-29a [74]. C66, a new analog of curcumin, has been reported to increase the expression of miR-200a, upregulate the level of Nrf2, reduce renal oxidative damage caused by DM, and play an important role in renal protection [75]. Currently, more clinical trials are needed to validate the efficacy of drugs based on miRNA induction, and drugs that induce the high expression of miRNAs with protective effects will provide us with another option for treating DKD.

4.2. MiRNA Mimics

MiRNA mimics are double-stranded RNA oligonucleotides that mimic endogenous miRNAs. They play a vital role in enhancing or rebuilding endogenous miRNAs [76].

MiRNA mimics with nucleotide sequences are similar to mature miRNAs and can be effectively delivered to cells through lipid reagents or electroporation and then bind to specific genes and to regulate the expression of miRNAs [77,78]. Many studies have focused on the potential application prospects of miRNA mimics in DKD. Cheng et al. reported that supplementation with miR-122-5p mimics could alleviate renal tubular injury and interstitial fibrosis in STZ-induced DKD model mice, reduce urinary creatinine levels, and exert renoprotective effects [79]. Bai et al. reported that the transfection of miRNA-20 mimics could inhibit high glucose-induced inflammation and apoptosis in renal tubular cells, alleviate high glucose-induced damage to renal tubular cells, and serve as a potential therapeutic target for DKD [80]. Clinical trials related to miRNA mimics are already underway, with MRG-201 (an LNA miRNA-29b mimic) inhibiting collagen expression and fibrosis development in skin incision wounds by restoring the levels of targeted miRNAs [81]. In addition, MRX34 (a miRNA-34a mimic) has also entered clinical trials for hepatocellular carcinoma, renal cell carcinoma, and melanoma [82]. Therefore, regulating or administering protective miRNA mimics could become a new approach for the treatment of DKD, with good research value and feasibility.

4.3. *Small-Molecule Inhibitors*

Some small molecules that inhibit or downregulate the expression of target miRNAs are also used in research on DKD. Paclitaxel (used for cancer chemotherapy), a mitotic inhibitor, has been shown to downregulate the expression of miR-192, thereby improving fibrosis in the residual kidney [83]. Hyperoside is the active component of Betula platyphylla flowers and can improve glomerulosclerosis in DKD by downregulating miR-21 to increase the expression of MMP-9 [84]. Wang et al. reported that Astragaloside IV (a bioactive saponin extracted from Astragalus root) could inhibit the expression of miR-21 and miR-192 in DKD models and alleviate renal fibrosis by acting on the TGF-β1/Smad pathway [85,86]. Jia et al. reported that icariin (dried stem and leaf extracts of epimedium) promoted the expression of GLP-1R by downregulating miR-192-5p and exerting renal protection by regulating glucose metabolism [87]. The above research suggests that high-throughput screening of drugs that can reduce the expression of miRNAs may be another option.

4.4. *MiRNA Sponges*

MiRNA sponges are RNA structures that contain multiple miRNA binding sites, that bind miRNAs and prevent them from performing their functions. Source mRNA targets are used to manipulate miRNAs at the cellular level [88]. By subcloning the miRNA binding site into a vector containing 50 and 30 stem loop elements of the U6 snRNA promoter, the sponge can be introduced into cells [89]. Research has shown that the expression of the miR-21 sponge inhibits endogenous miR-21, prevents high glucose-induced PTEN downregulation and Akt phosphorylation, and reduces the expression of fibronectin in the kidney [90].Moreover, circRNAs, which are covalently closed RNA loops produced by reverse-splicing pre-mRNAs, can also act as miRNA sponges [91]. Peng et al. reported that circRNA_010383, a natural sponge of miRNA-135a, inhibited the function of miR-135a by directly binding to miR-135a, leading to proteinuria and renal fibrosis in DKD [92]. Another type of circRNA, circ-AKT3, can act as a sponge of miR-296-3p to reduce the proliferation and fibrosis of mesangial cells in DKD [93], which provides a potential basis for the construction of additional miRNA sponges. At present, an increasing number of new strategies are being used to modify and reverse pathological changes in miRNA expression. In-depth research on miRNA sponges will undoubtedly promote the development of new therapeutic methods.

4.5. *Anti-miRNA Oligonucleotides*

Antisense oligonucleotide (ASO), also referred to as an antagomir, is a method that mainly blocks miRNA binding to target sequences by providing perfectly matched antisense nucleotide chains. Chemical modifications, such as 2′-O-methylation, 2′-Omethoxyethyl, 2′ fluoro-chemistries and locked nucleic acids (LNAs), can be used to improve the affinity of ASOs for target miRNAs and resistance to nuclease degradation in an effort to stably and effectively deliver them into cells [71,94,95]. Among these modifications, LNAs have the best affinity for binding to complementary RNA [96]. At present, relevant animal experiments have revealed the enormous potential of anti-miRNA oligonucleotides in the treatment of DKD. The injection of LNA-modified anti-miR-192 into diabetic mice, effectively inhibited the expression of miR-192 and downstream miRNAs (miR-216a, miR-217 and miR-200 families) and reduced the expression of collagen, TGF-β, fibronectin and p53, thereby alleviating renal fibrosis and proteinuria in DKD [72,97,98]. Antagomirs of miRNA-27a and miRNA-132 have also been reported to reduce collagen deposition, inhibit the progression of renal fibrosis, and alleviate the progression of DKD [99,100]. In addition, knocking down miR-29c with a specific 2′-O-methylation-modified ASO significantly reduced proteinuria and the accumulation of the renal mesangial matrix in DKD model mice, thereby reducing the incidence of DKD [101]. Notably, relevant clinical drugs have also entered the clinical trial stage. Miravirsen (LNA and phosphorothioate-modified antagomir targeting miR-122) and RG-012 (2′-o-methoxyethoxy-modified antagonist targeting miR-21) have shown efficacy in treating hepatitis C [102,103]. The targeted and specific regulation of miRNAs through anti-miRNA oligonucleotides may be a new approach for treating DKD.

The development of effective delivery methods and mechanisms targeting specific renal cell populations is an important challenge for successfully translating miRNAs into clinical applications. In addition, the combination of miRNAs with existing treatment methods such as ACEi and ARBs has provided new therapeutic potential, and further research is necessary to determine whether these combinations can exert synergistic effects [104].

5. Stem Cell Therapy

Stem cells are primitive undifferentiated cells with self-replication and multipotent differentiation potential [105]. Stem cells can be classified into embryonic stem cells, adult stem cells, and induced pluripotent stem cells based on their origin [106]. Adult stem cells can be further divided into endothelial progenitor cells, mesenchymal stem cells (MSCs), and osteogenic stem cells. Research has shown that stem cells can protect the kidneys through direct differentiation, paracrine effects, and other mechanisms. Stem cells can migrate to damaged tissues or organs, and in the treatment of kidney diseases, they can differentiate directly into intrinsic renal cells or secrete a series of bioactive molecules through paracrine mechanisms, regulate the microenvironment, inhibit inflammatory reactions, and alleviate kidney tissue damage [107,108]. In recent years, many scholars have explored the effectiveness and mechanism of stem cell therapy for DKD.

5.1. Mesenchymal Stem Cell Therapy

Common stem cell therapies mainly utilize MSCs, which not only have the ability to migrate to damaged tissue sites but also have strong immunosuppressive effects [105]. Therefore, the use of mesenchymal cells is expected to become an effective method for treating DKD. According to their isolation sources, MSCs can be derived from adipose tissue, bone marrow, placenta, or the umbilical cord [106].

5.1.1. Bone Marrow-Derived Mesenchymal Stem Cells

Bone marrow mesenchymal stem cells (BMSCs) can be obtained from bone marrow by isolating monocytes. In recent years, many scholars have explored the efficacy and mechanism of BMSCs in treating DKD. BMSCs can also exert renoprotective effects by regulating the levels of blood glucose. Lv et al. reported that BMSCs could inhibit glucose uptake mediated by GLUT1 and lower blood glucose levels, thereby suppressing oxidative stress, reducing urinary albumin excretion, and improving glomerulosclerosis [109].

Pan et al. transplanted BMSCs into diabetic tree shrews via intravenous infusion, which increased their insulin level and decreased their postprandial blood glucose, blood creatinine and urea nitrogen levels [110]. In addition, BMSCs alleviate DKD by regulating immune cells. BMSCs can significantly downregulate the expression of transcription factors necessary for the development of CD103+ DCs, inhibit DC maturation, reduce the expression of inflammatory factors and chemokines, inhibit CD8+ T-cell proliferation, and confer renal protection [111]. BMSCs promote macrophage differentiation into the M2 phenotype and alleviate the inflammatory response in DKD model mice by activating the transcription factor EB [112]. In addition, Lv et al. reported that the implantation of BMSCs could effectively reduce blood glucose and improve renal function in DKD mice and could also inhibit the TGF-β/Smad signaling pathway through the secretion of bone morphogenetic protein 7, improving glomerular fibrosis [113].

5.1.2. Umbilical Cord Mesenchymal Stem Cells

MSCs derived from the umbilical cord have lower immunogenicity, greater proliferation potential and differentiation ability, a faster self-renewal ability, and broader application prospects [114]. In STZ-induced diabetic mice, Li et al. reported that repeated injection of umbilical cord MSCs could inhibit TGF-β1-triggered myofibroblast differentiation through paracrine action and increase the levels of MMP2 and MMP9, thereby reducing the deposition of fibronectin and collagen I and alleviating proteinuria, glomerular injury, and fibrosis in DKD mouse models [115]. In addition to inhibiting fibrosis during the process of DKD, scholars have reported that MSCs in DKD could reduce cell death and delay disease progression through antiapoptotic effects [116]. Moreover, MSCs prevent the progression of DKD by inducing Arg1 expression in macrophages, which subsequently reduces mitochondrial dysfunction in tubular epithelial cells [117], to improve blood glucose control and anti-inflammatory activity [118].

5.1.3. Placenta-Derived Mesenchymal Stem Cells

Placenta-derived mesenchymal stem cells (PMSCs) are favored because of their accessibility and lack of associated ethical issues [119]. Wang et al. reported that PMSCs could be recruited into immune organs to promote Th17/Treg balance in the kidneys and blood of DKD rats by acting on the PD-1/PD-L1 pathway, which reduced the levels of proinflammatory cytokines (IL-17A and IL-1β) and improved renal function and pathological damage in DKD rats [120]. In addition, Han et al. reported that PMSCs could also regulate molecular pathways to exert renoprotective effects. Injecting PMSCs into DKD rats activated the expression of the SIRT1-PGC-1α-TFAM pathway to alleviate podocyte damage and mitochondrial autophagy inhibition [121].

5.1.4. Adipose-Derived Mesenchymal Stem Cells

Because of their abundance in vivo, adipose-derived mesenchymal stem cells (ADMSCs) have become an attractive cell source [122,123]. ADMSC transplantation may alleviate cell apoptosis and improve renal histology to alleviate DKD damage by activating klotho and inhibiting the Wnt/β-catenin pathway [124]. Yang et al. investigated the effect of ADMSCs combined with empagliflozin (EMPA) on DKD, and the results showed that ADMSC-EMPA combined therapy protected renal function and ultrastructural integrity in DKD mice better than monotherapy [125]. However, the retention and survival rates of ADMSCs are low in the kidneys. Therefore, scholars constructed ADMSC sheets to extend the survival time of ADMSCs in the kidneys, which allowed these cells to play a protective role in maintaining the renal tubular structure [126].

5.1.5. Mesenchymal Stem Cell-Derived Exosomes

Exosomes are extracellular vesicles with a diameter of 30 to 150 nanometers secreted by MSCs [127]. In kidney disease, in addition to their ability to function as stem cells themselves, MSC-derived exosomes can carry complex molecular cargo such as proteins, lipids, and nucleic acids (DNA, miRNA, and circRNA), to regulate cell apoptosis, proliferation, and immune response [108,128]. At present, extracellular vesicles from different sources of MSCs have been studied in DKD. Ebrahim et al. reported that BMSC-derived exosomes improved the autophagy of tubular cells in DKD mice by inhibiting the expression of rapamycin targets, reducing fibrosis, and improving kidney function [129]. The intravenous injection of extracellular vesicles derived from human umbilical cord MSCs could reduce blood glucose levels and improve insulin resistance in diabetic rats, indicating their potential application value in DKD [130]. Extracellular vesicles derived from ADMSCs have the potential to become intervention targets for DKD treatment by alleviating podocyte apoptosis, reducing the level of blood glucose in DKD mice, and inhibiting mesangial proliferation and renal fibrosis [131,132].

5.2. Other Types of Stem Cell Therapies

Induced pluripotent stem cells (iPSCs) are pluripotent cells formed by reprogramming terminally differentiated somatic cells with specific transcription factors, similar to human embryonic stem cells. Given the high differentiation ability of iPSCs, they can be induced to differentiate into renal tubular cells and podocytes [133]. Raikwar et al. generated endodermal cells from human iPSCs by treating them with activin A and induced them to differentiate into three-dimensional islet cell clusters through a series of growth factors. IPSCs reportedly play a vital role in improving the blood glucose level of diabetic mice [134]. Embryonic stem cells (ESCs) are a type of multipotent cell derived from the cell population within the blastocyst that can be induced into renal cells through a series of determined growth factors or inducers [135,136], indicating the potential for transplantation and regenerative therapy of damaged kidneys [108]. Currently, research on the treatment of DKD with iPSCs and ESCs remains in the early experimental stage. In addition, pancreatic Ngn3 stem cells can differentiate into mature beta cells within the pancreas, which can secrete insulin and maintain blood glucose stability in the body [137]. Intensive insulin therapy can reduce the occurrence of microalbuminuria, pancreatic stem cells may become the cell source for pancreatic islet formation [138,139], providing a new approach for the treatment of DM.

Stem cells have a wide range of sources, and autologous stem cells do not have immune rejection reactions, providing prospects for clinical applications. However, further research is needed to evaluate their therapeutic effects. The transplantation of stem cells will provide more options for the treatment of DKD.

6. Gut Microbiota-Targeted Therapy

The plethora of different types of bacteria in the intestine are collectively referred to as the gut microbiome. The disturbance of the intestinal flora plays an important role in the pathogenesis of DKD, which may be attributed to several factors [140]. For example, uremia in patients with CKD affects the composition and metabolism of the intestinal flora. Intestinal ecological imbalance may cause damage to the epithelial barrier, ultimately leading to increased exposure of the host to endotoxins [141]. Moreover, the large amount of proinflammatory and nephrotoxic substances produced by gut microbiota can trigger local inflammation in the kidneys, leading to a decrease in the eGFR [142]. Uremic toxins, such as p-cresyl sulfate, trimethylamine-N-oxide, and indoxyl sulfate, originate from microbial metabolism, and their accumulation exacerbates damage to the kidneys and cardiovascular system [143]. Intervening in the human gut microbiome is expected to become an effective treatment for alleviating DKD. Currently, many scholars are focusing on the role of regulating the gut microbiome to treat DKD, mainly with dietary strategies, probiotic and prebiotic treatment, and fecal microbiota transplantation (FMT) methods.

Diet is the main exogenous factor affecting the composition of the gut microbiome. Short-chain fatty acids (SCFAs), which are considered to play a key role in microbial host crosstalk, are the main metabolites produced by bacterial fermentation of dietary fiber. Research has shown that SCFAs can alleviate the inflammatory response in renal ischemia–reperfusion injury [144]. Li et al. reported that a high-fiber diet could improve the degree and probability of DKD development in a mouse model. Dietary fiber can prevent DKD by regulating the intestinal flora, enriching bacteria that produce SCFAs and increasing the production of SCFAs [145].

Probiotics and prebiotics are commonly used to regulate the gut microbiome. Probiotics are living microorganisms that synthesize multiple vitamins to improve the balance of host gut microbiota. Prebiotics, regulating the composition of gut microbiota and benefiting the human health, are ingredients of fermented food [146]. Many scholars have reported the positive effects of probiotic or prebiotic supplementation in CKD patients, including reduced uremic toxins and blood glucose levels, a restructured gut microbiome, and reductions in oxidative stress and inflammation [146,147]. In DKD patients, Dai et al. reported that probiotics could delay the progression of renal dysfunction, improve glucose and lipid metabolism, and reduce symptoms and oxidative stress [148].

Intestinal FMT is a process by which the gut microbiome of a healthy donor is transferred into a recipient to introduce or restore a stable gut microbiome. Previous studies have shown that FMT can prevent diabetic patients from gaining weight, reduce proteinuria and local intestinal inflammation, and improve insulin resistance [149]. Shang et al. reported that the level of uremic toxins in DKD mice significantly decreased and that DKD lesions gradually improved after the transplantation of feces from healthy mice [150]. In a mouse model of DKD, Bastos et al. reported that FMT could prevent weight gain, reduce albuminuria and local intestinal inflammation, and improve insulin resistance [149]. Because the mechanisms underlying the effects of FMT in humans are complex, the clinical applicability of FMT needs to be verified in clinical trials.

In general, the understanding of the relationship between the gut microbiome and DKD is in its infancy. It is still unclear which gut microbiota play a beneficial role and which play a harmful role in DKD. Further research is needed on the dosage, treatment duration, and long-term effectiveness of various bacterial colonies. Before gut microbiota-targeted therapy can be systematically used to treat or prevent DKD, additional research is needed [151].

7. Lifestyle Intervention

Except for the therapy methods discussed above, lifestyle management has also gradually gained attention in the treatment of DKD, including smoking cessation, exercise, weight loss and so on. Obesity can increase insulin resistance, activate inflammation and increase oxidative stress [152].Bolignano et al. found that weight loss could significantly reduce albuminuria and proteinuria in diabetic patients undergoing weight loss surgery [153].Therefore, weight loss is an important method to treat DKD. Physical activity is a basic component of lifestyle management in diabetic care. Research found that regular moderate and high-intensity physical activity is associated with a reduction in incidence rate and progression of DKD, as well as cardiovascular events and mortality. A comparative study in CKD patients (including DKD patients) suggested that exercise could reduce the decline in eGFR by approximately 0.5% annually for every 60-min increase in weekly activity duration [154].Smoking can lead to the progression and deterioration of DKD. Fedoroff et al. found that smoking is a risk factor for the progress of DKD, and the risk is increased with increased smoking dose [155]. Quitting smoking has been recommended in KDIGO for DKD patients [156]. Lifestyle intervention is also an important treatment method for DKD patients.

8. Conclusions

Several challenges exist in the molecular diagnosis and treatment of DKD. Firstly, the description of endpoints in DKD mainly include a decline in eGFR, the need for dialysis or transplant, and mortality. It is of great significance to explore more rational evaluation indicators of endpoints for DKD, such as proteinuria, to assist in evaluating the effectiveness of clinical interventions. Secondly, different phenotypes including proteinuric and non-proteinuric manifestations exist in DKD, and the specific diagnosis and treatment of the two different phenotypes requires further exploration. Thirdly, novel biomarkers that can not only predict the progression of DKD but also accurately reflect the response to current treatments are needed. However, novel biomarkers applied in clinical practice are still scarce and require further research.

Currently, although several novel drugs, such as RAS blockers, SGLT2 inhibitors, and MRAs, have shown clinical benefits for DKD, the current treatment options still have a limited ability to prevent the progression of kidney disease and reduce the risk of complications and death in DKD patients. Therefore, it is highly important to explore new treatment methods and optimize treatment strategies for controlling the progression of DKD. This review introduces new treatment strategies, including gene therapy, stem cell therapy, miRNA therapy, gut microbiota-targeted therapy and lifestyle intervention, which have the potential to usher in a new era of DKD management that improves angiopathy, kidney function, and survival rates (Figure 1). However, more multicenter, large-scale prospective clinical studies are urgently needed to verify the prognostic impact of these novel therapeutics on DKD patients. Hopefully, an increasing number of drugs will enter clinical trials to provide new ideas for the treatment and intervention of DKD and to inhibit the development of DKD into ESRD.

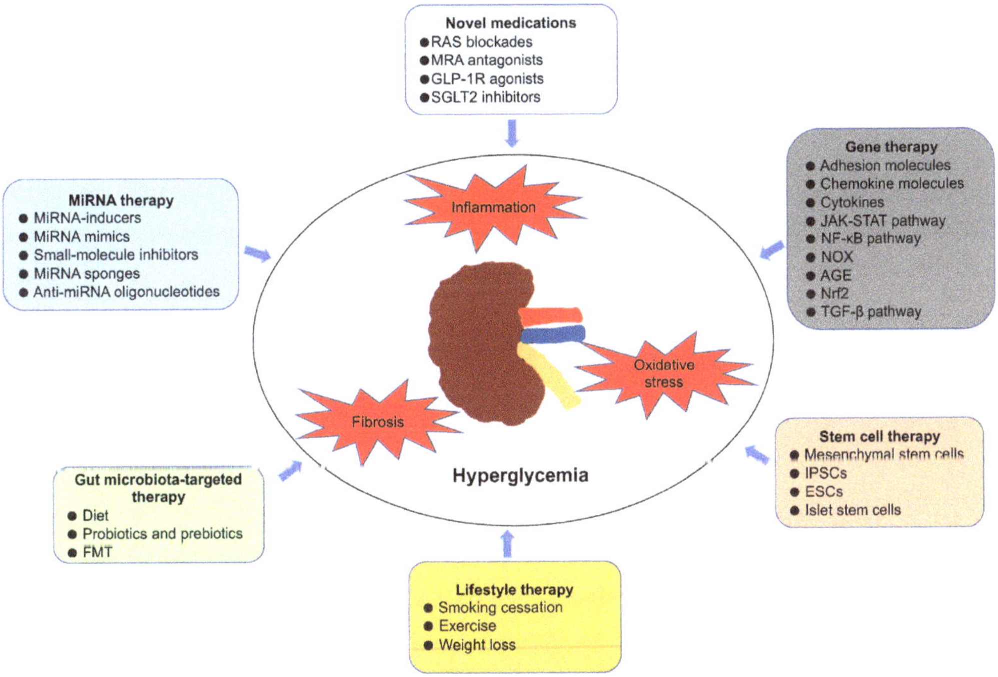

Figure 1. Current and potential targets for therapeutic interventions of DKD. Abbreviations: RAS: renin–angiotensin–aldosterone system; MRAs: mineralocorticoid system; SGLT2: sodium–glucose cotransporter 2; GLP-1R: glucagon-like peptide 1 receptor; JAK/STAT: Janus kinase/signal transducer and activator of transcription; NOX: nicotinamide adenine dinucleotide phosphate oxidases; AGE: advanced glycation end product; Nrf2: nuclear factor erythroid 2-related factor 2; IPSC: induced pluripotent stem cells; ESC: embryonic stem cell; FMT: fecal microbiota transplantation.

Author Contributions: Conceptualization, M.G. and C.Z.; writing—original draft preparation, M.G.; writing—review and editing, F.H. and C.Z.; validation M.G.; supervision, F.H. and C.Z. All authors have read and agreed to the published version of the manuscript.

Funding: This research was funded by the National Natural Science Foundation of China (82370728, 81974096, 81974097, 82170773, 82100729, 82200808, 82200841, 82300843, and 82300786), National Key Research and Development Program of China (2021YFC2500200), and Key Research and Development Program of Hubei Province (2023BCB034).

Institutional Review Board Statement: Not applicable.

Informed Consent Statement: Not applicable.

Data Availability Statement: Not applicable.

Conflicts of Interest: The authors declare no conflicts of interest.

References

1. Yuan, C.M.; Nee, R.; Ceckowski, K.A.; Knight, K.R.; Abbott, K.C. Diabetic nephropathy as the cause of end-stage kidney disease reported on the medical evidence form CMS2728 at a single center. *Clin. Kidney J.* **2017**, *10*, 257–262. [CrossRef] [PubMed]
2. Fried, L.F.; Folkerts, K.; Smeta, B.; Bowrin, K.D.; Mernagh, P.; Millier, A.; Kovesdy, C.P. Targeted literature review of the burden of illness in patients with chronic kidney disease and type 2 diabetes. *Am. J. Manag. Care* **2021**, *27*, S168–S177. [CrossRef] [PubMed]
3. Alicic, R.Z.; Rooney, M.T.; Tuttle, K.R. Diabetic Kidney Disease: Challenges, Progress, and Possibilities. *Clin. J. Am. Soc. Nephrol.* **2017**, *12*, 2032–2045. [CrossRef] [PubMed]
4. Tuttle, K.R.; Agarwal, R.; Alpers, C.E.; Bakris, G.L.; Brosius, F.C.; Kolkhof, P.; Uribarri, J. Molecular mechanisms and therapeutic targets for diabetic kidney disease. *Kidney Int.* **2022**, *102*, 248–260. [CrossRef]
5. Jager, K.J.; Kovesdy, C.; Langham, R.; Rosenberg, M.; Jha, V.; Zoccali, C. A single number for advocacy and communication-worldwide more than 850 million individuals have kidney diseases. *Nephrol. Dial. Transplant.* **2019**, *34*, 1803–1805. [CrossRef]
6. Golestaneh, L.; Alvarez, P.J.; Reaven, N.L.; Funk, S.E.; McGaughey, K.J.; Romero, A.; Brenner, M.S.; Onuigbo, M. All-cause costs increase exponentially with increased chronic kidney disease stage. *Am. J. Manag. Care* **2017**, *23*, S163–S172.
7. Folkerts, K.; Petruski-Ivleva, N.; Kelly, A.; Fried, L.; Blankenburg, M.; Gay, A.; Kovesdy, C.P. Annual health care resource utilization and cost among type 2 diabetes patients with newly recognized chronic kidney disease within a large U.S. administrative claims database. *J. Manag. Care Spec. Pharm.* **2020**, *26*, 1506–1516. [CrossRef]
8. Jung, C.Y.; Yoo, T.H. Pathophysiologic Mechanisms and Potential Biomarkers in Diabetic Kidney Disease. *Diabetes Metab. J.* **2022**, *46*, 181–197. [CrossRef]
9. Sawaf, H.; Thomas, G.; Taliercio, J.J.; Nakhoul, G.; Vachharajani, T.J.; Mehdi, A. Therapeutic Advances in Diabetic Nephropathy. *J. Clin. Med.* **2022**, *11*, 378. [CrossRef]
10. Ruggenenti, P.; Cravedi, P.; Remuzzi, G. The RAAS in the pathogenesis and treatment of diabetic nephropathy. *Nat. Rev. Nephrol.* **2010**, *6*, 319–330. [CrossRef]
11. Mohsen, M.; Elberry, A.A.; Mohamed Rabea, A.; Abdelrahim, M.E.A.; Hussein, R.R.S. Recent therapeutic targets in diabetic nephropathy. *Int. J. Clin. Pract.* **2021**, *75*, e14650. [CrossRef] [PubMed]
12. Siragy, H.M.; Carey, R.M. Role of the intrarenal renin-angiotensin-aldosterone system in chronic kidney disease. *Am. J. Nephrol.* **2010**, *31*, 541–550. [CrossRef] [PubMed]
13. Gupta, S.; Dominguez, M.; Golestaneh, L. Diabetic Kidney Disease: An Update. *Med. Clin. N. Am.* **2023**, *107*, 689–705. [CrossRef] [PubMed]
14. Kim, D.L.; Lee, S.E.; Kim, N.H. Renal Protection of Mineralocorticoid Receptor Antagonist, Finerenone, in Diabetic Kidney Disease. *Endocrinol. Metab.* **2023**, *38*, 43–55. [CrossRef] [PubMed]
15. Bakris, G.L.; Agarwal, R.; Anker, S.D.; Pitt, B.; Ruilope, L.M.; Rossing, P.; Kolkhof, P.; Nowack, C.; Schloemer, P.; Joseph, A.; et al. Effect of Finerenone on Chronic Kidney Disease Outcomes in Type 2 Diabetes. *N. Engl. J. Med.* **2020**, *383*, 2219–2229. [CrossRef]
16. Sattar, N.; Lee, M.M.Y.; Kristensen, S.L.; Branch, K.R.H.; Del Prato, S.; Khurmi, N.S.; Lam, C.S.P.; Lopes, R.D.; McMurray, J.J.V.; Pratley, R.E.; et al. Cardiovascular, mortality, and kidney outcomes with GLP-1 receptor agonists in patients with type 2 diabetes: A systematic review and meta-analysis of randomised trials. *Lancet Diabetes Endocrinol.* **2021**, *9*, 653–662. [CrossRef]
17. de Boer, I.H.; Khunti, K.; Sadusky, T.; Tuttle, K.R.; Neumiller, J.J.; Rhee, C.M.; Rosas, S.E.; Rossing, P.; Bakris, G. Diabetes Management in Chronic Kidney Disease: A Consensus Report by the American Diabetes Association (ADA) and Kidney Disease: Improving Global Outcomes (KDIGO). *Diabetes Care* **2022**, *45*, 3075–3090. [CrossRef]
18. Alicic, R.Z.; Neumiller, J.J.; Johnson, E.J.; Dieter, B.; Tuttle, K.R. Sodium-Glucose Cotransporter 2 Inhibition and Diabetic Kidney Disease. *Diabetes* **2019**, *68*, 248–257. [CrossRef]
19. Yamazaki, T.; Mimura, I.; Tanaka, T.; Nangaku, M. Treatment of Diabetic Kidney Disease: Current and Future. *Diabetes Metab. J.* **2021**, *45*, 11–26. [CrossRef]
20. Heerspink, H.J.L.; Perco, P.; Mulder, S.; Leierer, J.; Hansen, M.K.; Heinzel, A.; Mayer, G. Canagliflozin reduces inflammation and fibrosis biomarkers: A potential mechanism of action for beneficial effects of SGLT2 inhibitors in diabetic kidney disease. *Diabetologia* **2019**, *62*, 1154–1166. [CrossRef]
21. Zinman, B.; Wanner, C.; Lachin, J.M.; Fitchett, D.; Bluhmki, E.; Hantel, S.; Mattheus, M.; Devins, T.; Johansen, O.E.; Woerle, H.J.; et al. Empagliflozin, Cardiovascular Outcomes, and Mortality in Type 2 Diabetes. *N. Engl. J. Med.* **2015**, *373*, 2117–2128. [CrossRef] [PubMed]

22. Wanner, C.; Inzucchi, S.E.; Lachin, J.M.; Fitchett, D.; von Eynatten, M.; Mattheus, M.; Johansen, O.E.; Woerle, H.J.; Broedl, U.C.; Zinman, B. Empagliflozin and Progression of Kidney Disease in Type 2 Diabetes. *N. Engl. J. Med.* **2016**, *375*, 323–334. [CrossRef] [PubMed]
23. Neal, B.; Perkovic, V.; Mahaffey, K.W.; de Zeeuw, D.; Fulcher, G.; Erondu, N.; Shaw, W.; Law, G.; Desai, M.; Matthews, D.R. Canagliflozin and Cardiovascular and Renal Events in Type 2 Diabetes. *N. Engl. J. Med.* **2017**, *377*, 644–657. [CrossRef] [PubMed]
24. Wiviott, S.D.; Raz, I.; Bonaca, M.P.; Mosenzon, O.; Kato, E.T.; Cahn, A.; Silverman, M.G.; Zelniker, T.A.; Kuder, J.F.; Murphy, S.A.; et al. Dapagliflozin and Cardiovascular Outcomes in Type 2 Diabetes. *N. Engl. J. Med.* **2019**, *380*, 347–357. [CrossRef] [PubMed]
25. Zelniker, T.A.; Wiviott, S.D.; Raz, I.; Im, K.; Goodrich, E.L.; Bonaca, M.P.; Mosenzon, O.; Kato, E.T.; Cahn, A.; Furtado, R.H.M.; et al. SGLT2 inhibitors for primary and secondary prevention of cardiovascular and renal outcomes in type 2 diabetes: A systematic review and meta-analysis of cardiovascular outcome trials. *Lancet* **2019**, *393*, 31–39. [CrossRef]
26. Donate-Correa, J.; Luis-Rodríguez, D.; Martín-Núñez, E.; Tagua, V.G.; Hernández-Carballo, C.; Ferri, C.; Rodríguez-Rodríguez, A.E.; Mora-Fernández, C.; Navarro-González, J.F. Inflammatory Targets in Diabetic Nephropathy. *J. Clin. Med.* **2020**, *9*, 458. [CrossRef]
27. Navarro-González, J.F.; Mora-Fernández, C.; Muros de Fuentes, M.; García-Pérez, J. Inflammatory molecules and pathways in the pathogenesis of diabetic nephropathy. *Nat. Rev. Nephrol.* **2011**, *7*, 327–340. [CrossRef]
28. Rayego-Mateos, S.; Rodrigues-Diez, R.R.; Fernandez-Fernandez, B.; Mora-Fernández, C.; Marchant, V.; Donate-Correa, J.; Navarro-González, J.F.; Ortiz, A.; Ruiz-Ortega, M. Targeting inflammation to treat diabetic kidney disease: The road to 2030. *Kidney Int.* **2023**, *103*, 282–296. [CrossRef]
29. Li, H.Y.; Lin, H.A.; Nien, F.J.; Wu, V.C.; Jiang, Y.D.; Chang, T.J.; Kao, H.L.; Lin, M.S.; Wei, J.N.; Lin, C.H.; et al. Serum Vascular Adhesion Protein-1 Predicts End-Stage Renal Disease in Patients with Type 2 Diabetes. *PLoS ONE* **2016**, *11*, e0147981. [CrossRef]
30. Qian, Y.; Li, S.; Ye, S.; Chen, Y.; Zhai, Z.; Chen, K.; Yang, G. Renoprotective effect of rosiglitazone through the suppression of renal intercellular adhesion molecule-1 expression in streptozotocin-induced diabetic rats. *J. Endocrinol. Investig.* **2008**, *31*, 1069–1074. [CrossRef]
31. Rubio-Guerra, A.F.; Vargas-Robles, H.; Lozano Nuevo, J.J.; Escalante-Acosta, B.A. Correlation between circulating adhesion molecule levels and albuminuria in type-2 diabetic hypertensive patients. *Kidney Blood Press. Res.* **2009**, *32*, 106–109. [CrossRef] [PubMed]
32. de Zeeuw, D.; Renfurm, R.W.; Bakris, G.; Rossing, P.; Perkovic, V.; Hou, F.F.; Nangaku, M.; Sharma, K.; Heerspink, H.J.L.; Garcia-Hernandez, A.; et al. Efficacy of a novel inhibitor of vascular adhesion protein-1 in reducing albuminuria in patients with diabetic kidney disease (ALBUM): A randomised, placebo-controlled, phase 2 trial. *Lancet Diabetes Endocrinol.* **2018**, *6*, 925–933. [CrossRef] [PubMed]
33. Wada, J.; Makino, H. Inflammation and the pathogenesis of diabetic nephropathy. *Clin. Sci.* **2013**, *124*, 139–152. [CrossRef] [PubMed]
34. Giunti, S.; Barutta, F.; Perin, P.C.; Gruden, G. Targeting the MCP-1/CCR2 System in diabetic kidney disease. *Curr. Vasc. Pharmacol.* **2010**, *8*, 849–860. [CrossRef] [PubMed]
35. Menne, J.; Eulberg, D.; Beyer, D.; Baumann, M.; Saudek, F.; Valkusz, Z.; Więcek, A.; Haller, H. C-C motif-ligand 2 inhibition with emapticap pegol (NOX-E36) in type 2 diabetic patients with albuminuria. *Nephrol. Dial. Transplant.* **2017**, *32*, 307–315. [CrossRef]
36. Sayyed, S.G.; Hägele, H.; Kulkarni, O.P.; Endlich, K.; Segerer, S.; Eulberg, D.; Klussmann, S.; Anders, H.J. Podocytes produce homeostatic chemokine stromal cell-derived factor-1/CXCL12, which contributes to glomerulosclerosis, podocyte loss and albuminuria in a mouse model of type 2 diabetes. *Diabetologia* **2009**, *52*, 2445–2454. [CrossRef]
37. Moreno, J.A.; Gomez Guerrero, C.; Mas, S.; Sanz, A.B.; Lorenzo, O.; Ruiz-Ortega, M.; Opazo, L.; Mezzano, S.; Egido, J. Targeting inflammation in diabetic nephropathy: A tale of hope. *Expert Opin. Investig. Drugs* **2018**, *27*, 917–930. [CrossRef]
38. Luo, C.; Li, T.; Zhang, C.; Chen, Q.; Li, Z.; Liu, J.; Wang, Y. Therapeutic effect of alprostadil in diabetic nephropathy: Possible roles of angiopoietin-2 and IL-18. *Cell Physiol. Biochem.* **2014**, *34*, 916–928. [CrossRef]
39. Yang, S.M.; Ka, S.M.; Wu, H.L.; Yeh, Y.C.; Kuo, C.H.; Hua, K.F.; Shi, G.Y.; Hung, Y.J.; Hsiao, F.C.; Yang, S.S.; et al. Thrombomodulin domain 1 ameliorates diabetic nephropathy in mice via anti-NF-κB/NLRP3 inflammasome-mediated inflammation, enhancement of NRF2 antioxidant activity and inhibition of apoptosis. *Diabetologia* **2014**, *57*, 424–434. [CrossRef]
40. Moriwaki, Y.; Inokuchi, T.; Yamamoto, A.; Ka, T.; Tsutsumi, Z.; Takahashi, S.; Yamamoto, T. Effect of TNF-alpha inhibition on urinary albumin excretion in experimental diabetic rats. *Acta Diabetol.* **2007**, *44*, 215–218. [CrossRef]
41. Cavelti-Weder, C.; Babians-Brunner, A.; Keller, C.; Stahel, M.A.; Kurz-Levin, M.; Zayed, H.; Solinger, A.M.; Mandrup-Poulsen, T.; Dinarello, C.A.; Donath, M.Y. Effects of gevokizumab on glycemia and inflammatory markers in type 2 diabetes. *Diabetes Care* **2012**, *35*, 1654–1662. [CrossRef] [PubMed]
42. Ridker, P.M.; MacFadyen, J.G.; Glynn, R.J.; Koenig, W.; Libby, P.; Everett, B.M.; Lefkowitz, M.; Thuren, T.; Cornel, J.H. Inhibition of Interleukin-1β by Canakinumab and Cardiovascular Outcomes in Patients With Chronic Kidney Disease. *J. Am. Coll. Cardiol.* **2018**, *71*, 2405–2414. [CrossRef] [PubMed]
43. Banes, A.K.; Shaw, S.; Jenkins, J.; Redd, H.; Amiri, F.; Pollock, D.M.; Marrero, M.B. Angiotensin II blockade prevents hyperglycemia-induced activation of JAK and STAT proteins in diabetic rat kidney glomeruli. *Am. J. Physiol. Ren. Physiol.* **2004**, *286*, F653–F659. [CrossRef] [PubMed]
44. Cetkovic-Cvrlje, M.; Dragt, A.L.; Vassilev, A.; Liu, X.P.; Uckun, F.M. Targeting JAK3 with JANEX-1 for prevention of autoimmune type 1 diabetes in NOD mice. *Clin. Immunol.* **2003**, *106*, 213–225. [CrossRef]

45. Said, E.; Zaitone, S.A.; Eldosoky, M.; Elsherbiny, N.M. Nifuroxazide, a STAT3 inhibitor, mitigates inflammatory burden and protects against diabetes-induced nephropathy in rats. *Chem. Biol. Interact.* **2018**, *281*, 111–120. [CrossRef]
46. Tuttle, K.R.; Brosius, F.C., 3rd; Adler, S.G.; Kretzler, M.; Mehta, R.L.; Tumlin, J.A.; Tanaka, Y.; Haneda, M.; Liu, J.; Silk, M.E.; et al. JAK1/JAK2 inhibition by baricitinib in diabetic kidney disease: Results from a Phase 2 randomized controlled clinical trial. *Nephrol. Dial. Transplant.* **2018**, *33*, 1950–1959. [CrossRef]
47. Foresto-Neto, O.; Albino, A.H.; Arias, S.C.A.; Faustino, V.D.; Zambom, F.F.F.; Cenedeze, M.A.; Elias, R.M.; Malheiros, D.; Camara, N.O.S.; Fujihara, C.K.; et al. NF-κB System Is Chronically Activated and Promotes Glomerular Injury in Experimental Type 1 Diabetic Kidney Disease. *Front. Physiol.* **2020**, *11*, 84. [CrossRef]
48. Kim, J.E.; Lee, M.H.; Nam, D.H.; Song, H.K.; Kang, Y.S.; Lee, J.E.; Kim, H.W.; Cha, J.J.; Hyun, Y.Y.; Han, S.Y.; et al. Celastrol, an NF-κB inhibitor, improves insulin resistance and attenuates renal injury in db/db mice. *PLoS ONE* **2013**, *8*, e62068. [CrossRef]
49. Borgohain, M.P.; Lahkar, M.; Ahmed, S.; Chowdhury, L.; Kumar, S.; Pant, R.; Choubey, A. Small Molecule Inhibiting Nuclear Factor-kB Ameliorates Oxidative Stress and Suppresses Renal Inflammation in Early Stage of Alloxan-Induced Diabetic Nephropathy in Rat. *Basic Clin. Pharmacol. Toxicol.* **2017**, *120*, 442–449. [CrossRef]
50. Kolati, S.R.; Kasala, E.R.; Bodduluru, L.N.; Mahareddy, J.R.; Uppulapu, S.K.; Gogoi, R.; Barua, C.C.; Lahkar, M. BAY 11-7082 ameliorates diabetic nephropathy by attenuating hyperglycemia-mediated oxidative stress and renal inflammation via NF-κB pathway. *Environ. Toxicol. Pharmacol.* **2015**, *39*, 690–699. [CrossRef]
51. Navarro-González, J.F.; Mora-Fernández, C.; Muros de Fuentes, M.; Chahin, J.; Méndez, M.L.; Gallego, E.; Macía, M.; del Castillo, N.; Rivero, A.; Getino, M.A.; et al. Effect of pentoxifylline on renal function and urinary albumin excretion in patients with diabetic kidney disease: The PREDIAN trial. *J. Am. Soc. Nephrol.* **2015**, *26*, 220–229. [CrossRef] [PubMed]
52. Ma, X.; Ma, J.; Leng, T.; Yuan, Z.; Hu, T.; Liu, Q.; Shen, T. Advances in oxidative stress in pathogenesis of diabetic kidney disease and efficacy of TCM intervention. *Ren. Fail.* **2023**, *45*, 2146512. [CrossRef] [PubMed]
53. Mazzieri, A.; Porcellati, F.; Timio, F.; Reboldi, G. Molecular Targets of Novel Therapeutics for Diabetic Kidney Disease: A New Era of Nephroprotection. *Int. J. Mol. Sci.* **2024**, *25*, 3969. [CrossRef] [PubMed]
54. Shi, Q.; Lee, D.Y.; Féliers, D.; Abboud, H.E.; Bhat, M.A.; Gorin, Y. Interplay between RNA-binding protein HuR and Nox4 as a novel therapeutic target in diabetic kidney disease. *Mol. Metab.* **2020**, *36*, 100968. [CrossRef] [PubMed]
55. Jha, J.C.; Gray, S.P.; Barit, D.; Okabe, J.; El-Osta, A.; Namikoshi, T.; Thallas-Bonke, V.; Wingler, K.; Szyndralewiez, C.; Heitz, F.; et al. Genetic targeting or pharmacologic inhibition of NADPH oxidase nox4 provides renoprotection in long-term diabetic nephropathy. *J. Am. Soc. Nephrol.* **2014**, *25*, 1237–1254. [CrossRef] [PubMed]
56. Kwon, G.; Uddin, M.J.; Lee, G.; Jiang, S.; Cho, A.; Lee, J.H.; Lee, S.R.; Bae, Y.S.; Moon, S.H.; Lee, S.J.; et al. A novel pan-Nox inhibitor, APX-115, protects kidney injury in streptozotocin-induced diabetic mice: Possible role of peroxisomal and mitochondrial biogenesis. *Oncotarget* **2017**, *8*, 74217–74232. [CrossRef] [PubMed]
57. Østergaard, J.A.; Cooper, M.E.; Jandeleit-Dahm, K.A.M. Targeting oxidative stress and anti-oxidant defence in diabetic kidney disease. *J. Nephrol.* **2020**, *33*, 917–929. [CrossRef]
58. Ahmed, S.M.; Luo, L.; Namani, A.; Wang, X.J.; Tang, X. Nrf2 signaling pathway: Pivotal roles in inflammation. *Biochim. Biophys. Acta Mol. Basis Dis.* **2017**, *1863*, 585–597. [CrossRef]
59. Kovac, S.; Angelova, P.R.; Holmström, K.M.; Zhang, Y.; Dinkova-Kostova, A.T.; Abramov, A.Y. Nrf2 regulates ROS production by mitochondria and NADPH oxidase. *Biochim. Biophys. Acta* **2015**, *1850*, 794–801. [CrossRef]
60. Zheng, H.; Whitman, S.A.; Wu, W.; Wondrak, G.T.; Wong, P.K.; Fang, D.; Zhang, D.D. Therapeutic potential of Nrf2 activators in streptozotocin-induced diabetic nephropathy. *Diabetes* **2011**, *60*, 3055–3066. [CrossRef]
61. Wang, N.; Zhang, C. Oxidative Stress: A Culprit in the Progression of Diabetic Kidney Disease. *Antioxidants* **2024**, *13*, 455. [CrossRef] [PubMed]
62. Zhao, L.; Zou, Y.; Liu, F. Transforming Growth Factor-Beta1 in Diabetic Kidney Disease. *Front. Cell Dev. Biol.* **2020**, *8*, 187. [CrossRef] [PubMed]
63. RamachandraRao, S.P.; Zhu, Y.; Ravasi, T.; McGowan, T.A.; Toh, I.; Dunn, S.R.; Okada, S.; Shaw, M.A.; Sharma, K. Pirfenidone is renoprotective in diabetic kidney disease. *J. Am. Soc. Nephrol.* **2009**, *20*, 1765–1775. [CrossRef] [PubMed]
64. Benigni, A.; Zoja, C.; Campana, M.; Corna, D.; Sangalli, F.; Rottoli, D.; Gagliardini, E.; Conti, S.; Ledbetter, S.; Remuzzi, G. Beneficial effect of TGFbeta antagonism in treating diabetic nephropathy depends on when treatment is started. *Nephron Exp. Nephrol.* **2006**, *104*, e158–e168. [CrossRef] [PubMed]
65. Sharma, K.; Ix, J.H.; Mathew, A.V.; Cho, M.; Pflueger, A.; Dunn, S.R.; Francos, B.; Sharma, S.; Falkner, B.; McGowan, T.A.; et al. Pirfenidone for diabetic nephropathy. *J. Am. Soc. Nephrol.* **2011**, *22*, 1144–1151. [CrossRef]
66. Tang, J.; Liu, F.; Cooper, M.E.; Chai, Z. Renal fibrosis as a hallmark of diabetic kidney disease: Potential role of targeting transforming growth factor-beta (TGF-β) and related molecules. *Expert Opin. Ther. Targets* **2022**, *26*, 721–738. [CrossRef]
67. Cerqueira, D.M.; Tayeb, M.; Ho, J. MicroRNAs in kidney development and disease. *JCI Insight* **2022**, *7*, 158277. [CrossRef]
68. Friedman, R.C.; Farh, K.K.; Burge, C.B.; Bartel, D.P. Most mammalian mRNAs are conserved targets of microRNAs. *Genome Res.* **2009**, *19*, 92–105. [CrossRef]
69. Palazzo, A.F.; Lee, E.S. Non-coding RNA: What is functional and what is junk? *Front. Genet.* **2015**, *6*, 2. [CrossRef]
70. Zahari Sham, S.Y.; Abdullah, M.; Osman, M.; Seow, H.F. An insight of dysregulation of microRNAs in the pathogenesis of diabetic kidney disease. *Malays. J. Pathol.* **2022**, *44*, 187–201.

71. Yarahmadi, A.; Shahrokhi, S.Z.; Mostafavi-Pour, Z.; Azarpira, N. MicroRNAs in diabetic nephropathy: From molecular mechanisms to new therapeutic targets of treatment. *Biochem. Pharmacol.* **2021**, *189*, 114301. [CrossRef] [PubMed]
72. Kato, M.; Natarajan, R. MicroRNAs in diabetic nephropathy: Functions, biomarkers, and therapeutic targets. *Ann. N. Y. Acad. Sci.* **2015**, *1353*, 72–88. [CrossRef] [PubMed]
73. Han, F.; Wang, S.; Chang, Y.; Li, C.; Yang, J.; Han, Z.; Chang, B.; Sun, B.; Chen, L. Triptolide prevents extracellular matrix accumulation in experimental diabetic kidney disease by targeting microRNA-137/Notch1 pathway. *J. Cell Physiol.* **2018**, *233*, 2225–2237. [CrossRef] [PubMed]
74. Kanasaki, K.; Shi, S.; Kanasaki, M.; He, J.; Nagai, T.; Nakamura, Y.; Ishigaki, Y.; Kitada, M.; Srivastava, S.P.; Koya, D. Linagliptin-mediated DPP-4 inhibition ameliorates kidney fibrosis in streptozotocin-induced diabetic mice by inhibiting endothelial-to-mesenchymal transition in a therapeutic regimen. *Diabetes* **2014**, *63*, 2120–2131. [CrossRef]
75. Wu, H.; Kong, L.; Tan, Y.; Epstein, P.N.; Zeng, J.; Gu, J.; Liang, G.; Kong, M.; Chen, X.; Miao, L.; et al. C66 ameliorates diabetic nephropathy in mice by both upregulating NRF2 function via increase in miR-200a and inhibiting miR-21. *Diabetologia* **2016**, *59*, 1558–1568. [CrossRef]
76. Chau, B.N.; Xin, C.; Hartner, J.; Ren, S.; Castano, A.P.; Linn, G.; Li, J.; Tran, P.T.; Kaimal, V.; Huang, X.; et al. MicroRNA-21 promotes fibrosis of the kidney by silencing metabolic pathways. *Sci. Transl. Med.* **2012**, *4*, 121ra118. [CrossRef]
77. Kaadt, E.; Alsing, S.; Cecchi, C.R.; Damgaard, C.K.; Corydon, T.J.; Aagaard, L. Efficient Knockdown and Lack of Passenger Strand Activity by Dicer-Independent shRNAs Expressed from Pol II-Driven MicroRNA Scaffolds. *Mol. Ther. Nucleic Acids* **2019**, *14*, 318–328. [CrossRef]
78. Ling, H.; Fabbri, M.; Calin, G.A. MicroRNAs and other non-coding RNAs as targets for anticancer drug development. *Nat. Rev. Drug Discov.* **2013**, *12*, 847–865. [CrossRef]
79. Cheng, L.; Qiu, X.; He, L.; Liu, L. MicroRNA-122-5p ameliorates tubular injury in diabetic nephropathy via FIH-1/HIF-1α pathway. *Ren. Fail.* **2022**, *44*, 293–303. [CrossRef]
80. Bai, Y.; Li, H.; Dong, J. Up-regulation of miR-20a weakens inflammation and apoptosis in high-glucose-induced renal tubular cell mediating diabetic kidney disease by repressing CXCL8 expression. *Arch. Physiol. Biochem.* **2022**, *128*, 1603–1610. [CrossRef]
81. Gallant-Behm, C.L.; Piper, J.; Lynch, J.M.; Seto, A.G.; Hong, S.J.; Mustoe, T.A.; Maari, C.; Pestano, L.A.; Dalby, C.M.; Jackson, A.L.; et al. A MicroRNA-29 Mimic (Remlarsen) Represses Extracellular Matrix Expression and Fibroplasia in the Skin. *J. Investig. Dermatol.* **2019**, *139*, 1073–1081. [CrossRef] [PubMed]
82. Ling, H.; Girnita, L.; Buda, O.; Calin, G.A. Non-coding RNAs: The cancer genome dark matter that matters! *Clin. Chem. Lab. Med.* **2017**, *55*, 705–714. [CrossRef] [PubMed]
83. Sun, L.; Zhang, D.; Liu, F.; Xiang, X.; Ling, G.; Xiao, L.; Liu, Y.; Zhu, X.; Zhan, M.; Yang, Y.; et al. Low-dose paclitaxel ameliorates fibrosis in the remnant kidney model by down-regulating miR-192. *J. Pathol.* **2011**, *225*, 364–377. [CrossRef] [PubMed]
84. Zhang, L.; He, S.; Yang, F.; Yu, H.; Xie, W.; Dai, Q.; Zhang, D.; Liu, X.; Zhou, S.; Zhang, K. Hyperoside ameliorates glomerulosclerosis in diabetic nephropathy by downregulating miR-21. *Can. J. Physiol. Pharmacol.* **2016**, *94*, 1249–1256. [CrossRef] [PubMed]
85. Mao, Q.; Chen, C.; Liang, H.; Zhong, S.; Cheng, X.; Li, L. Astragaloside IV inhibits excessive mesangial cell proliferation and renal fibrosis caused by diabetic nephropathy via modulation of the TGF-β1/Smad/miR-192 signaling pathway. *Exp. Ther. Med.* **2019**, *18*, 3053–3061. [CrossRef]
86. Wang, X.; Gao, Y.; Tian, N.; Zou, D.; Shi, Y.; Zhang, N. Astragaloside IV improves renal function and fibrosis via inhibition of miR-21-induced podocyte dedifferentiation and mesangial cell activation in diabetic mice. *Drug Des. Dev. Ther.* **2018**, *12*, 2431–2442. [CrossRef]
87. Jia, Z.; Wang, K.; Zhang, Y.; Duan, Y.; Xiao, K.; Liu, S.; Ding, X. Icariin Ameliorates Diabetic Renal Tubulointerstitial Fibrosis by Restoring Autophagy via Regulation of the miR-192-5p/GLP-1R Pathway. *Front. Pharmacol.* **2021**, *12*, 720387. [CrossRef]
88. Diener, C.; Keller, A.; Meese, E. Emerging concepts of miRNA therapeutics: From cells to clinic. *Trends Genet.* **2022**, *38*, 613–626. [CrossRef]
89. Alvarez, M.L.; DiStefano, J.K. Towards microRNA-based therapeutics for diabetic nephropathy. *Diabetologia* **2013**, *56*, 444–456. [CrossRef]
90. Dey, N.; Das, F.; Mariappan, M.M.; Mandal, C.C.; Ghosh-Choudhury, N.; Kasinath, B.S.; Choudhury, G.G. MicroRNA-21 orchestrates high glucose-induced signals to TOR complex 1, resulting in renal cell pathology in diabetes. *J. Biol. Chem.* **2011**, *286*, 25586–25603. [CrossRef]
91. Liu, M.; Zhao, J. Circular RNAs in Diabetic Nephropathy: Updates and Perspectives. *Aging Dis.* **2022**, *13*, 1365–1380. [CrossRef] [PubMed]
92. Peng, F.; Gong, W.; Li, S.; Yin, B.; Zhao, C.; Liu, W.; Chen, X.; Luo, C.; Huang, Q.; Chen, T.; et al. circRNA_010383 Acts as a Sponge for miR-135a, and Its Downregulated Expression Contributes to Renal Fibrosis in Diabetic Nephropathy. *Diabetes* **2021**, *70*, 603–615. [CrossRef] [PubMed]
93. Tang, B.; Li, W.; Ji, T.T.; Li, X.Y.; Qu, X.; Feng, L.; Bai, S. Circ-AKT3 inhibits the accumulation of extracellular matrix of mesangial cells in diabetic nephropathy via modulating miR-296-3p/E-cadherin signals. *J. Cell Mol. Med.* **2020**, *24*, 8779–8788. [CrossRef] [PubMed]
94. Kato, M. Noncoding RNAs as therapeutic targets in early stage diabetic kidney disease. *Kidney Res. Clin. Pract.* **2018**, *37*, 197–209. [CrossRef]

95. Lennox, K.A.; Behlke, M.A. Chemical modification and design of anti-miRNA oligonucleotides. *Gene Ther.* **2011**, *18*, 1111–1120. [CrossRef]
96. Lima, J.F.; Cerqueira, L.; Figueiredo, C.; Oliveira, C.; Azevedo, N.F. Anti-miRNA oligonucleotides: A comprehensive guide for design. *RNA Biol.* **2018**, *15*, 338–352. [CrossRef]
97. Putta, S.; Lanting, L.; Sun, G.; Lawson, G.; Kato, M.; Natarajan, R. Inhibiting microRNA-192 ameliorates renal fibrosis in diabetic nephropathy. *J. Am. Soc. Nephrol.* **2012**, *23*, 458–469. [CrossRef]
98. Deshpande, S.D.; Putta, S.; Wang, M.; Lai, J.Y.; Bitzer, M.; Nelson, R.G.; Lanting, L.L.; Kato, M.; Natarajan, R. Transforming growth factor-β-induced cross talk between p53 and a microRNA in the pathogenesis of diabetic nephropathy. *Diabetes* **2013**, *62*, 3151–3162. [CrossRef]
99. Zhang, Y.; Zhao, S.; Wu, D.; Liu, X.; Shi, M.; Wang, Y.; Zhang, F.; Ding, J.; Xiao, Y.; Guo, B. MicroRNA-22 Promotes Renal Tubulointerstitial Fibrosis by Targeting PTEN and Suppressing Autophagy in Diabetic Nephropathy. *J. Diabetes Res.* **2018**, *2018*, 4728645. [CrossRef]
100. Bijkerk, R.; de Bruin, R.G.; van Solingen, C.; van Gils, J.M.; Duijs, J.M.; van der Veer, E.P.; Rabelink, T.J.; Humphreys, B.D.; van Zonneveld, A.J. Silencing of microRNA-132 reduces renal fibrosis by selectively inhibiting myofibroblast proliferation. *Kidney Int.* **2016**, *89*, 1268–1280. [CrossRef]
101. Long, J.; Wang, Y.; Wang, W.; Chang, B.H.; Danesh, F.R. MicroRNA-29c is a signature microRNA under high glucose conditions that targets Sprouty homolog 1, and its in vivo knockdown prevents progression of diabetic nephropathy. *J. Biol. Chem.* **2011**, *286*, 11837–11848. [CrossRef] [PubMed]
102. Lindow, M.; Kauppinen, S. Discovering the first microRNA-targeted drug. *J. Cell Biol.* **2012**, *199*, 407–412. [CrossRef] [PubMed]
103. van der Ree, M.H.; de Vree, J.M.; Stelma, F.; Willemse, S.; van der Valk, M.; Rietdijk, S.; Molenkamp, R.; Schinkel, J.; van Nuenen, A.C.; Beuers, U.; et al. Safety, tolerability, and antiviral effect of RG-101 in patients with chronic hepatitis C: A phase 1B, double-blind, randomised controlled trial. *Lancet* **2017**, *389*, 709–717. [CrossRef] [PubMed]
104. Hagiwara, S.; Gohda, T.; Kantharidis, P.; Okabe, J.; Murakoshi, M.; Suzuki, Y. Potential of Modulating Aldosterone Signaling and Mineralocorticoid Receptor with microRNAs to Attenuate Diabetic Kidney Disease. *Int. J. Mol. Sci.* **2024**, *25*, 869. [CrossRef] [PubMed]
105. Xu, N.; Liu, J.; Li, X. Therapeutic role of mesenchymal stem cells (MSCs) in diabetic kidney disease (DKD). *Endocr. J.* **2022**, *69*, 1159–1172. [CrossRef]
106. Habiba, U.E.; Khan, N.; Greene, D.L.; Shamim, S.; Umer, A. The therapeutic effect of mesenchymal stem cells in diabetic kidney disease. *J. Mol. Med.* **2024**, *102*, 537–570. [CrossRef]
107. Hickson, L.J.; Abedalqader, T.; Ben-Bernard, G.; Mondy, J.M.; Bian, X.; Conley, S.M.; Zhu, X.; Herrmann, S.M.; Kukla, A.; Lorenz, E.C.; et al. A systematic review and meta-analysis of cell-based interventions in experimental diabetic kidney disease. *Stem Cells Transl. Med.* **2021**, *10*, 1304–1319. [CrossRef]
108. Wang, Y.; Jin, M.; Cheng, C.K.; Li, Q. Tubular injury in diabetic kidney disease: Molecular mechanisms and potential therapeutic perspectives. *Front. Endocrinol.* **2023**, *14*, 1238927. [CrossRef]
109. Lv, S.; Cheng, J.; Sun, A.; Li, J.; Wang, W.; Guan, G.; Liu, G.; Su, M. Mesenchymal stem cells transplantation ameliorates glomerular injury in streptozotocin-induced diabetic nephropathy in rats via inhibiting oxidative stress. *Diabetes Res. Clin. Pract.* **2014**, *104*, 143–154. [CrossRef]
110. Pan, X.H.; Yang, X.Y.; Yao, X.; Sun, X.M.; Zhu, L.; Wang, J.X.; Pang, R.Q.; Cai, X.M.; Dai, J.J.; Ruan, G.P. Bone-marrow mesenchymal stem cell transplantation to treat diabetic nephropathy in tree shrews. *Cell Biochem. Funct.* **2014**, *32*, 453–463. [CrossRef]
111. Zhang, F.; Wang, C.; Wen, X.; Chen, Y.; Mao, R.; Cui, D.; Li, L.; Liu, J.; Chen, Y.; Cheng, J.; et al. Mesenchymal stem cells alleviate rat diabetic nephropathy by suppressing CD103(+) DCs-mediated CD8(+) T cell responses. *J. Cell Mol. Med.* **2020**, *24*, 5817–5831. [CrossRef] [PubMed]
112. Yuan, Y.; Li, L.; Zhu, L.; Liu, F.; Tang, X.; Liao, G.; Liu, J.; Cheng, J.; Chen, Y.; Lu, Y. Mesenchymal stem cells elicit macrophages into M2 phenotype via improving transcription factor EB-mediated autophagy to alleviate diabetic nephropathy. *Stem Cells* **2020**, *38*, 639–652. [CrossRef] [PubMed]
113. Lv, S.; Liu, G.; Sun, A.; Wang, J.; Cheng, J.; Wang, W.; Liu, X.; Nie, H.; Guan, G. Mesenchymal stem cells ameliorate diabetic glomerular fibrosis in vivo and in vitro by inhibiting TGF-β signalling via secretion of bone morphogenetic protein 7. *Diabetes Vasc. Dis. Res.* **2014**, *11*, 251–261. [CrossRef] [PubMed]
114. Li, T.; Xia, M.; Gao, Y.; Chen, Y.; Xu, Y. Human umbilical cord mesenchymal stem cells: An overview of their potential in cell-based therapy. *Expert Opin. Biol. Ther.* **2015**, *15*, 1293–1306. [CrossRef] [PubMed]
115. Li, H.; Rong, P.; Ma, X.; Nie, W.; Chen, Y.; Zhang, J.; Dong, Q.; Yang, M.; Wang, W. Mouse Umbilical Cord Mesenchymal Stem Cell Paracrine Alleviates Renal Fibrosis in Diabetic Nephropathy by Reducing Myofibroblast Transdifferentiation and Cell Proliferation and Upregulating MMPs in Mesangial Cells. *J. Diabetes Res.* **2020**, *2020*, 3847171. [CrossRef]
116. Zheng, S.; Zhang, K.; Zhang, Y.; He, J.; Ouyang, Y.; Lang, R.; Ao, C.; Jiang, Y.; Xiao, H.; Li, Y.; et al. Human Umbilical Cord Mesenchymal Stem Cells Inhibit Pyroptosis of Renal Tubular Epithelial Cells through miR-342-3p/Caspase1 Signaling Pathway in Diabetic Nephropathy. *Stem Cells Int.* **2023**, *2023*, 5584894. [CrossRef]
117. Lee, S.E.; Jang, J.E.; Kim, H.S.; Jung, M.K.; Ko, M.S.; Kim, M.O.; Park, H.S.; Oh, W.; Choi, S.J.; Jin, H.J.; et al. Mesenchymal stem cells prevent the progression of diabetic nephropathy by improving mitochondrial function in tubular epithelial cells. *Exp. Mol. Med.* **2019**, *51*, 1–14. [CrossRef]

118. An, X.; Liao, G.; Chen, Y.; Luo, A.; Liu, J.; Yuan, Y.; Li, L.; Yang, L.; Wang, H.; Liu, F.; et al. Intervention for early diabetic nephropathy by mesenchymal stem cells in a preclinical nonhuman primate model. *Stem Cell Res. Ther.* **2019**, *10*, 363. [CrossRef]
119. Wang, J.; Zeng, X.X.; Cai, W.; Han, Z.B.; Zhu, L.Y.; Liu, J.Y.; Xu, J.X. Safety and Efficacy of Placenta-Derived Mesenchymal Stem Cell Treatment for Diabetic Patients with Critical Limb Ischemia: A Pilot Study. *Exp. Clin. Endocrinol. Diabetes* **2021**, *129*, 542–548. [CrossRef]
120. Wang, J.; Liu, H.; Yue, G.; Deng, Y.; Cai, W.; Xu, J. Human placenta-derived mesenchymal stem cells ameliorate diabetic kidney disease by modulating the T helper 17 cell/regulatory T-cell balance through the programmed death 1/programmed death-ligand 1 pathway. *Diabetes Obes. Metab.* **2024**, *26*, 32–45. [CrossRef]
121. Han, X.; Wang, J.; Li, R.; Huang, M.; Yue, G.; Guan, L.; Deng, Y.; Cai, W.; Xu, J. Placental Mesenchymal Stem Cells Alleviate Podocyte Injury in Diabetic Kidney Disease by Modulating Mitophagy via the SIRT1-PGC-1alpha-TFAM Pathway. *Int. J. Mol. Sci.* **2023**, *24*, 4696. [CrossRef] [PubMed]
122. Lotfy, A.; Salama, M.; Zahran, F.; Jones, E.; Badawy, A.; Sobh, M. Characterization of mesenchymal stem cells derived from rat bone marrow and adipose tissue: A comparative study. *Int. J. Stem Cells* **2014**, *7*, 135–142. [CrossRef]
123. Kern, S.; Eichler, H.; Stoeve, J.; Klüter, H.; Bieback, K. Comparative analysis of mesenchymal stem cells from bone marrow, umbilical cord blood, or adipose tissue. *Stem Cells* **2006**, *24*, 1294–1301. [CrossRef] [PubMed]
124. Ni, W.; Fang, Y.; Xie, L.; Liu, X.; Shan, W.; Zeng, R.; Liu, J.; Liu, X. Adipose-Derived Mesenchymal Stem Cells Transplantation Alleviates Renal Injury in Streptozotocin-Induced Diabetic Nephropathy. *J. Histochem. Cytochem.* **2015**, *63*, 842–853. [CrossRef] [PubMed]
125. Yang, C.C.; Chen, Y.L.; Sung, P.H.; Chiang, J.Y.; Chen, C.H.; Li, Y.C.; Yip, H.K. Repeated administration of adipose-derived mesenchymal stem cells added on beneficial effects of empagliflozin on protecting renal function in diabetic kidney disease rat. *Biomed. J.* **2024**, *47*, 100613. [CrossRef]
126. Takemura, S.; Shimizu, T.; Oka, M.; Sekiya, S.; Babazono, T. Transplantation of adipose-derived mesenchymal stem cell sheets directly into the kidney suppresses the progression of renal injury in a diabetic nephropathy rat model. *J. Diabetes Investig.* **2020**, *11*, 545–553. [CrossRef]
127. Zhang, W.; Bai, X.; Zhao, B.; Li, Y.; Zhang, Y.; Li, Z.; Wang, X.; Luo, L.; Han, F.; Zhang, J.; et al. Cell-free therapy based on adipose tissue stem cell-derived exosomes promotes wound healing via the PI3K/Akt signaling pathway. *Exp. Cell Res.* **2018**, *370*, 333–342. [CrossRef]
128. Bochon, B.; Kozubska, M.; Surygała, G.; Witkowska, A.; Kuźniewicz, R.; Grzeszczak, W.; Wystrychowski, G. Mesenchymal Stem Cells-Potential Applications in Kidney Diseases. *Int. J. Mol. Sci.* **2019**, *20*, 2462. [CrossRef]
129. Ebrahim, N.; Ahmed, I.A.; Hussien, N.I.; Dessouky, A.A.; Farid, A.S.; Elshazly, A.M.; Mostafa, O.; Gazzar, W.B.E.; Sorour, S.M.; Seleem, Y.; et al. Mesenchymal Stem Cell-Derived Exosomes Ameliorated Diabetic Nephropathy by Autophagy Induction through the mTOR Signaling Pathway. *Cells* **2018**, *7*, 226. [CrossRef]
130. Sun, Y.; Shi, H.; Yin, S.; Ji, C.; Zhang, X.; Zhang, B.; Wu, P.; Shi, Y.; Mao, F.; Yan, Y.; et al. Human Mesenchymal Stem Cell Derived Exosomes Alleviate Type 2 Diabetes Mellitus by Reversing Peripheral Insulin Resistance and Relieving β-Cell Destruction. *ACS Nano* **2018**, *12*, 7613–7628. [CrossRef]
131. Jin, J.; Shi, Y.; Gong, J.; Zhao, L.; Li, Y.; He, Q.; Huang, H. Exosome secreted from adipose-derived stem cells attenuates diabetic nephropathy by promoting autophagy flux and inhibiting apoptosis in podocyte. *Stem Cell Res. Ther.* **2019**, *10*, 95. [CrossRef] [PubMed]
132. Hao, Y.; Miao, J.; Liu, W.; Cai, K.; Huang, X.; Peng, L. Mesenchymal Stem Cell-Derived Exosomes Carry MicroRNA-125a to Protect Against Diabetic Nephropathy by Targeting Histone Deacetylase 1 and Downregulating Endothelin-1. *Diabetes Metab. Syndr. Obes.* **2021**, *14*, 1405–1418. [CrossRef] [PubMed]
133. Tögel, F.; Weiss, K.; Yang, Y.; Hu, Z.; Zhang, P.; Westenfelder, C. Vasculotropic, paracrine actions of infused mesenchymal stem cells are important to the recovery from acute kidney injury. *Am. J. Physiol. Ren. Physiol.* **2007**, *292*, F1626–F1635. [CrossRef] [PubMed]
134. Raikwar, S.P.; Kim, E.M.; Sivitz, W.I.; Allamargot, C.; Thedens, D.R.; Zavazava, N. Human iPS cell-derived insulin producing cells form vascularized organoids under the kidney capsules of diabetic mice. *PLoS ONE* **2015**, *10*, e0116582. [CrossRef] [PubMed]
135. Kim, D.; Dressler, G.R. Nephrogenic factors promote differentiation of mouse embryonic stem cells into renal epithelia. *J. Am. Soc. Nephrol.* **2005**, *16*, 3527–3534. [CrossRef]
136. Narayanan, K.; Schumacher, K.M.; Tasnim, F.; Kandasamy, K.; Schumacher, A.; Ni, M.; Gao, S.; Gopalan, B.; Zink, D.; Ying, J.Y. Human embryonic stem cells differentiate into functional renal proximal tubular-like cells. *Kidney Int.* **2013**, *83*, 593–603. [CrossRef]
137. Gribben, C.; Lambert, C.; Messal, H.A.; Hubber, E.L.; Rackham, C.; Evans, I.; Heimberg, H.; Jones, P.; Sancho, R.; Behrens, A. Ductal Ngn3-expressing progenitors contribute to adult β cell neogenesis in the pancreas. *Cell Stem Cell* **2021**, *28*, 2000–2008.e2004. [CrossRef]
138. Tan, J.; Liu, L.; Li, B.; Xie, Q.; Sun, J.; Pu, H.; Zhang, L. Pancreatic stem cells differentiate into insulin-secreting cells on fibroblast-modified PLGA membranes. *Mater. Sci. Eng. C Mater. Biol. Appl.* **2019**, *97*, 593–601. [CrossRef]
139. Rajput, R.; Sinha, B.; Majumdar, S.; Shunmugavelu, M.; Bajaj, S. Consensus statement on insulin therapy in chronic kidney disease. *Diabetes Res. Clin. Pract.* **2017**, *127*, 10–20. [CrossRef]
140. Evenepoel, P.; Poesen, R.; Meijers, B. The gut-kidney axis. *Pediatr. Nephrol.* **2017**, *32*, 2005–2014. [CrossRef]

141. Vaziri, N.D. CKD impairs barrier function and alters microbial flora of the intestine: A major link to inflammation and uremic toxicity. *Curr. Opin. Nephrol. Hypertens.* **2012**, *21*, 587–592. [CrossRef] [PubMed]
142. Anders, H.J.; Andersen, K.; Stecher, B. The intestinal microbiota, a leaky gut, and abnormal immunity in kidney disease. *Kidney Int.* **2013**, *83*, 1010–1016. [CrossRef] [PubMed]
143. Heianza, Y.; Ma, W.; Manson, J.E.; Rexrode, K.M.; Qi, L. Gut Microbiota Metabolites and Risk of Major Adverse Cardiovascular Disease Events and Death: A Systematic Review and Meta-Analysis of Prospective Studies. *J. Am. Heart Assoc.* **2017**, *6*, e004947. [CrossRef] [PubMed]
144. Andrade-Oliveira, V.; Amano, M.T.; Correa-Costa, M.; Castoldi, A.; Felizardo, R.J.; de Almeida, D.C.; Bassi, E.J.; Moraes-Vieira, P.M.; Hiyane, M.I.; Rodas, A.C.; et al. Gut Bacteria Products Prevent AKI Induced by Ischemia-Reperfusion. *J. Am. Soc. Nephrol.* **2015**, *26*, 1877–1888. [CrossRef] [PubMed]
145. Li, Y.J.; Chen, X.; Kwan, T.K.; Loh, Y.W.; Singer, J.; Liu, Y.; Ma, J.; Tan, J.; Macia, L.; Mackay, C.R.; et al. Dietary Fiber Protects against Diabetic Nephropathy through Short-Chain Fatty Acid-Mediated Activation of G Protein-Coupled Receptors GPR43 and GPR109A. *J. Am. Soc. Nephrol.* **2020**, *31*, 1267–1281. [CrossRef]
146. Wu, X.; Zhao, L.; Zhang, Y.; Li, K.; Yang, J. The role and mechanism of the gut microbiota in the development and treatment of diabetic kidney disease. *Front. Physiol.* **2023**, *14*, 1166685. [CrossRef]
147. Lv, Q.; Li, Z.; Sui, A.; Yang, X.; Han, Y.; Yao, R. The role and mechanisms of gut microbiota in diabetic nephropathy, diabetic retinopathy and cardiovascular diseases. *Front. Microbiol.* **2022**, *13*, 977187. [CrossRef]
148. Dai, Y.; Quan, J.; Xiong, L.; Luo, Y.; Yi, B. Probiotics improve renal function, glucose, lipids, inflammation and oxidative stress in diabetic kidney disease: A systematic review and meta-analysis. *Ren. Fail.* **2022**, *44*, 862–880. [CrossRef]
149. Bastos, R.M.C.; Simplício-Filho, A.; Sávio-Silva, C.; Oliveira, L.F.V.; Cruz, G.N.F.; Sousa, E.H.; Noronha, I.L.; Mangueira, C.L.P.; Quaglierini-Ribeiro, H.; Josefi-Rocha, G.R.; et al. Fecal Microbiota Transplant in a Pre-Clinical Model of Type 2 Diabetes Mellitus, Obesity and Diabetic Kidney Disease. *Int. J. Mol. Sci.* **2022**, *23*, 3842. [CrossRef]
150. Shang, J.; Cui, W.; Guo, R.; Zhang, Y.; Wang, P.; Yu, W.; Zheng, X.; Wang, T.; Dong, Y.; Zhao, J.; et al. The harmful intestinal microbial community accumulates during DKD exacerbation and microbiome-metabolome combined validation in a mouse model. *Front. Endocrinol.* **2022**, *13*, 964389. [CrossRef]
151. Lin, J.R.; Wang, Z.T.; Sun, J.J.; Yang, Y.Y.; Li, X.X.; Wang, X.R.; Shi, Y.; Zhu, Y.Y.; Wang, R.T.; Wang, M.N.; et al. Gut microbiota and diabetic kidney diseases: Pathogenesis and therapeutic perspectives. *World J. Diabetes* **2022**, *13*, 308–318. [CrossRef] [PubMed]
152. Wang, N.; Zhang, C. Recent Advances in the Management of Diabetic Kidney Disease: Slowing Progression. *Int. J. Mol. Sci.* **2024**, *25*, 3086. [CrossRef] [PubMed]
153. Bolignano, D.; Zoccali, C. Effects of weight loss on renal function in obese CKD patients: A systematic review. *Nephrol. Dial. Transplant.* **2013**, *28* (Suppl. 4), iv82–iv98. [CrossRef] [PubMed]
154. Robinson-Cohen, C.; Littman, A.J.; Duncan, G.E.; Weiss, N.S.; Sachs, M.C.; Ruzinski, J.; Kundzins, J.; Rock, D.; de Boer, I.H.; Ikizler, T.A.; et al. Physical activity and change in estimated GFR among persons with CKD. *J. Am. Soc. Nephrol.* **2014**, *25*, 399–406. [CrossRef] [PubMed]
155. Feodoroff, M.; Harjutsalo, V.; Forsblom, C.; Thorn, L.; Wadén, J.; Tolonen, N.; Lithovius, R.; Groop, P.H. Smoking and progression of diabetic nephropathy in patients with type 1 diabetes. *Acta Diabetol.* **2016**, *53*, 525–533. [CrossRef]
156. Kidney Disease: Improving Global Outcomes (KDIGO) Diabetes Work Group. KDIGO 2022 Clinical Practice Guideline for Diabetes Management in Chronic Kidney Disease. *Kidney Int.* **2022**, *102*, S1–S127. [CrossRef]

International Journal of
Molecular Sciences

MDPI

Review

Endocrine-Disrupting Chemicals and the Development of Diabetes Mellitus Type 1: A 5-Year Systematic Review

Georgia-Nektaria Keskesiadou [1], Sophia Tsokkou [1,2], Ioannis Konstantinidis [1], Maria-Nefeli Georgaki [1,3], Antonia Sioga [1,2], Theodora Papamitsou [1,2,*] and Sofia Karachrysafi [1,2]

1 Research Team "Histologistas", Interinstitutional Postgraduate Program "Health and Environmental Factors", Department of Medicine, Faculty of Health Sciences, Aristotle University of Thessaloniki, 54124 Thessaloniki, Greece; gwgwkeske@gmail.com (G.-N.K.); stsokkou@auth.gr (S.T.); ikonsc@auth.gr (I.K.); mgeorgaki@cheng.auth.gr (M.-N.G.); asioga@auth.gr (A.S.); sofia_karachrysafi@outlook.com (S.K.)

2 Laboratory of Histology-Embryology, Department of Medicine, Faculty of Health Sciences, Aristotle University of Thessaloniki, 54124 Thessaloniki, Greece

3 Environmental Engineering Laboratory, Department of Chemical Engineering, Aristotle University of Thessaloniki, 54124 Thessaloniki, Greece

* Correspondence: thpapami@auth.gr

Abstract: Introduction: According to the Institute of Environmental Sciences, endocrine-disrupting chemicals (EDCs) are "natural or human-made chemicals that may mimic, block, or interfere with the body's hormones, associated with a wide array of health issues", mainly in the endocrine system. Recent studies have discussed the potential contribution of EDCs as risk factors leading to diabetes mellitus type 1 (T1DM), through various cellular and molecular pathways. Purpose: The purpose of this study was to investigate the correlation between the EDCs and the development of T1DM. Methodology: Thus, a 5-year systematic review was conducted to bring light to this research question. Using the meta-analysis and systematic review guideline protocol, a PRISMA flow diagram was constructed and, using the keywords (diabetes mellitus type 1) AND (endocrine-disrupting chemicals) in the databases PubMed, Scopus and ScienceDirect, the relevant data was collected and extracted into tables. Quality assessment tools were employed to evaluate the quality of the content of each article retrieved. Results: Based on the data collected and extracted from both human and animal studies, an association was found between T1DM and certain EDCs, such as bisphenol A (BPA), bisphenol S (BPS), persistent organic pollutants (POPs), phthalates and dioxins. Moreover, based on the quality assessments performed, using the Newcastle–Ottawa Scale and ARRIVE quality assessment tool, the articles were considered of high quality and thus eligible to justify the correlation of the EDCs and the development of T1DM. Conclusion: Based on the above study, the correlation can be justified; however, additional studies can be made focusing mainly on humans to understand further the pathophysiologic mechanism involved in this association.

Keywords: Endocrine Disrupting Chemicals (EDCs); Diabetes Mellitus Type 1 (T1DM); phthalates; Persistent Organic Pollutants (POPs); bisphenol A (BPA)

1. Introduction

1.1. Endocrine Disrupting Chemicals

Endocrine-disrupting chemicals (EDCs) are substances present in the environment such as soil, air and water and in food sources. They can also be found in manufactured and cosmetic care products [1]. They are found in their natural form or are human-produced and have the tendency to mimic, inhibit and intervene in the natural pathway of the human body's hormones and thus lead to a wide range of health pathological alterations [2]. Generally, EDCs can act in three main ways. They can act by mimicking human hormones and inhibiting the natural purpose of a hormone in the physiologic pathway. The second way of interfering is either by increasing or decreasing the hormones by affecting their metabolic and storing processes. Thirdly, ECDs can affect the level of sensitivity the human body has to the various types of hormones. These kinds of alterations can lead to several adverse effects on the health of an individual, such as diabetes mellitus (DM), obesity and cardiovascular diseases (CVD), as well as growth restrictions and neurological and cognitive disabilities [1,2]. EDCs are thoroughly distributed in our daily life and are found in daily products such as cosmetics, seafoods, pesticides and packaging of children's toys [2].

1.2. Diabetes Mellitus

Diabetes mellitus (DM) is a group of metabolic diseases characterized by chronic hyperglycemia due to disturbances in insulin secretion, insulin action in the form of resistance, or a combination of these two. Inadequate insulin secretion and/or reduced tissue response to insulin results in insufficient action of insulin on the targeted tissues, leading to disruptions of carbohydrate, fat and protein metabolism. Both decreased insulin secretion and insulin action may coexist in the same patient. While the etiology of diabetes is heterogeneous, most cases of DM can be classified into two broad etiopathogenetic categories. The first type is T1DM, which is mainly characterized by insulin secretion deficiency, while T2DM results from a combination of resistance to insulin action and the insufficient compensatory insulin secretion response. While T1DM remains the most common form of diabetes in young people in many populations, especially those with a European background, T2DM is becoming an increasingly important public health concern worldwide, among children as well as in people with severe obesity [1,2].

1.2.1. Diabetes Mellitus Type 1

More specifically, T1DM, also known as juvenile diabetes or insulin-dependent diabetes, has been of particular concern in recent years, as the number of cases is ever increasing, and is the most common type in children and adolescents. The Atlas of Diabetes (7th edition) reports that its global prevalence is estimated at 415 million (8.8%), which is projected to increase to 642 million over the next 25 years. In India, there are about 69.2 million people with diabetes, but this figure is expected to reach 123.5 million by 2040 [3]. The reason for the increase in the number of cases is still unclear, but there are several risk factors that lead to predisposition to the development of T1DM, including environmental factors, genetic factors, epigenetic factors, lifestyle and complications during pregnancy, such as preeclampsia and infections [4,5].

1.2.2. Diabetes Mellitus Type 1 and EDCs (Figure 1)

Research that is based on the environmental factors playing a role in the etiology of pancreatic islet autoimmunity focus mainly on the impact of viruses, inappropriate early infant nutritional diet, high childhood weight gain, vitamin D deficiency, gut microbiome and endocrine-disrupting chemicals (EDCs) [6].

EDCs are a group of exogenous compounds with high heterogeneity that can be found naturally in living organisms or are synthesized industrially in food and consumer products [6,7]. They can influence the development and function of beta cells or immune system genes, promoting autoimmunity and increasing susceptibility to autoimmune attack as -individual chemical agents or as chemical mixtures. However, there are only a few conflicting human studies showing the possible role of exposure to EDCs in the pathogenesis of T1DM. In addition, the presence of a familial form of mild diabetes during adolescence should raise the suspicion of monogenic diabetes, which is responsible for 1% to 6% of cases of childhood DM.

Furthermore, there is increasing evidence indicating a connection between exposure to endocrine-disrupting chemicals (EDCs) and the development of various pancreatic diseases, including type 1 diabetes mellitus (T1DM). This correlation is linked to the aryl hydrocarbon receptor (AHR), which acts as a transcription factor that is activated by ligands. The AHR plays a critical role in regulating essential cellular and molecular processes, such as the metabolism of foreign substances, immune responses and the formation of cancer. Notably, in the developing pancreas of the embryo and in the early stages of life, a variety of endocrine-disrupting substances (EDCs) function as agonists or antagonists of AHR, causing dysregulation of critical cellular and molecular pathways and disturbance of the endocrine system [8].

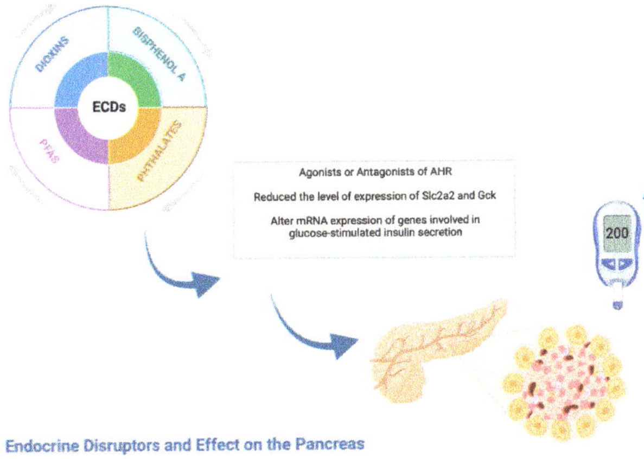

Figure 1. Endocrine disruptors and effect on the pancreas, created by BioRender.

1.2.3. Epidemiology

According to the International Diabetes Federation (IDF), 1 in 10 adults have DM worldwide, while 3 in 4 people with DM live in low-income countries and 1 in 2 patients with DM are not diagnosed. T1DM affects 1.2 million children worldwide. Moreover, according to the IDF, there is an exponential increase in cases, especially in regions of the world that are of lower socioeconomic status [9].

1.2.4. Stages of Diabetes Mellitus Type 1

T1DM is characterized by three stages. More precisely, at stage 1, the presymptomatic stage, there are autoantibodies against islets (two or more types), and normal blood glucose. The immune system at this stage has already begun to attack the beta cells of the pancreas; however, blood sugar remains normal and symptoms do not yet appear. Subsequently, in stage 2, in addition to the presence of autoantibodies there is abnormal glucose tolerance, usually being presymptomatic, due to increasing beta cell loss/destruction. For both stages (stage 1 and stage 2) the risk of developing T1DM approaches its peak. Finally, in stage 3, the clinical diagnosis of T1DM takes place, as blood glucose is now above diagnostic limits, while at the same time the common symptoms of polyuria, polydipsia, weight loss and fatigue appear. Clinical research supports the necessity of early diagnosis of T1DM before stage 3, as the sooner the diagnosis is made, the sooner the intervention will be made, reducing the occurrence of long-term complications [5–7].

1.2.5. Diagnostic Criteria of T1DM (Figure 2)

According to the World Health Organization (WHO), the presence of even a random value of fasting plasma glucose $\geq$200 mg/dL in combination with the characteristic clinical symptoms of diabetes make the diagnosis definitive. In addition, a fasting glucose value of $\geq$126 mg/dL, as long as fasting lasts at least 8 h before the test, as well as the absence of symptoms confirmation with a second test or a second sample, can result in a diagnosis of diabetes. Another method employed to diagnose diabetes in recent years is glycosylated hemoglobin A1c (HbA1c), a widely used marker for chronic glycemia that measures non-enzymatic glycation of hemoglobin and reflects average blood glucose levels over a period of 2 to 3 months. If it is at an HbA1c level of ≥6.5% (according to an estimated average glucose of 140 mg/dL), then the diagnosis is made again. Finally, there is the oral glucose tolerance test (OGTT), which most of the time—especially for the diagnosis of T1D in children and adolescents—is not necessary, but in cases of asymptomatic hyperglycemia and investigation of abnormal glucose metabolism outside T1D the test is useful and should be applied when appropriate. OGTT is performed after loading with 1.75 g of glucose per kg of body weight (BS), with a maximum dose of 75 g of glucose. The diagnosis, therefore, is definitive when the blood glucose value two hours after this test is ≥200 mg/dL [1,2].

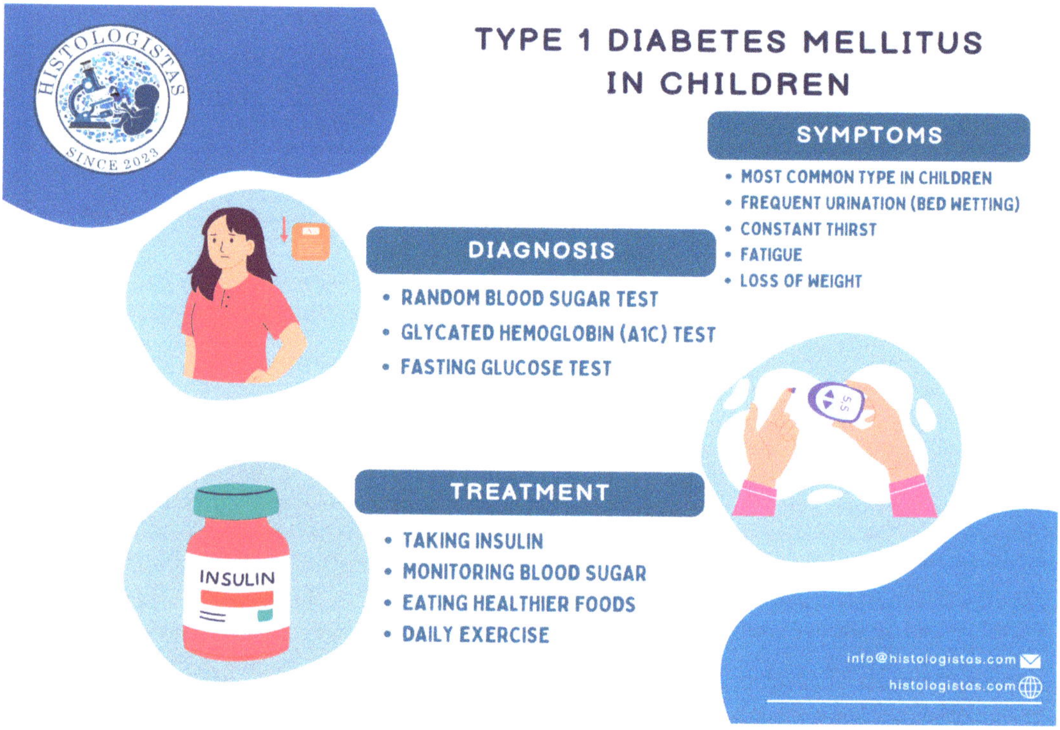

Figure 2. Symptoms, diagnostic criteria and treatment of T1DM.

1.2.6. Basic Treatment Options for T1DM (Figure 2)

Regarding basic treatment options for T1DM, the initiation of subcutaneous insulin is of utmost importance, in combination with intensive blood glucose monitoring, adoption of a healthy diet and daily exercise. Many regimens of insulin injection intensity have been proposed, with the main notion being the utilization of both a long-acting insulin (basal insulin) and a rapid-acting insulin before meals [1,2].

2. Methods

2.1. Purpose of Study

The purpose of this study was to investigate the correlation between the development of T1DM and EDCs. Thus, a 5-year systematic review was constructed to answer the question raised.

2.2. PICO Table–PRISMA Flow Diagram–Table–Quality Assessment

The Population, Intervention, Control and Outcome (PICO) table was created to facilitate and clarify the question more clearly. In addition, a PRISMA flow diagram was produced using the following keywords in the code: (diabetes mellitus type 1) AND (endocrine disruptors). Then the databases PubMed, Scopus and Science Direct were searched with the following criteria: duration in the last 5 years (2019–2024), the articles had to be Reviews, Research, Articles and Case Reports, written in English and have open access. Tables were then constructed with the most crucial information retrieved (Cohort Studies, Basic Research, Literature Review) and quality assessments were performed (Newcastle–Ottawa Scale, ARRIVE quality assessment tool) to check the accuracy of the papers. The Newcastle–Ottawa Scale assessment is for cohort studies, with a scale of 0 (lowest) to 7 (highest). The second rating scale used was the ARRIVE scale, which is specific to animal studies and ranges from 1 to 10.

3. Results

3.1. PICO Table

According to the PICO, the study population consisted of children, adolescents, and young adults (10 to 22 years old), the risk factor was endocrine disruptors, as an incidence of T1DM and as a result endocrine disruptors as risk factors for developing T1DM (Table 1).

Table 1. Population, Intervention, Comparator, Outcome.

PICO	
P	Children and Adolescents and Young Adults (10–22 years old)
I	Endocrine-Disrupting Chemicals
C	Diabetes Mellitus Type 1 (T1DM)
O	Endocrine Disrupting Chemicals as Risk Factors of Diabetes Mellitus Type 1 (T1DM)

3.2. PRISMA Flow Diagram

After the PRISMA flow diagram was created, the following results were found. A total of 17,947 articles were found using the keywords (diabetes mellitus type 1) AND (endocrine disruptors). Specifically, 37 articles were found in PubMed, 205 in Scopus and 17,705 in Science Direct. After using the criteria, which are: in the last 5 years (2019–2024), the articles had to be Reviews, Studies, Articles and Case Reports, written in English and have open access, the automatic data engine removed 16,413 articles, leaving 1534 articles remaining. These articles were studied, and 14 articles were removed as duplicates, leaving 1520. From these, a total of 1505 articles were removed, specifically 189 articles were removed due to low correlation based on title, 634 based on abstract, 671 based on total text and, finally, 11 since only abstracts were found. A total of 16 articles were selected from the databases. An additional 6 articles from external sources were added to the total number, bringing the final number to 22 articles (Figure 3). Three tables were then created that included the most important information, more specifically, the names of the authors, the chronology and type of articles, the groups with the highest correlation with T1DM, the types of endocrine disruptors and their correlation with T1DM (Tables 2–4).

3.3. Quality Assessment

3.3.1. Newcastle–Ottawa Scale (NOS)

Firstly, the Newcastle–Ottawa Scale (NOS) was used to evaluate four nonrandomized trials, of which two of the four, namely Salo HM, 2019 [10] and Lee, I, 2021 [11] had a score of 7, which is the highest score, followed by the Bresson SE 2019 [12] study, with a score of 6, and, finally, the Duforun study, P, 2023 [13] with a score of 5. The last two studies were on the middle of the scale, but the overall score of all four papers can support the credibility of the overall content (Table 5).

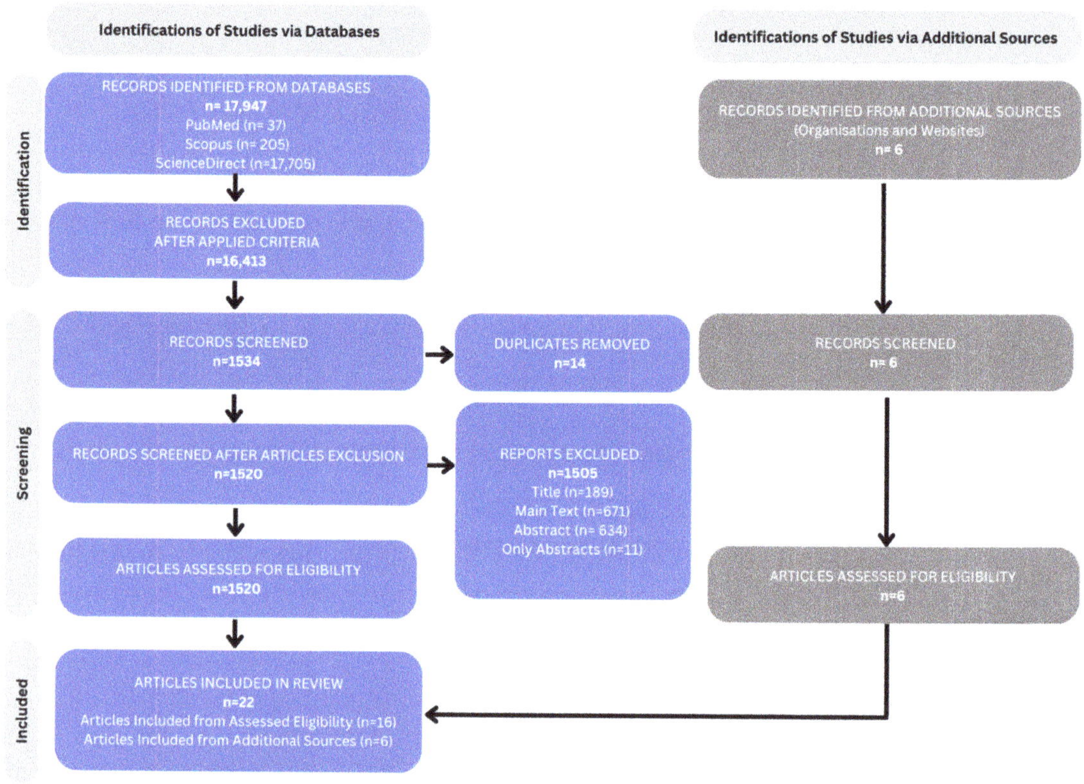

Figure 3. PRISMA flow diagram.

3.3.2. ARRIVE Quality Assessment Scale

In addition, the ARRIVE scale used for animal studies showed the following results. Xu's study, J, 2019 [14] had the highest score of 9, while the remaining 4 studies had scores of 7, which is on the medium to high scale (Table 6).

Table 2. Cohort studies.

Authors	Study Type	Type of ECDS	ECDS Correlation with T1DM	Number of Population	Method of Assessment	Risk Factors	Sex	Age
Salo HM, 2019 [10]	Cohort	1. Persistent organic pollutants (POPs) 2. dioxins 3. polychlorinated biphenyls (PCBs) 4. Pesticides 5. Brominated flame-retardants	Could not observe any definite associations between increased exposure to chemical pollutants at birth, at 12 or at 48 months of age, and risk of β-cell autoimmunity. The current work indicates that prenatal or early childhood exposure to POPs, including PFASs, is not an apparent risk factor for later β-cell autoimmunity. In 48-month-old children, PFDA was above the LOQ in 34% of the autoantibody-negative children, in 63% of the autoantibody-positive children, and in 88% of the children diagnosed with type 1 diabetes. PFDA has been demonstrated to interfere with the function of thyroid hormones in in vitro studies, and endocrine disruption is an interesting mode of action of PFDA in biological systems.	136	1. FINDIA pilot study 2. DIABIMMUNE study	Autoantibody positive cases, HLA risk genotype, Breastfeeding, Formula feeding	XX, XY	from newborns to 6 years old
Bresson SE, 2019, [12]	Cohort	1. Polychlorinated biphenyls (PCBs) 2. Organochlorine pesticides 3. POPs	There was a significant association between T1D/IS and several POPs including p, p′-DDE, trans-nonachlor and PCB-153. PCB-153 and p, p′-DDE reduce insulin production and secretion in β-cells. Interestingly, PCB-153 but not p, p′-DDE, reduced the expression level of Slc2a2 and Gck. PCB-153 and p, p′-DDE alter mRNA expression of genes involved in glucose-stimulated insulin secretion.	442	SEARCH Case-Control (SEARCH-CC)	BMI, Insulin Sensitivity, Insulin Resistance	XX, XY	10 to 22 years
Dufour, P, 2023 [13]	Cohort	7 phthalate metabolites, 4 parabens, 7 bisphenols, benzophenone 3 and triclosan were measured in urine, while 15 organochlorine pesticides, 4 polychlorinated biphenyls (PCBs) and 7 perfluoroalkyl substances 1. Phthalate metabolites 2. PCBs 3. Bisphenols 4. PCB 5. Miscellaneous, OCP, PFAS	This work investigated the link between the exposure to some environmental pollutants, from persistent organic pollutants to some non-persistent plasticizers and antimicrobials, and thyroid disorders in type 1 diabetes children. Associations between the levels (in serum or urine) of some PFASs, PCBs, phthalates and bisphenols and thyroid hormone were highlighted, suggesting an impact of these pollutants on the thyroid function in this population suspected to be particularly vulnerable toward endocrine disruption.	54	Urine and Blood Analysis	-	XX, XY	3 to 18 years

Table 2. *Cont.*

Authors	Study Type	Type of ECDS	ECDS Correlation with T1DM	Number of Population	Method of Assessment	Risk Factors	Sex	Age
Lee, I, 2021 [11]	Cohort	Phthalates and bisphenol A (BPA), parabens	BPA in urine was associated with higher rates of DM	3787	1. Korean National Environmental Health Survey (KoNEHS) 2. Concentrations of phthalate metabolites, BPA, and parabens were measured in the spot urine samples using liquid–liquid extraction and ultraperformance liquid chromatography mass spectrometry separation, followed by electrospray ionization and tandem mass spectrometry 3. Urinary dilution, in addition to two traditional methods (i.e., Cr adjustment and SG adjustment), a CAS was used	-	XX, XY	19 and older

Table 3. Basic Research–Animal Studies.

Authors	Type of Study	Type of ECDS	ECDS Correlation with T1DM	Number of Population	Method of Assessment	Sex	Age
Xu J, 2019 [14]	Basic Research Study	Bisphenol A (BPA)	Not statistically significant, but a shift towards accelerated T1D development was observed	30	BW and non-fasting BGLs were measured every 1–2 weeks. Accu-Chek Diabetes monitoring kit (Roche Diagnostics, Indianapolis, IN, USA) or Contour Blood Glucose Meter (Ascensia Diabetes Care, Parsippany, NJ, USA) were used to measure BGLs from a small sample of venous blood (tail nick).	XX	8 to 12 weeks old
McDonough CM, 2022 [15]	Basic Research Study	Bisphenol S (BPS)	Significant adverse effect	12	Body weight, blood glucose measurement, diabetic incidence, GTT and ITT, behaviour test, Y-maze test, Flow Cytometric Analysis	XX, XY	10 weeks
Sinioja T, 2022 [16]	Basic Research Study	Persistent organic pollutants (POPs), organochlorides, organobromides, and per- and polyfluoroalkyl substances (PFAS)	Risk of T1D	29	Folch procedure for Lipidomic Analysis	XX, XY	8 to 10 weeks old
Xu J, 2019 [17]	Basic Research Study	Bisphenol A (BPA)	Sex plays an important role in BPA altering T1D risk. BPA accelerated T1D development in adult NOD females but delayed T1D development in male mice.	-	Tolerance tests and insulin measurement, antibody measurement, cytokine/chemokine measurement, GMB, bioinformatics and metabolomics and bioinformatics analysis	XX, XY	8 to 12 weeks old
Xu J, 2019 [18]	Basic Research Study	Bisphenol S (BPS)	PS exposure had dose-related protective effects on T1D in females. This suggests that BPS uses different mechanisms from BPA to alter glucose homeostasis and T1D	-	Tolerance tests and insulin measurement, antibody measurement, cytokine/chemokine measurement, GMB, bioinformatics and metabolomics and bioinformatics analysis		8 to 15 weeks old

Table 4. Literature reviews.

Authors	Study Type	Type of ECDS	ECDS Correlation with T1DM	Doi
Melissa and Cairro, 2023 [19]	Review	Phthalates	There is a possible link between the exposure to phthalates and the development of DM. Phthalates may lead to insulin resistance and consequent diabetes mellitus through oxidative stress, the activation of different hormone receptors (PPAR and ER), and impaired inflammatory factors.	10.3390/metabo13060746
Prediere et al., 2020 [20]	Review	Chemical substances (bisphenol A; pesticides; phthalates; polychlorinated biphenyls; polyfluorinated substances)	They could affect the development and the function of the immune system or of the β-cells function, promoting autoimmunity and increasing the susceptibility to autoimmune attack. However, the studies are few and demonstrated contradictory results, according to gender and age.	10.3390/ijms21082937
HInault et al., 2023 [21]	Review	Organochlorine (OCs) pesticides: dichloro-diphenyl-trichloroethane (DDT) and its metabolites (chlordane) and industrial chemicals: polychlorinated biphenyls (PCBs), polybrominated diphenyl ethers (PBDE), dioxins (TCDD: 2,3,7,8-tétrachlorodibenzo-p-dioxine), and per- and polyfluoroalkyl substances (PFAs), bisphenol A (BPA)	It remains difficult to form a comprehensive view on the causal relationship between EDCs and diabetes (both T1DM and T2DM), and further experiments are required.	10.3390/ijms24054537
Khali et al., 2023 [22]	Review	Ambient air pollution, persistent organic pollutants (POPs), metals (bisphenol A [BPA]), phthalates, polybrominated diphenyl ethers (PBDEs)	The research has shown inconsistent results regarding direct pathogenesis, since these diseases are multifactorial. Many studies attempted to determine the impact of isolated atmospheric compounds but did not taking into consideration their compounding effects, and others used smaller sample sizes.	10.3390/ijms24108870

Table 4. *Cont.*

Authors	Study Type	Type of ECDS	ECDS Correlation with T1DM	Doi
Heo and Kim, 2021 [23]	Review	Ambient air pollution (PM, NO_2, and NOx)	Altered immune response, oxidative stress, neuroinflammation, inadequate placental development, and epigenetic modulation are some of the underlying factors that have been identified. However, it is difficult to demonstrate causality.	10.6065/apem.2142132.066
Jiang et al., 2023 [24]	Review	bisphenol A (BPA)	BPA exposure is associated with target organ damage in DM and may exacerbate the progression of some chronic complications of DM.	10.1016/j.heliyon. 2023.e16340
Ibarra et al., 2020 [25]	Review	bisphenol A (BPA) & phthalates	There is association with a wide range of reproductive, metabolic and neurological diseases, as well as hormone-related cancers.	10.1016/j.envpol.2020.116380

Table 5. Newcastle–Ottawa Quality Assesment Scale–Cohort Studies.

	Selection				Comparability	Outcome		
Source	Representativeness of the Sample	Sample Size	Non–Respondents	Ascertainment of the Exposure	The Subjects in Different Outcome Groups Are Comparable	Assessment of the Outcome	Statistical Test	Quality
Salo HM, 2019 [10]	1	1	1	1	1	1	1	7
Bresson SE, 2019, [11]	1	1	1	1	0	1	1	6
Lee, I, 2021 [12]	1	1	1	1	1	1	1	7
Dufour, P, 2023 [13]	1	1	1	1	0	0	1	5

Methodological quality according to total score: <5: low quality, 5–7: moderate quality, >7: high quality.

Table 6. ARRIVE assessment scale—animal studies.

Source	Study Design	Sample Size	Inclusion and Exclusion Criteria	Randomisation	Blinding	Outcome Measures	Statistical Methods	Experimental Animals	Experimental Procedures	Results	Quality
Xu J, 2019 [14]	+	+	+	+	?	+	+	+	+	+	9
McDonough CM, 2022 [15]	+	+	?	?	?	+	+	+	+	+	7
Sinioja T, 2022 [16]	+	+	-	+	?	?	+	+	+	+	7
Xu J, 2019 [17]	+	-	-	+	?	+	+	+	+	+	7
Xu J, 2019 [18]	+	-	-	+	?	+	+	+	+	+	7

"?" = uncertain, not stated; "+" = positive answer to assessment questions; "-"= negative answer to assessment questions.

4. Discussion

Based on the systematic research conducted, data collected using the PRISMA guidelines and PICO model, the following results can be stated from the information summarized in the three tables that, respectively, represent cohorts carried out with human participants (Table 2), basic research studies conducted with animals (Table 3) and, finally, literature and systematic reviews (Table 4).

4.1. Correlation between ECDs and T1DM Based on Human Cohort Studies

Firstly, two major retrospective cohort studies discussed the association of polychlorinated biphenyls (PCBs), organochlorine pesticides and persistent organic pollutants (POPs) with the descriptive results of the studies mentioned in Table 2. In Bresson SE et al., 2019, where 442 people, both males and females, of age 10 to 20 years old were tested according to the SEARCH Case-Control protocol [11], there was a significant association between T1DM/IS occurrence and several POPs, including p, p′-DDE, trans-nonachlor and PCB-153. PCB-153 and p, p′-DDE were found to reduce insulin production and secretion in pancreatic β-cells. Interestingly, PCB-153 but not p, p′-DDE reduced the level of expression of *Slc2a2* and *Gck*, but both were confirmed to alter mRNA expression of genes involved in glucose-stimulated insulin secretion. Additionally, they identified increased BMI, reduced insulin sensitivity and insulin resistance as potential risk factors for ECDs to cause T1DM.

However, Salo HM et al., 2019 [10], where 136 children, both male and female, from newborn to 6 years old were studied according to FINDIA pilot study and DIABIMMUNE study protocols [1], could not observe any definite associations between increased exposure to chemical pollutants at birth, or at 12 or at 48 months of age, and risk of β-cell autoimmunity and, thus, the development of T1DM. Prenatal or early childhood exposure to POPs, including PFASs, is not an apparent risk factor for later β-cell autoimmunity, but in 48-month-old children, PFDA was above the LOQ in 34% of the autoantibody-negative children, in 63% of the autoantibody-positive children, and in 88% of the children diagnosed with type 1 diabetes. PFDA has been demonstrated to interfere with the function of thyroid hormones in in vitro studies, and endocrine disruption is an interesting mode of action of PFDA in biological systems. Autoantibody positive cases, HLA risk genotype, breastfeeding and formula feeding were observed as latent risk factors.

Secondly, phthalates, bisphenol A (BPA) and parabens were investigated as impending ECDs in the pathophysiology of T1DM in Lee I. et al., 2021 [12]. They studied 3787 participants older than 19 years old, both male and female, utilizing the Korean National Environmental Health Survey (KoNEHS), concentrations of phthalate metabolites, BPA, and parabens that were measured in spot urine samples using liquid–liquid extraction and ultraperformance liquid chromatography mass spectrometry separation, followed by electrospray ionization and tandem mass spectrometry and urinary dilution, in addition to two traditional methods (i.e., Cr and SG adjustment). This cohort concluded that increased BPA in urine was associated with higher rates of T1DM.

Moreover, Dufour P. et al., 2023 [13] analyzed blood and urine samples of 54 children aged 3 to 18 years old, both male and female, in an attempt to investigate the link between the exposure to many environmental pollutants, from persistent organic pollutants to some non-persistent plasticizers and antimicrobials [i.e., 7 phthalate metabolites, 4 parabens, 7 bisphenols, benzophenone 3, triclosan, 15 organochlorine pesticides, 4 polychlorinated biphenyls (PCBs) and 7 perfluoroalkyl substances] and thyroid disorders in T1DM children. Associations between the levels (in serum or urine) of some PFASs, PCBs, phthalates and bisphenols and thyroid hormone levels were highlighted, suggesting an impact of these pollutants on thyroid function in this population, suspected to be particularly vulnerable toward endocrine disruption, but there was no direct implication that they correlate with the development of T1DM.

4.2. Correlation between ECDs and T1DM Based on Basic Research Animal Studies

Furthermore, four different study protocols by Xu J. et al., 2019 [14,16–18], using non-obese diabetic (NOD) mice, male and/or female, aged 8 to 12 weeks old, rummaged the association of bisphenol A (BPA) and bisphenol S (BPS) in T1DM occurrence. In the first experiment, BW and non-fasting BGLs were measured every 1–2 weeks using an Accu-Chek Diabetes monitoring kit (Roche Diagnostics, Indianapolis, IN, USA) or Contour Blood Glucose Meter (Ascensia Diabetes Care, Parsippany, NJ, USA) from a small sample of venous blood (tail nick), and it revealed a not statistically significant correlation, but a shift towards accelerated T1DM development was observed. The other two studies assessed NOD mice with tolerance tests and insulin, antibody, cytokine/chemokine measurement, GMB, bioinformatics and metabolomics analysis and verified that sex plays an important role in BPA altering T1DM risk, as BPA accelerated T1DM development in adult NOD females, but delayed male mice from T1DM development, and, also, that BPS exposure had dose-related protective effects on T1DM in females, suggesting that BPS uses different mechanisms from BPA to alter glucose homeostasis and T1DM. In addition, McDonough CM et al., 2022 [15] assessed 10-week-old NOD mice using body weight, blood glucose measurement, diabetic incidence, GTT and ITT, behavior tests, the Y-maze test and Flow Cytometric Analysis to substantiate that BPS has a significant adverse effect on T1DM occurrence. The last animal study to examine the correlation between ECDs and T1DM development risk was by Sinioja T et al., 2022 [16], where they attested that persistent organic pollutants (POPs), organochlorides, organobromides and per- and poly-fluoroalkyl substances (PFAS) are risk factors for the development of T1DM, using 29 8- to 10-week-old NOD mice and assessing them with the Folch procedure for Lipidomic Analysis.

4.3. Correlation between ECDs and T1DM Based on the Literature and Systematic Reviews

Many reviews tried to support the association between different disruptors and the development of T1DM. Firstly, Melissa and Cairro, 2023 [19] stated that there is a possible link between the exposure to phthalates and the development of DM, as they may lead to insulin resistance and consequent DM through oxidative stress, the activation of different hormone receptors (PPAR and ER) and impaired inflammatory factors. Secondly, Ibarra et al., 2020 [25] asserted that there is an association of bisphenol A (BPA) and phthalates with a wide range of reproductive, metabolic and neurological diseases, as well as hormone-related cancers, while Jiang et al., 2023 [24] declared that BPA exposure is associated with target organ damage in DM and may exacerbate the progression of some chronic complications of DM. Moreover, Prediere et al., 2020 [20] averred that many chemical substances (Table 4) could affect the development and the function of the immune system or of the pancreatic β-cells, promoting autoimmunity and increasing the susceptibility to autoimmune assault, but the studies cited are few and demonstrated contradictory results, according to gender and age. Furthermore, Heo and Kim, 2021 [23] testified that altered immune response, oxidative stress, neuroinflammation, inadequate placental development and epigenetic modulation are some of the underlying risk factors that have been identified for the pathophysiology of ambient air pollution (i.e., PM, NO_2 and NOx) in the development of T1DM. Additionally, Khali et al., 2023 [22] expressed that current research has shown inconsistent results with regard to direct pathogenesis, due to the fact that DM is multifactorial and many studies attempted to determine the impact of isolated atmospheric compounds, such as ambient air pollution, persistent organic pollutants (POPs) and metals, i.e., bisphenol A (BPA), phthalates and polybrominated-diphenyl-ethers (PBDEs), but did not take into consideration their compounding effects. Finally, Hinault et al., 2023 [21] affirmed that it remains difficult to form a comprehensive view on the causal relationship between EDCs and diabetes (both T1DM and T2DM) and that further experiments are required.

5. Conclusions

Based on the data examined, it can be concluded that stronger association was found between persistent organic pollutants (POPs), phthalates and bisphenols, mostly BPA, and the risk for development of T1DM, mainly through reduced insulin sensitivity, autoimmunity of pancreatic β-cells, oxidative and inflammatory stress and epigenetic modulation. It is obvious that further research is required to bear out the important role of ECDs in the pathophysiology of DM in general and, specifically, T1DM; studies need to consider the compound and accumulating effects of many disruptors at a time. An important setback to this analysis was the limited data found compared to the wide variety of ECDs described; however, based on the data collected and the quality assessments performed, the studies found are credible and valid enough to support the correlation and linkage between endocrine disruptors and T1DM.

Author Contributions: Conceptualization, G.-N.K. and S.K.; methodology, S.T.; software, M.-N.G.; validation, G.-N.K., S.T. and I.K.; formal analysis, G.-N.K.; investigation, G.-N.K.; resources, S.T.; data curation, S.T.; writing—original draft preparation, S.T. and I.K.; writing—review and editing, S.T. and I.K.; visualization, S.T.; supervision, A.S. and T.P.; project administration, S.K.; funding acquisition, T.P. All authors have read and agreed to the published version of the manuscript.

Funding: This research received no external funding.

Conflicts of Interest: The authors declare no conflict of interest.

References

1. Endocrine Society. *Endocrine-Disrupting Chemicals (EDCs) [Internet]*; Endocrine Society: Washington, DC, USA, 2023. Available online: https://www.endocrine.org/patient-engagement/endocrine-library/edcs (accessed on 11 August 2024).
2. National Institute of Environmental Health Sciences. *Endocrine Disruptors [Internet]*; NIEHS: Durham, NC, USA, 2023. Available online: https://www.niehs.nih.gov/health/topics/agents/endocrine (accessed on 11 August 2024).
3. *Wiley Online Library [Internet]*; Wiley: Hoboken, NJ, USA, 2024. Available online: https://onlinelibrary.wiley.com/ (accessed on 12 June 2024).
4. Lucier, J. Type 1 Diabetes [Internet]. StatPearls. 2023. Available online: https://www.ncbi.nlm.nih.gov/books/NBK507713/#:~:text=Type%201%20diabetes%20mellitus%20(T1D,an%20automated%20insulin%20delivery%20system (accessed on 12 June 2024).
5. Diabetes Mellitus in Children and Adolescents—Pediatrics [Internet]. MSD Manual Professional Edition. Available online: https://www.msdmanuals.com/professional/pediatrics/endocrine-disorders-in-children/diabetes-mellitus-in-children-and-adolescents# (accessed on 12 June 2024).
6. BridgetChapple. Ketones and Diabetes [Internet]. Diabetes UK. Available online: https://www.diabetes.org.uk/guide-to-diabetes/managing-your-diabetes/ketones-and-diabetes#:~:text=Ketones%20are%20a%20type%20of,without%20it%20being%20a%20problem (accessed on 12 June 2024).
7. Diabetic Ketoacidosis (DKA)—Endocrine and Metabolic Disorders [Internet]. MSD Manual Professional Edition. Available online: https://www.msdmanuals.com/professional/endocrine-and-metabolic-disorders/diabetes-mellitus-and-disorders-of-carbohydrate-metabolism/diabetic-ketoacidosis-dka (accessed on 12 June 2024).
8. Kim, K. The Role of Endocrine Disruption Chemical-Regulated Aryl Hydrocarbon Receptor Activity in the Pathogenesis of Pancreatic Diseases and Cancer. *Int. J. Mol. Sci.* **2024**, *25*, 3818. [CrossRef] [PubMed] [PubMed Central]
9. Pandol, S.J. Pancreatic Embryology and Development [Internet]. The Exocrine Pancreas. 1970. Available online: https://www.ncbi.nlm.nih.gov/books/NBK54135/ (accessed on 12 June 2024).
10. Salo, H.M.; Koponen, J.; Kiviranta, H.; Rantakokko, P.; Honkanen, J.; Härkönen, T.; Ilonen, J.; Virtanen, S.M.; Tillmann, V.; Knip, M.; et al. No evidence of the role of early chemical exposure in the development of β-cell autoimmunity. *Environ. Sci. Pollut. Res. Int.* **2019**, *26*, 1370–1378. [CrossRef] [PubMed]
11. Lee, I.; Park, Y.J.; Kim, M.J.; Kim, S.; Choi, S.; Park, J.; Cho, Y.H.; Hong, S.; Yoo, J.; Park, H.; et al. Associations of urinary concentrations of phthalate metabolites, bisphenol A, and parabens with obesity and diabetes mellitus in a Korean adult population: Korean National Environmental Health Survey (KoNEHS) 2015–2017. *Environ. Int.* **2021**, *146*, 106227. [CrossRef] [PubMed]
12. Bresson, S.E.; Isom, S.; Jensen, E.T.; Huber, S.; Oulhote, Y.; Rigdon, J.; Lovato, J.; Liese, A.D.; Pihoker, C.; Dabelea, D.; et al. Associations between persistent organic pollutants and type 1 diabetes in youth. *Environ. Int.* **2022**, *163*, 107175. [CrossRef] [PubMed]
13. Dufour, P.; Pirard, C.; Lebrethon, M.C.; Charlier, C. Associations between endocrine disruptor contamination and thyroid hormone homeostasis in Belgian type 1 diabetic children. *Int. Arch. Occup. Environ. Health* **2023**, *96*, 869–881. [CrossRef] [PubMed]
14. Xu, J.; Huang, G.; Nagy, T.; Guo, T.L. Bisphenol A alteration of type 1 diabetes in non-obese diabetic (NOD) female mice is dependent on window of exposure. *Arch. Toxicol.* **2019**, *93*, 1083–1093. [CrossRef] [PubMed]

15. McDonough, C.M.; Xu, J.; Guo, T.L. Behavioral changes and hyperglycemia in NODEF mice following bisphenol S exposure are affected by diets. *Neurotoxicology* **2021**, *85*, 209–221. [CrossRef] [PubMed]
16. Sinioja, T.; Bodin, J.; Duberg, D.; Dirven, H.; Berntsen, H.F.; Zimmer, K.; Nygaard, U.C.; Orešič, M.; Hyötyläinen, T. Exposure to persistent organic pollutants alters the serum metabolome in non-obese diabetic mice. *Metabolomics* **2022**, *18*, 87. [CrossRef] [PubMed]
17. Xu, J.; Huang, G.; Guo, T.L. Sex-dependent effects of bisphenol A on type 1 diabetes development in non-obese diabetic (NOD) mice. *Arch. Toxicol.* **2019**, *93*, 997–1008. [CrossRef] [PubMed]
18. Xu, J.; Huang, G.; Guo, T.L. Bisphenol S modulates type 1 diabetes development in non-obese diabetic (NOD) mice with diet- and sex-related effects. *Toxics* **2019**, *7*, 35. [CrossRef] [PubMed]
19. Mariana, M.; Cairrao, E. The relationship between phthalates and diabetes: A review. *Metabolites* **2023**, *13*, 746. [CrossRef] [PubMed]
20. Predieri, B.; Bruzzi, P.; Bigi, E.; Ciancia, S.; Madeo, S.F.; Lucaccioni, L.; Iughetti, L. Endocrine disrupting chemicals and type 1 diabetes. *Int. J. Mol. Sci.* **2020**, *21*, 2937. [CrossRef] [PubMed]
21. Hinault, C.; Caroli-Bosc, P.; Bost, F.; Chevalier, N. Critical overview on endocrine disruptors in diabetes mellitus. *Int. J. Mol. Sci.* **2023**, *24*, 4537. [CrossRef] [PubMed]
22. Khalil, W.J.; Akeblersane, M.; Khan, A.S.; Moin, A.S.M.; Butler, A.E. Environmental pollution and the risk of developing metabolic disorders: Obesity and diabetes. *Int. J. Mol. Sci.* **2023**, *24*, 8870. [CrossRef] [PubMed]
23. Heo, Y.J.; Kim, H.S. Ambient air pollution and endocrinologic disorders in childhood. *Ann. Pediatr. Endocrinol. Metab.* **2021**, *26*, 158–170. [CrossRef] [PubMed]
24. Jiang, W.; Ding, K.; Huang, W.; Xu, F.; Lei, M.; Yue, R. Potential effects of bisphenol A on diabetes mellitus and its chronic complications: A narrative review. *Heliyon* **2023**, *9*, e16340. [CrossRef] [PubMed]
25. Martínez-Ibarra, A.; Martínez-Razo, L.D.; MacDonald-Ramos, K.; Morales-Pacheco, M.; Vázquez-Martínez, E.R.; López-López, M.; Dorantes, M.R.; Cerbón, M. Multisystemic alterations in humans induced by bisphenol A and phthalates: Experimental, epidemiological and clinical studies reveal the need to change health policies. *Environ Pollut.* **2021**, *271*, 116380. [CrossRef] [PubMed]

International Journal of
Molecular Sciences

MDPI

Review

Mesenchymal Stem Cell Therapy: Therapeutic Opportunities and Challenges for Diabetic Kidney Disease

Jia Cheng and Chun Zhang *

Department of Nephrology, Union Hospital, Tongji Medical College, Huazhong University of Science and Technology, Wuhan 430000, China; u201712217@hust.edu.cn
* Correspondence: drzhangchun@hust.edu.cn

Abstract: Diabetic kidney disease (DKD) is the leading cause of end-stage renal disease (ESRD), which severely affects the quality of patients' lives. However, the current therapeutic approaches can only postpone its progression to ESRD. It is therefore imperative to develop a novel therapeutic strategy for renal injury in DKD, with the objective of restoring renal function and reversing the process of ESRD. In recent years, the potential of mesenchymal stem cell (MSC) therapy for DKD has garnered increasing attention within the scientific community. Preclinical research on MSC therapy has yielded promising results, and the safety of MSC treatment in vivo has been substantiated in clinical studies. An increasing body of evidence suggests that MSC therapy has significant potential for the treatment of DKD. This article reviews the existing research on MSCs and their derived exosomes in treating DKD and analyzes the underlying mechanism of MSC-based therapy for DKD. Additionally, we discuss the potential of combining MSC therapy with conventional pharmacological treatments, along with the constraints and prospects of MSC therapy for DKD. We hope this review can provide a precise and comprehensive understanding of MSCs for the treatment of DKD.

Keywords: mesenchymal stem cell; diabetic kidney disease; mesenchymal stem cell therapy

1. Introduction

Diabetic kidney disease (DKD) is the most common microvascular complication of diabetes and the main cause of end stage renal disease (ESRD) in the majority of countries. Patients with DKD frequently lack efficacious treatment options in the terminal stages and are therefore reliant on hemodialysis or kidney transplantation for survival. The pathogenesis of DKD is a complex process whereby chronic persistent hyperglycemia may lead to nephron damage through immune–inflammatory processes, mitochondrial dysfunction, and oxidative stress [1]. The renal pathological manifestations of DKD patients include mesangial dilation, collagen deposition, basement membrane thickening, podocyte loss and hypertrophy, albuminuria, tubular epithelial atrophy, activated myofibroblasts, matrix accumulation, and inflammatory cell influx [2]. Conventional treatments for DKD are confined to regulating blood pressure (BP), glucose, and lipids, which can merely defer but not wholly reinstate normal renal functioning. It is, therefore, imperative to identify a novel treatment for DKD.

Mesenchymal stem cells (MSCs) represent a specific type of undifferentiated stem cell, offering distinctive advantages such as self-renewal capacity, multipotent differentiation potential, and low immunogenicity [3,4]. Furthermore, MSCs possess immunomodulatory [3,5], antioxidant [6], antiapoptotic [7], mitochondrial quality control [8,9], and robust tissue regeneration capabilities [10], rendering them highly versatile in the domain of regenerative medicine. Additionally, MSCs can be sourced from diverse locations, such as bone marrow (BM-MSCs) [11], adipose tissue (ADSCs) [12], umbilical cord (UC-MSCs) [13], Wharton's jelly (WJMSCs) [14], placenta from full-term pregnancies (PMSCs) [15], and human exfoliated deciduous teeth (SHED) [16] (Figure 1), making them accessible and ethically acceptable for use in research and clinical applications. MSCs can transport intracellular substances, including nucleic acids, proteins, and mitochondria (Mt), through secreting exosomes to damaged cells, thereby regulating the pathological and physiological processes of damaged cells [17,18]. The method of using MSCs and their secreted derivatives (such as exosomes) to treat diseases is collectively called MSC therapy.

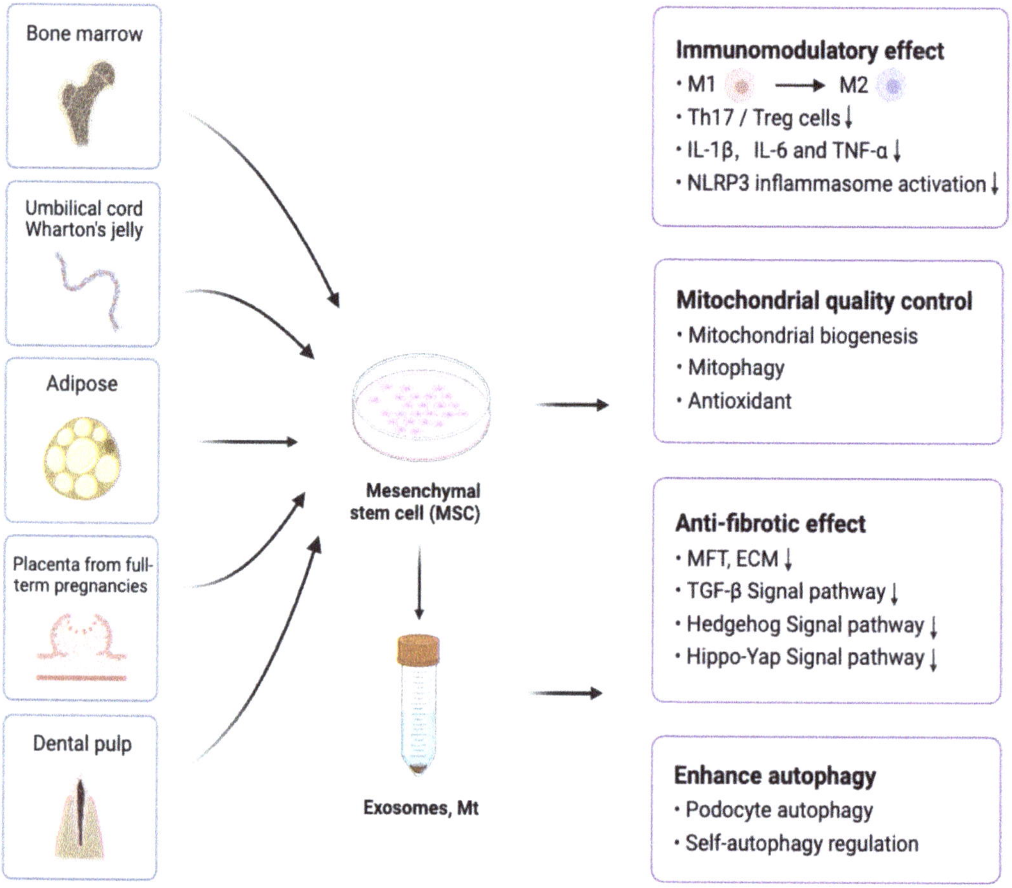

Figure 1. The therapeutic effects of MSCs and their derivatives from different sources in the treatment of DKD. MSCs can be isolated from bone marrow, the umbilical cord, the Wharton's jelly, adipose tissue, placenta and dental pulp. MSCs and their derivatives have anti-inflammatory and anti-fibrotic effects, control the targeted cells mitochondrial quality, as well as enhance autophagy to protect the kidney from diabetic injury.

MSC therapy has recently garnered significant interest mainly due to its potential for restoring damaged nephrons. There have been notable advancements in the research of MSC therapy in the treatment of DKD. This article provides a summary of the research on MSC therapy in DKD, introduces the potential mechanisms of MSC therapy in DKD treatment, and further analyzes the possibility of combining MSC therapy with traditional pharmacological treatments for DKD. It also considers the prospects and challenges of MSC therapy in treating DKD. This study aims to elucidate the value of MSC therapy in the treatment of DKD and provide new insights into the treatment of DKD.

2. The Use of MSC Therapy in DKD

2.1. Preclinical Study

The therapeutic potential of MSCs in DKD was initially found because exogenous human bone marrow pluripotent stromal cells (hMSCs) significantly differentiated into endothelial cells in the glomeruli of immunodeficiency NOD/scid diabetic mice [19]. The hMSCs were administered to mice via intracardiac infusion, and subsequent analysis revealed a minor degree of aggregation within the glomeruli. The degree of mesangial thickening was found to be diminished in mice that had been infused with hMSCs, as was the infiltration of macrophages [19]. The author speculated that transplanted hMSCs either prevented pathological changes in the glomeruli or enhanced their regeneration to improve renal lesions in diabetes patients [19]. Subsequently, it was established that MSCs derived from disparate species and sources exhibited therapeutic efficacy in the amelioration of the experimental model of DKD. Concurrently, it was demonstrated that MSCs exert their effects through paracrine pathways [17], specifically by delivering their derivatives such as exosomes [18,20]. Furthermore, it is hypothesized that the mitochondria in MSCs exert a significant influence on the enhancement of mitochondrial quality in damaged renal cells. Derivatives secreted from MSCs are more readily obtainable and storable, with minimal ethical constraints [21], and thus they are regarded as a promising novel therapy for treating kidney injury [22]. The potential of MSC therapy from various sources to treat DKD has been the subject of extensive investigation in recent times. A summary of the preclinical studies (Table 1) of MSC therapy for DKD in the past five years based on species and sources is presented. The studies primarily concentrate on the remission of DKD by BM-MSCs, ADSCs, and UC-MSCs, with a particular focus on UC-MSCs. In contrast to BM-MSCs, the source of UC-MSCs is more readily accessible than bone marrow, obviating the necessity for invasive procedures and significantly reducing the discomfort experienced by patients. Moreover, UC-MSCs exhibit a greater capacity for proliferation [23]. Furthermore, UC-MSCs can be stored at birth and are derived from autologous cells, which exhibit reduced immunogenicity and markedly enhance the safety of mesenchymal therapy. Recently, it was found that MSCs from dental pulp have a similar role in alleviating the progression of DKD and can also be used as a source of materials for future MSC therapy. It is anticipated that in the future, further tissues with robust proliferative and differentiative capabilities will be identified as sources of MSCs.

Table 1. Preclinical studies on the treatment of DKD with MSCs and their derived exosomes in the past five years.

Type of MSCs	Subjects	Method of DKD Induction	Administration	Frequency and Dose	Effect	Reference
Mouse BM-MSCs	Male BALB/c mice	i.p. of 150 mg/kg STZ	Tail vein injection	Once every two weeks, three times after the onset 5×10^5	MSCs reprogram Mφ into M2 via improvement of the lysosome–autophagy pathway and mitochondrial bioenergetics with transcription factor EB activation.	[24]
	Male mice BTBR.Cg-Lep^{ob}/WiscJ	-	Intraperitoneal injection	8th and 10th weeks of age 1×10^6	MSC-treated animals exhibited lighter renal pathological impairment, upregulation of mitochondria-related survival genes, and a decrease in autophagy hyperactivation and apoptosis.	[25]

Table 1. *Cont.*

Type of MSCs	Subjects	Method of DKD Induction	Administration	Frequency and Dose	Effect	Reference
Rat BM-MSCs	Male C57BL/6 mice male SD rats	i.p. of 150 mg/kg STZ (diabetic mice) tail vein injection of 55 mg/kg of STZ (diabetic rats)	Tail vein injection	8 w and 10 w after the onset 1×10^4	Mt transfer from BM-MSCs to damaged PTECs, injection of BM-MSC-derived isolated Mt in renal capsule share the same effect.	[26]
	Male C57BL/6 mice	Not provided	Tail vein injection	Not provided	MSCs protect DKD kidneys by regulating M6 A methylation through Smad2/3/WTAP/ENO1.	[27]
	Male SD rats	i.p. of 55 mg/kg STZ	Tail vein injection	Once a week for 6 weeks 1×10^7	Show immunosuppression of CD8 T-cell proliferation and activation mediated by CD103 DCs.	[28]
	Male SD rats	i.p. of 65 mg/kg STZ	Tail vein injection	4×10^6	Improve renal function and collagen accumulation; inhibit inflammatory and fibrotic cytokines by downregulating TLR-4/NF-κB expression.	[29]
	Male SD rats	i.p. of 60 mg/kg STZ	Tail vein injection	Once a week for two continuous weeks 5×10^6	MSCs suppress progression of diabetic nephropathy (DN) pathogenesis through LXA4 by targeting TGF-β/Smad signals and pro-inflammatory cytokines in DN.	[30]
	GECs	30 mmol/L D-glucose	Co-culture	5×10^5	Rejuvenate damaged GEs via Mt transfer.	[31]
BM-MSC-Exos	Male SD rats	i.p. of 35 mg/kg STZ	Tail vein injection	Once a week for 12 weeks 100 μg	Increase apoptosis; decrease GLU. Scr, BUN.	[32]
Rat BM-MSCs-derived exosomes	Male albino rats	i.p. of 60 mg/kg STZ	Tail vein injection	Two injections of exosomes 100 μg/kg/dose	Ameliorate diabetic nephropathy by autophagy induction through the mTOR signaling pathway.	[33]
Human BM-MSCs-derived extracellular vesicles (EVs)	Male NSG mice	i.p. of 37 mg/kg STZ for 4 days	Intravenous injection	Once a week for 4 weeks 1×10^{10} particles	Revert the progression of glomerular and interstitial fibrosis.	[34]
Rat ADSCs	Male SD rats	8 weeks of high-fat diet (HFD) and a single dose of 25 mg/kg STZ	Tail vein injection	Once a week for 24 weeks 3×10^6	Reduce blood glucose and insulin demand, reduce the expression of SLGT2 of PTEC and reduce kidney damage and inflammation.	[35]
Mouse ADSCs-EVs	C57BL/KsJ db/db	-	Tail vein injection	Once a week for 12 weeks -	miR-26a-5p delivered by ADSC-derived EVs suppress glomerular podocyte apoptosis and protect against DN by regulating TLR4.	[36]
Mouse ADSCs-Exos	C57BL/KsJ db/db	-	Tail vein injection	Once a week for 12 weeks -	Reduce proteinuria, Scr, blood urea nitrogen (BUN), and podocyte apoptosis; miR-486 of ADSCs-Exo promote autophagy flux.	[37]

Table 1. *Cont.*

Type of MSCs	Subjects	Method of DKD Induction	Administration	Frequency and Dose	Effect	Reference
Rat ADSCs-exos	Male SD rats	i.p. of 60 mg/kg STZ	Tail vein injection	50 μg exosomes t twice a week for 3 weeks	Decrease levels of blood glucose, serum creatinine (Scr), 24 h urinary protein, UACR, and kidney weight/body weight, and they suppressed mesangial hyperplasia and kidney fibrosis, which is related to miR-125a.	[38]
Human ADSCs-EVs	Mouse podocyte clone 5 (MPC5)	Normal glucose (NG, 5.5 mM), NG + mannitol (5.5 mM glucose +24.5 mM MA), HG (30 mM glucose)	-	25 μg/mL	EV-derived miR-15b-5p could protect MPC5 cell apoptosis and inflammation via downregulation of the VEGF/PDK4 axis.	[39]
Rat ADSCs Sheet	Spontaneously Diabetic Torii (SDT) fatty rat	-	Femoral vein injection or renal capsule transplantation	6×10^6	Transplantation of adipose-derived mesenchymal stem cell sheets directly into the kidney improves transplantation efficiency and suppresses inflammation and renal injury progression.	[40]
Human UC-MSCs	Male CD1 mice	i.p. of 80 mg/kg STZ	-	Thrice every 4 weeks after the onset 5×10^5	Promote the expression of Arg1 in macrophages to promote M2 polarization and improve mitochondrial function of renal tubular epithelial cells.	[41]
	Male SD rats	i.p. of 60 mg/kg STZ	Tail vein injection	Once a week for two consecutive weeks 2×10^6	Reduce urinary total protein, UACR, Scr, and BUN, improve renal pathological abnormalities, promote the expression of antiapoptotic protein Bcl-xl, and activate the apoptotic pathway.	[42]
	Male SD rats	i.p. of 60 mg/kg STZ	Tail vein injection	2×10^6	Ameliorate functional parameters, improve renal pathological changes, reduce the levels of pro-inflammatory cytokines (IL-6, IL-1β, and TNF-α) and profibrotic factor (TGF-β) in the kidney and blood.	[43]
	Male SD rats	i.p. of 60 mg/kg STZ	Tail vein injection	Twice; group 1: weeks 7 and 8; group 2: weeks 9 and 10 2×10^6	UC-MSC-derived miR-146a-5p restores renal function in DN rats by facilitating M2 macrophage polarization by targeting TRAF6.	[44]
	Male SD rats	HFD and i.p. of 35 mg/kg STZ	Tail vein injection	Three times every 10 days 2×10^6	Attenuate renal oxidative damage and apoptosis, activate Nrf2.	[45]

Table 1. *Cont.*

Type of MSCs	Subjects	Method of DKD Induction	Administration	Frequency and Dose	Effect	Reference
Human UC-MSCs	Male SD rats	5/6 nephrectomy, followed by intraperitoneal administration of aminoguanidine (180 mg/kg) and streptozotocin (30 mg/kg)	Intrarenal arterial injection (IRA)	Day 21 after CKD induction 2.1×10^5	IRA injection of xenogeneic MSCs was safe and effectively protected the residual renal function and architectural integrity.	[46]
	Male SD rats	i.p. of 50 mg/kg STZ	Intravenous injection	Once a week for 4 weeks Low-dose hucMSCs-treated group (MSC-L):5 $\times 10^6$; high-dose hucMSCs-treated group (MSC-H):1 $\times 10^7$	Improve cell viability, wound healing, and senescence of the high glucose-damaged rat podocytes through a paracrine action mode; activate autophagy and attenuate senescence through the AMPK/mTOR pathway.	[47]
	Male SD rats	HFD for 4 w, followed by i.p. of 30 mg/kg STZ	Tail vein injection	Once per week for four consecutive weeks 1×10^7	HUC-MSCs downregulated the expression of IGF1/IGF1R in the renal tissue of diabetic rats, inhibited the activity of the target genes *CHK2* and *p53*, reduced apoptosis, and improved diabetic nephropathy.	[48]
	Male C57BL/6 mice	i.p. of 60 mg/kg STZ	Tail vein injection	8 w or 16 w after the onset 5×10^5	Reduce uACR and improve multiple glomerular and renal interstitial abnormalities; reduce circulating TGF-β1 and restore intrarenal autophagy; reduce early inflammation of the disease.	[49]
Mouse UC-MSCs	Male BALB/C mice	i.p. of 150 mg/kg STZ	Tail vein injection	Weekly for 4 weeks 1×10^4	Alleviate albuminuria, glomerulus injury, and fibrosis by inhibiting TGF-β1-triggered MFT and cell proliferation mediated by PI3K/Akt and MAPK signaling pathways and elevating the levels of MMP2 and MMP9.	[50]
Human UC-MSCs-exos	Male C57BL/KsJ-db/db	-	Tail vein injection	MSCs: every week for 6 weeks 1×10^6 MSC-exos: twice a week for 6 weeks 10 mg/kg bw	MSC-exos could inhibit high glucose-induced apoptosis and EMT through miR-424-5p targeting of YAP1.	[51]
Human UC-MSCs-exos	Male db/db mice	-	Tail vein injection	3 times in the first week, and then twice a week for the next 3 weeks 100 μg	MSCs-Exos attenuated the expression of inflammatory factors in podocytes and DN mice in vivo and in vitro, inhibited the activation of the NLRP3 signaling pathway, and improved renal injury. MiR-22-3p may play a role in the anti-inflammatory effect of MSCs-Exo.	[52]

Table 1. *Cont.*

Type of MSCs	Subjects	Method of DKD Induction	Administration	Frequency and Dose	Effect	Reference
Human UC-MSCs-exos	Male C57BL/6J mice; male db/db mice	HFD for 6 weeks, followed by i.p. of 100 mg/kg STZ for three days	Not provided	3 times in the first week, and then twice a week for the next 3 weeks 100 μg	Alleviate the inflammatory response; inhibit the activation of NOD2 signaling pathway; prevent apoptosis; increase cell viability in podocytes.	[53]
Human UC-MSCs-EVs	Male SD rats	HFD and i.p. of 35 mg/kg STZ	Tail vein injection	10 mg/kg	Protein 14-3-3ζ in hucMSC-sEVs promotes YAP cytoplasmic retention instead of entering the nucleus, enhancing the level of autophagy in the cytoplasm to remove the excessive YAP protein.	[54]
Human UC-MSCs-EVs	Male C57BLKS/J db/db	-	Intravenous injection	Two times a week from 8 to 18 weeks old 100 ug	Attenuate renal fibrosis and inflammation; MiR-23a-3p and its target Krüppel-like factor 3 (KLF3) inhibit high glucose (HG)-induced STAT3 signaling pathway; miR-23a-3p is packaged into MSC sEV by RNA-binding motif protein X-linked (RBMX).	[55]
Human UC-MSCs/UC-MSCs-exos	Male SD rats	i.p. of 60 mg/kg STZ	Tail vein injection	Twice UC-MSCs: 2×10^6 UC-MSC-exo: 100 μg/kg	Attenuate kidney damage; inhibit EMT and renal fibrosis; decrease SMO expression targeting Hedgehog/SMO pathway.	[56]
Human PMSCs	Male SD rats	i.p. of 60 mg/kg STZ	Tail vein injection	1×10^6	Reverse podocyte injury and mitophagy.	[57]
	Male SD rats	i.p. of 60 mg/kg STZ	Tail vein injection	Once a week for three weeks 1×10^6	Regulate TH17/Treg through systemic immune regulation by upregulating PD-1; enhance the autophagy level of DKD rat kidney and podocyte by upregulating the expression of SIRT1 and FOXO1.	[58,59]
SHED Human BM-MSCs	Male Goto-Kakizaki (GK) rats	HFD for 2–4 weeks	Tail vein injection	4×10^6	Attenuate hyperglycemia, hyperlipidemia, increased urinary albumin excretion, ECM accumulation, and a fractional mesangial area.	[60]
Human Amniotic MSCs	Male SD rats	i.p. of 55 mg/kg STZ	Penile vein injection	2×10^6	Increase the expression of ARF and decrease blood glucose, 24-h urinary protein, Scr, urea, kidney injury molecule-1 (KIM-1), and renal injury index.	[61]

2.2. MSC Therapies Toward Clinical Translation

Currently, a considerable number of clinical studies are investigating the potential of MSC therapy as a treatment for diabetes mellitus (DM). Nevertheless, only two studies have concentrated on the particular issue of kidney damage associated with diabetes and DKD. The clinical application of MSC therapy in the treatment of DKD has made some progress. The two clinical studies primarily explored the safety of bone marrow-derived MSCs (ORBCEL-M) and mesenchymal precursor cells (MPCs, rexlemestrocel-L) in vivo, despite the evidence of adverse effects, in the treatment of DKD [62,63]. Nevertheless, the observation period of rexlemestrocel-L is relatively short, and it cannot be excluded that more rare security incidents may occur after long-term observation. The improvement in renal function reported in the two clinical studies was not statistically significant due to the relatively small sample sizes, although there was a discernible improvement in renal function.

Indeed, the absence of consistent and standardized methodologies to characterize its safety and efficacy represents a significant obstacle to the advancement of MSC therapy toward clinical application. The sources of MSCs are diverse, and there are notable discrepancies in the administration doses employed across disparate studies. It is difficult to determine the optimal efficacious dose in individual clinical studies. Secondly, neither study detected the pharmacokinetics of MSCs in vivo, which is meaningful for us in order to explore effective and safe doses for clinical applications. Indeed, when conducting experiments on cell products, it is essential to confirm the pharmacokinetics of cells within the human body. This will facilitate a deeper comprehension of the metabolism of MSCs in vivo and enable more accurate determination of safe and efficacious doses. The latest research provides some insights into the circulation dynamics of MSCs by detecting the SRY gene in peripheral blood [64]. Furthermore, the relatively small number of subjects included in the two clinical studies, the significant differences in baseline levels observed, and the inconsistent endpoints of observation may have contributed to the lack of significant statistical significance. Given the limited research results on the clinical efficacy, optimal treatment timing, and regimen of MSC therapy for DKD, it is imperative to explore reasonable and safe infusion doses to reach a consensus and standardize clinical trial observation indicators.

MSC therapy in the treatment of DKD has achieved remarkable results in vitro and in vivo. Nevertheless, the proof of human studies has been relatively limited in recent years. There are objective limitations to the translation of animal experiments into human applications. Currently, commonly used DKD animal models include pharmacological induction models (STZ injection) and genetic models (db/db mice). These models cannot well reflect the pathophysiological processes of human DKD. The exact progression of kidney disease may vary greatly from humans, including timing, severity, and specific histopathological features. The timeline of disease progression in animal models is often accelerated compared to humans, leading to a misunderstanding of the long-term effects of interventions. The existing literature has employed unilateral nephrectomy in db/db mouse models as a means of standardizing the onset time of DKD [65]. In addition, human behaviors such as dietary habits, physical activity, and lifestyle choices play a key role in the development of DKD, which cannot be accurately replicated in animal models. It appears that MSCs do not offer a distinct advantage in the treatment of human diabetes nephropathy. There are many challenges in clinical practice, such as heterogeneity in subject renal function and urinary albumin levels, short follow-up observation time, inconsistent biomarker observations, comorbidities, and different drug treatment regimens. These issues collectively contribute to its uncertain clinical value. In the future, it may be necessary to design studies with a larger sample size, that are multicenter, with a longer follow-up time, and with more unified observation endpoints to explore the overall impact of MSC therapy on DKD.

3. The Mechanism of MSC Therapy in Treating DKD

3.1. *Immunomodulatory Effect*

MSCs have been demonstrated to possess potent immune regulatory functions. They are capable of ameliorating the local immune microenvironment of the kidney and systemic immune function through direct contact and the secretion of cytokines.

MSCs may play an essential role in suppressing the renal local immune response of DKD by reducing inflammatory cell infiltration and inflammatory response. Studies have indicated that MSCs can suppress the expression of pro-inflammatory factors including IL-1β, IL-6, and TNF-α in the kidney of DKD rats through local immune regulation. Furthermore, the infiltration of inflammatory cells in the kidney tissue of DKD rats treated with

MSCs was diminished [28,66], as evidenced by a reduction in CD103 dendritic cells (DCs) and CD8 T cells in the kidney of DKD rats treated with MSC [28]. Additionally, studies have demonstrated that MSC treatment reduces macrophage infiltration by inhibiting the expression of monocyte chemoattractant protein-1 (MCP-1) [66]. The anti-inflammatory effect of M2 macrophages is of pivotal importance in the prevention and treatment of inflammatory diseases. BM-MSCs, UC-MSCs, and their derived exosomes and microRNAs have been demonstrated to enhance macrophage M2 polarization, augment the anti-inflammatory response, and mitigate kidney damage resulting from DKD inflammation [24,41,44,67,68]. MSCs may promote the transformation of macrophages to the M2 phenotype by enhancing the expression of the M2 macrophage marker Arg1 [41], which exerts beneficial effects on mitochondrial dysfunction.

Interestingly, AD-MSCs were demonstrated to increase similar expression of Arg1 and M1 macrophage marker iNOS, resulting in a null effect on mitochondrial dysfunction [41]. The precise reason for this discrepancy in outcomes between the administration of MSCs derived from different sources is currently unclear. The primary immunomodulatory effects of MSC in the kidney are achieved by regulating innate immune cell infiltration, which attenuates local inflammatory responses.

MSCs may also exert therapeutic effects through systemic immune regulation. A single injection of human umbilical cord-derived mesenchymal stem cells (hUC-MSCs) was observed to result in a decrease in serum IL-6, TNF-α, and TGF-β1 levels in the early stage of DKD (8 weeks after STZ injection). In comparison, no change was noted in serum IL-6 and TNF-α levels in the late stage (16 weeks after STZ injection) [19]. This suggests that systemic immune suppression of MSCs is more likely to play a role in the early stage of DKD. The imbalance of TH17/Treg cells plays a vital role in the pathogenesis of diabetes nephropathy. Specifically, the proportion of TH17 cells in the peripheral blood of patients with diabetes nephropathy tends to increase while the proportion of Treg cells decreases. This imbalance may lead to excessive inflammatory response, further exacerbating kidney damage. The infusion of human placenta-derived mesenchymal stem cells (PMSCs) has been demonstrated to promote Th17/Treg balance in the kidneys and blood of DKD rats while simultaneously reducing the levels of pro-inflammatory cytokines (IL-17A and IL-1 β) [58]. It is postulated that MSC may regulate TH17/Treg-related DKD through systemic immune regulation, thereby reducing inflammatory factors in the body and kidneys and weakening the inflammatory response.

In addition, MSCs can inhibit various inflammatory pathways. Human umbilical cord mesenchymal stem cells (UC-MSCs), bone marrow mesenchymal stem cells (BMSCs) [29], and exosomes derived from adipose-derived mesenchymal stem cells (ADMSCs) containing microRNA-26a-5p [36] are beneficial for podocytes and diabetic nephropathy under HG by inhibiting Toll-like receptor (TLR) signaling and suppressing inflammation. MSC-derived exosomes miR-22–3p improve diabetes nephropathy through the NLRP3 signaling pathway [52].

3.2. Mitochondrial Quality Control

Many sophisticated quality control mechanisms have been found within mitochondria to counteract stress and maintain the integrity and functionality of organelles, including mitochondrial DNA (mtDNA) repair, mitochondrial dynamics (fusion and fission), mitochondrial autophagy, and mitochondrial biogenesis. The dysregulation of mitochondrial quality control mechanisms may induce mitochondrial damage and dysfunction, leading to cell death, tissue damage, and potential organ failure. DKD is frequently observed to occur alongside a reduction in mitochondrial number and an accompanying impairment in functionality. As a therapeutic method with exogenous restoration of mitochondrial quality control function [69], MSCs are essential in restoring mitochondrial quality in DKD.

It is well-known that MSCs can alleviate high glucose-induced renal mitochondrial dysfunction. Recent studies have shown that MSCs transfer their mitochondria to damaged cells through tunnel nanotubes to improve the mitochondrial quality of target cells. This process involves promoting mitochondrial biogenesis and mitophagy of target cells, which improves the metabolic capacity and antistress ability of cells. Specifically, under co-culture of MSCs and macrophages in vitro, mitochondrial transfer from the former to the latter occurs via tunneling nanotubes (TNT), and high glucose levels can facilitate this process [70]. The mitochondrial function of macrophages was improved by extracting the Mt of MSC and transferring them to macrophages, which may be caused by the activation of PGC-1α-mediated mitochondrial biogenesis and PGC-1α/TFEB-mediated autophagy in macrophages [70]. In vivo and in vitro studies have demonstrated that MSCs can transfer their Mt into diabetes-damaged proximal tubular epithelial cells (PTEC) and glomerular endothelial cells (GECs). MSCs could restore the structure of renal tubules and inhibit cell apoptosis and ROS production by regulating Mt-related factors such as Bcl-2, Bax, and PGC-1 α [26,41]. Moreover, glomeruli's apoptosis, proliferation, and mitochondrial function have also been improved [31]. The quality of the mitochondria in podocytes plays an integral role in ensuring the proper functioning of these cells. MSCs can maintain the normal structure and function of podocytes by regulating podocyte mitophagy. PMSCs can significantly improve kidney injury and reduce podocyte damage in DKD rats by regulating the SIRT1-PGC-1α-TFAM pathway and enhancing PINK1/Parkin-mediated podocyte mitophagy [57]. Nevertheless, it remains uncertain whether MSCs transfer their mitochondria to podocytes to facilitate podocyte mitophagy and enhance the mitochondrial quality of podocytes. It is currently thought that there are four potential pathways through which Mt can be transferred from MSCs to damaged cells. These include transfer through tunneling nanotubes, gap junctions, cell fusion, microvesicles, and direct uptake of isolated mitochondria [71]. Recent studies have revealed that MSCs may transfer mitochondria to recipient cells through a three-dimensional pathway, as mitochondrial transfer was found in adipose-derived MSCs and breast cancer cells even when the tunneling nanotube (TNT) was blocked [72]. MSCs do not seem to promote the metabolism of recipient cells by exogenously supplementing functional mitochondria. The latest research findings indicate that even dysfunctional mitochondria can trigger mitochondrial autophagy in receptor cells through the typical PINK1/Parkin signaling pathway after internalization and rapidly degrade in autophagosomes. To summarize, exogenous mitochondria serve only as a trigger for autophagy in recipient cells. If it could be demonstrated that mitochondria can be transferred to kidney cells without the involvement of MSCs, this approach would be more ethically sound in clinical applications.

In conclusion, the role of MSCs in alleviating DKD in diabetes may be attributed, at least in part, to the fact that MSCs restore normal cell function by delivering mitochondria to kidney cells or by promoting kidney cells to activate endogenous mitochondrial quality control mechanisms.

3.3. Antifibrotic Effect

Fibrosis and epithelial–mesenchymal transition are typical pathological changes associated with DKD, which can result in severe glomerulosclerosis and impaired filtration function. Cytokines, hyperglycemia, and advanced glycation end products (AGEs) have been demonstrated to promote myofibroblast transdifferentiation (MFT) and the profibrotic phenotype of nonfibroblast kidney cells [73]. Renal fibrosis is typified by augmented deposition of the extracellular matrix (ECM), with the degree of renal fibrosis as a predictor of the severity and probability of adverse outcomes in patients [74]. Antifibrotic therapy is of considerable significance in the postponement of the progression of DKD to ESRD.

Multiple MSCs and their derived exosomes have been recognized to play an important role in the delay and reversal of renal fibrosis in diabetic kidney disease by reducing the production of ECM proteins and promoting their degradation [29,34,43,50,55,56,75]. BMSCs reduce the protein expression of plasminogen activator inhibitor-1 (PAI-1) and decrease the accumulation of ECM via inhibiting the TGF-β1/Smad3 pathway, thereby balancing the fibrinolytic system [75]. The paracrine of UC-MSC has been demonstrated to alleviate renal fibrosis in diabetic nephropathy, which is related to the inhibition of the TGF-β1 and Hedgehog signaling pathway. TGF-β-triggered MFT and cell proliferation are mediated by the PI3K/Akt and MAPK signaling pathways. Additionally, the levels of MMP2 and MMP9 increase, resulting in a reduction in fibronectin and type I collagen deposition [50]. Moreover, US-MSCs and their derived exosomes can also inhibit the expression of the crucial protein SMO in Hedgehog signaling, thereby hindering EMT and reducing DN renal fibrosis [56]. It is worth noting that inflammation is also closely related to fibrosis. As mentioned above, the pro-inflammatory factors secreted by macrophages facilitate the MFT and promote the deposition of ECM.

Single-cell RNA transcriptomics has revealed how MSCs-derived small extracellular vesicles (MSCs-EVs) combat DKD-induced fibrosis. It was observed that the administration of MSC sEV induced an increase in the phosphorylation of YAP at serine 381 and serine 127, concomitant with a reduction in the overall protein level of YAP [76], which CK1δ and β-TRCP, the main regulators of YAP degradation, may mediate. When siRNA is used to knock down CK1δ and β-TRCP in MSC sEV, the antifibrosis effect of MSC sEV is attenuated, indicating that MSC sEV exerts its renal protective effect by delivering CK1δ and β-TRCP to the renal tissue [76].

3.4. *Enhance Autophagy*

Autophagy represents a highly conserved lysosomal degradation pathway that removes protein aggregates and damaged organelles to maintain cellular homeostasis. Autophagy is an important stress response mechanism, and its dysfunction has been linked to the pathogenesis of a wide range of diseases. Emerging evidence shows that in individuals with diabetes, an impairment in autophagy may contribute to the development of renal glomerular and tubulointerstitial lesions [77].

MSCs can increase autophagy levels in targeted cells to help maintain normal physiological processes. MSCs can influence the autophagy of immune cells implicated in injury-induced inflammation, thereby reducing their survival, proliferation, and function and facilitating the resolution of inflammation [78]. MSCs have restored renal autophagy throughout the DKD process [49]. MSC-derived exosomes improve diabetes nephropathy by relieving the autophagy attenuation induced by the mTOR signaling pathway [33,37]. Furthermore, MSC-derived exosomes facilitate the degradation of YAP protein in podocytes by enhancing autophagy, thereby reducing its entry into the nucleus and exerting antifibrotic effects [54]. PMSCs alleviate kidney damage in DKD by promoting SIRT1/FOXO1 to encourage autophagy of podocytes [59]. The utilization of the autophagy inhibitors chloroquine and 3-MA (3-methylpurine) resulted in the reversal of the PMSC-mediated enhancement in glucose and lipid metabolism and renal function and the reduction of podocyte damage in DKD [33,59]. Interestingly, the regulation of autophagy in MSCs affects their regenerative therapeutic potential [79].

4. The Signaling Pathways Involved in MSC Treatment of DKD

In the past year, there have been many studies that explored the signal pathways involved in the treatment of DKD with MSCs. These studies discussed in detail the multiple signal pathways that MSCs may involve in the treatment of DKD. We summarize the signal pathways that MSC therapy may participate in Figure 2, hoping to provide a new perspective for clarifying the specific mechanism of MSCs in DKD.

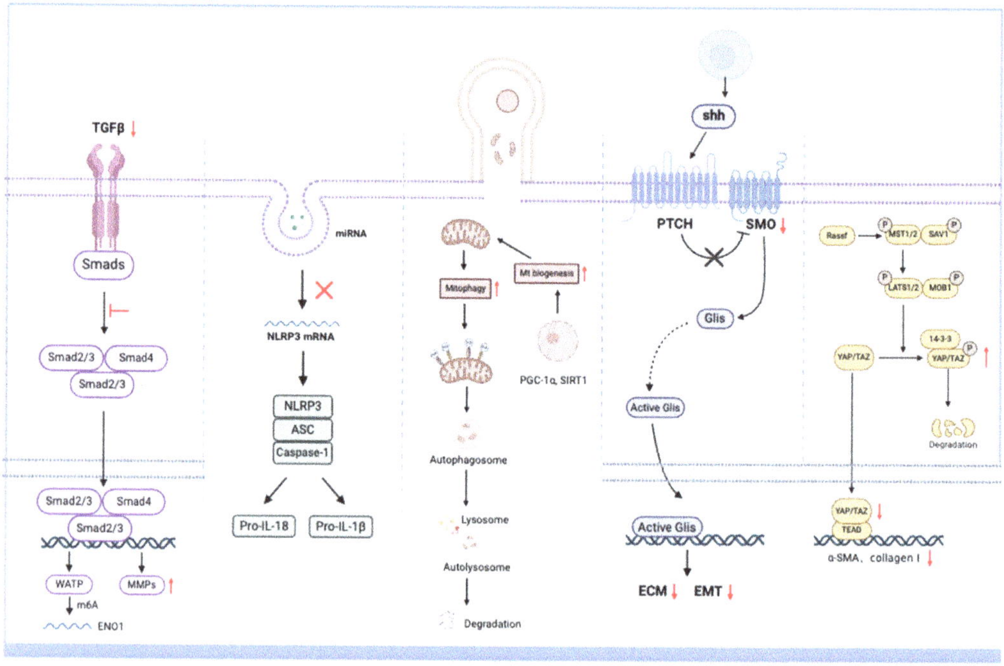

Figure 2. The signal pathways of MSCs in the treatment of DKD. MSCs reduce the expression of TGF-β 1 to inhibit the formation of SMAD complexes, suppress epithelial to mesenchymal transition (EMT), and promote fibrinolysis. MSCs deliver miRNA to receptor cells by secreting exosomes, inhibiting NLPR3 mRNA translation and reducing the release of pro-inflammatory cytokines. MSCs transport mitochondria to receptor cells to activate mitochondrial biogenesis and mitophagy, improving the quality of receptor cell mitochondria. MSCs inhibit the Hedgehog/SMO signaling pathway to reduce the expression of extracellular matrix (ECM) proteins. MSCs inhibit the Hippo/YAP signaling pathway by promoting phosphorylation of YAP protein and reducing ECM protein.

5. MSC Therapy and Traditional Drug Treatment

Previously, the primary objective in the prevention and treatment of DKD was the strict control of blood glucose and blood pressure. The most widely accepted drugs with renal protective effects are renin–angiotensin–aldosterone system (RAAS) inhibitors such as angiotensin-converting enzyme inhibitors (ACEIs) and angiotensin II receptor blockers (ARBs) and novel hypoglycemics including sodium–glucose co-transporter 2 (SGLT-2) inhibitors and the glucagon-like peptide-1 (GLP-1) receptor agonist. Recently, the new third-generation selective mineralocorticoid receptor antagonist Finerenone, which has great benefits in reducing urinary protein and delaying the risk of end-stage renal disease progression in DKD patients, was developed [80]. However, these symptomatic treatments can only delay kidney damage, with limited therapeutic effects for advanced DKD or ESRD. The loss of nephrons is irreversible. MSC therapy differs from traditional drug therapy because it has enormous therapeutic potential in promoting cell regeneration and restoring normal kidney function. DKD patients often have severe renal dysfunction when seeking medical assistance, and in this sense, MSC therapy has the potential for broader application.

In addition, combining MSC therapy with traditional pharmaceuticals has demonstrated enhanced therapeutic efficacy in the DKD model compared to using either treatment alone. Scientific investigation has established that the conjunction of ADMSCs with the GLP-1 receptor agonist exenatide can markedly enhance renal functionality and mitigate structural alterations by restoring equilibrium in inflammatory, fibrotic, and apoptotic markers [81]. The combination of ADMSCs and the SGLT-2 inhibitor empagliflozin (EMPA) demonstrated superior efficacy in restoring renal function, attenuating renal tubular epithelial cell injury, and reducing podocyte loss in a rat model of DKD compared to a single treatment. The authors posited that ADMSC augmented the renoprotective impact of EMPA by promoting antiapoptotic, anti-inflammatory, and antioxidative stress and restoration of mitochondrial autophagy functions without affecting the hypoglycemic effect of EMPA [82]. UC-MSCs combined with irbesartan significantly exerted better protective effects on glomerular podocyte injury and renal function compared to the administration of MSCs or irbesartan alone [83]. Pre-treating MSCs with drugs and then injecting the treated MSCs into the body also offered certain advantages. Compared with simple MSC treatment, MSCs modified with angiotensin-converting enzyme 2 (ACE2) had better performance in reducing albuminuria and improving glomerulosclerosis. In vitro and in vivo, MSCs-ACE2 is more beneficial than using MSCs alone in reducing Ang II and increasing Ang1-7, thereby inhibiting the adverse effects of Ang II accumulation by downregulating ECM and suppressing the transforming growth factor (TGF-β)/Smad pathway [84]. ACE2 therapy-modified MSCs have additional benefits for the progression of diabetes nephropathy (DN) by inhibiting renal RAS activation and reducing glomerular fibrosis. Nevertheless, the precise nature of the interaction between MSCs and diabetes drugs remains uncertain. In the double-blind control group (placebo) trial of allogeneic BM-MPCs for patients with renal dysfunction due to type 2 diabetes, the subjects also took antidiabetic drugs (including insulin, hypoglycemic traditional medicines such as metformin, and sulfonylureas) while receiving BM-MPCs [62]. However, these subjects did not demonstrate a superior therapeutic effect. Further scientific investigation and larger sample sizes are required to elucidate the impact of MSCs on the efficacy of other hypoglycemics and antihypertensive drugs.

6. Challenges and Prospects

Although the beneficial role of MSC therapy in DKD has been widely accepted, some potential issues still need to be improved in its clinical application and promotion.

The safety of MSCs in DKD has been preliminarily verified. To date, few studies have reported adverse events associated with the use of MSCs in DKD. Nevertheless, the safety of this approach remains a topic of debate, given the limited scope of clinical research and the relatively short observation period. A recent study has revealed that MSCs may undergo chromosomal abnormalities even during the early passages of their growth cycle and may develop into malignant tumors if transplanted within the body [85]. DKD-related factors (including hyperglycemia and uremia) may alter the differentiation potential of MSCs, which limits their application. While RNA sequencing indicates that the angiogenic and repair potential of MSCs derived from adipose tissue in DKD subjects can be preserved [86] and that diabetic microenvironment preconditioning of ADSCs has been demonstrated to enhance their antidiabetic, anti-long-term complications and anti-inflammatory effects in type 2 diabetic rats [87], alterations in the transcriptome associated with angiogenesis have been observed [86,88]. Furthermore, a randomized controlled trial yielded evidence that the duration of type 2 diabetes mellitus and obesity influence the efficacy of autologous transferred BM-MSCs [89]. Therefore, the microenvironment of diabetes may result in poor implantation and limited differentiation of stem cells, differentiation into undesired cell lineages, and malignant transformation or genetic distortion of stem cells.

In addition, the MSCs currently used in clinical studies are all administered in a single dose, with limited efficacy. The potential for multiple injections in the future cannot be ruled out, but this may result in additional adverse reactions, including sensitization reactions caused by repeated administrations, local complications (hematoma formation, local infection), and vascular obstruction (dyspnea, oliguria, myocardial infarction, venous thromboembolism events). In the future, if it is demonstrated that repeated in vivo injection of MSCs is safe and well tolerated, repeated administration may prove to be a more efficacious approach and may be suitable for assessing long-term effects on renal function.

Given the potential tumorigenicity and ethical concerns associated with the use of MSCs in clinical practice, there are now more recommendations for cell-free therapies. Cell-free therapy is a biological treatment that primarily uses extracellular matrixs, cytokines, biomolecules, or other biomolecules to treat diseases without the injection or use of live cells. This therapy aims to utilize the natural repair mechanisms within the body or provide bioactive ingredients to promote tissue regeneration and repair. In the treatment of DKD, it has been found that exosomes and mitochondria, two cell products, may have similar effects with applied cells.

As summarized in Table 1, the potential of MSC-exos as a cell-free treatment strategy for DKD has been demonstrated to hold negotiable promises. MSC-exos can avoid many safety issues of MSCs, not only reducing unexpected differentiation of cells at risk of tumor formation but also avoiding secondary ischemic injury caused by vascular coagulation Furthermore, studies have shown that the pretransplantation of exogenous mitochondria into endothelial cells (ECs) induces transient cell protection through mitochondrial autophagy, thereby enabling the implantation of ECs without the necessity of MSC support [69]. These strategies offer novel approaches to transplanting or regenerating DKD glomerular endothelial cells. A detailed exploration of the mechanism of MSCs in the treatment of DKD is likely to facilitate the development of new MSC treatments in the future. Specifically, the extraction and direct input of active ingredients of MSCs can effectively circumvent the potential risks associated with MSC infusion. It is imperative that future studies undertake a comprehensive evaluation of the safety of MSCs in the treatment of DN and that strategies to enhance their safety be explored, including autologous MSC transplantation, immunosuppressive therapy, and gene-editing techniques.

A further challenge is that the biodistribution of MSCs and exosomes after in vivo injection suggested that the aggregation of MSCs in the kidney was limited and their retention time was short, which diminished the therapeutic potential of MSCs in kidney diseases. At present, the most commonly used in vivo MSC injection method is intravenous injection. Still, fluorescence tracing shows that MSCs injected into the tail vein of rats have little or no renal homing. The PMSCs injected into the tail vein of mice mainly accumulated in the thymus and spleen and rarely in the kidney after 24 h in male SD DKD rats [58]. The concentration of human UC-MSCs-EVs in the kidney of db/db mice reached a maximum at 12 h post-tail vein injection, after which the distribution of the EVs in the kidneys decreased, with the majority accumulating in the liver [55]. It is essential to enhance the renal targeted delivery efficiency of MSCs and their derivatives to optimize the utilization rate of MSC therapy [40]. Some scholars have proposed that the direct implantation of mesenchymal cell sheets into the renal capsule can increase the local MSC infiltration rate and demonstrate superior therapeutic efficacy compared to the tail vein injection of MSCs. Stromal cell-derived factor-1 (SDF-1) plays a crucial role in MSC migration, involving activation, mobilization, homing, and retention [90], which may be related to poor homing in DN treatment. In early DN rats, the expression of SDF-1 in the kidneys is weak, which may be the reason for the poor homing effect of transplanted MSCs. The ultrasound-targeted disruption of microbubbles loaded with SDF-1 has the potential to increase the level of SDF-1 in targeted renal tissue, thereby promoting the homing of MSCs to early DN kidneys. However, the impact of this approach on the therapeutic effect on the kidneys has yet to be validated [91–93]. Recent studies have explored the combination of MSCs with kidney-targeted materials. Fe3O4-coated polydopamine nanoparticles (NP) internalized MSCs, which enhanced the homing of MSCs transplanted alone in the kidney tissue of DKD mice, improved the accuracy of MSCs homing to the kidneys, and also enhanced the renal function compared to MSCs transplanted alone [94]. It was also reported that the use of a hydrogel containing arginine glycine aspartate (RGD) peptide increased the renal retention and stability of MSC-EVs, had an excellent rescue effect on renal function in the mouse AKI model, reduced histopathological damage and renal tubular damage, and promoted cell proliferation in the early stage of AKI [95]. There is potential for improving the repair efficiency of DKD-damaged cells and enhancing the fibrosis process.

Furthermore, three-dimensional (3D) cultured MSCs have been shown to have higher cell survival rates and enhanced anti-inflammatory effects, which may be due to enhanced autophagy of MSCs [79]. In the future, the development of new composite applications of MSCs and related derivatives and materials, as well as the pretreatment of MSCs through various methods, are expected to make new progress in the treatment of DKD.

Despite the potential risks associated with the application of MSCs, their regenerative function in nephrons is irreplaceable, enabling the reversal of damage in advanced kidney disease. In addition, EVs secreted by MSCs have shown great potential as drug carriers in recent years [96–98]. Extracellular vesicles offer a promising approach to drug delivery, combining the advantages of cells and nanotechnology in drug delivery. For example, EVs can enhance the stability of drugs, and they have a natural targeting ability based on donor cells when delivering drugs. EVs are nanosized molecules with cell surface substances, so they have biological solid barrier permeability and can selectively penetrate tissue damage. Therefore, EVs will be a promising drug delivery system for treating DKD.

7. Conclusions

The management of DKD is a major challenge as there are currently no strategies available to regenerate lost nephron function. The pathological process of DKD is highly complicated, with a variety of contributing factors that collectively impede the development of effective therapies and preclude a complete cure. Therefore, it is important to gain a deeper understanding of the pathological mechanism to facilitate the development of more effective treatments for DKD. The vulnerability and limited regenerative capacity of the nephron significantly impede the recovery of renal function. As a regenerative medicine therapy, MSCs have brought new hope for the treatment of DKD, but various problems derived from them still need to be addressed. Cell-free therapy may be the choice in the future. Exosomes and mitochondria derived from MSCs have great potential in improving renal function. At the same time, the combined application of MSCs and other therapies including MSCs and exosomes as carriers of targeted delivery drugs, MSCs to enhance the efficacy of other therapeutic drugs, and MSCs combined with targeted kidney materials to enhance the efficacy of MSC therapy are worthy of attention. We need to optimize MSC therapy to improve the therapeutic effect in DKD, and further investigation is necessary to elucidate the precise function of MSCs in DKD. Overcoming current challenges will provide a comprehensive rationale for the clinical application of MSC therapy in the treatment of DKD, offering a promising avenue for the clinical management of DKD.

Author Contributions: Conceptualization, C.Z.; Writing—original draft preparation and editing, J.C.; Supervision, C.Z.; Funding acquisition, C.Z. All authors have read and agreed to the published version of the manuscript.

Funding: This work is funded by the National Natural Science Foundation of China (82370728, 81974096, 82170773, 81974097, 82100729, 82200808, 82200841, 82300843, and 82300786), the National Key Research and Development Program of China (2021YFC2500200), and the Key Research and Development Program of Hubei Province (2023BCB034).

Conflicts of Interest: The authors declare no conflicts of interest.

References

1. Ricciardi, C.A.; Gnudi, L. Kidney disease in diabetes: From mechanisms to clinical presentation and treatment strategies. *Metabolism* **2021**, *124*, 154890. [CrossRef] [PubMed]
2. Reidy, K.; Kang, H.M.; Hostetter, T.; Susztak, K. Molecular mechanisms of diabetic kidney disease. *J. Clin. Investig.* **2014**, *124*, 2333–2340. [CrossRef] [PubMed]
3. Uccelli, A.; Moretta, L.; Pistoia, V. Mesenchymal stem cells in health and disease. *Nat. Rev. Immunol.* **2008**, *8*, 726–736. [CrossRef]
4. Hoang, D.M.; Pham, P.T.; Bach, T.Q.; Ngo, A.T.L.; Nguyen, Q.T.; Phan, T.T.K.; Nguyen, G.H.; Le, P.T.T.; Hoang, V.T.; Forsyth, N.R.; et al. Stem cell-based therapy for human diseases. *Signal Transduct. Target. Ther.* **2022**, *7*, 272. [CrossRef] [PubMed]
5. Keating, A. How do mesenchymal stromal cells suppress T cells? *Cell Stem Cell* **2008**, *2*, 106–108. [CrossRef] [PubMed]

6. Wang, T.; Jian, Z.; Baskys, A.; Yang, J.; Li, J.; Guo, H.; Hei, Y.; Xian, P.; He, Z.; Li, Z.; et al. MSC-derived exosomes protect against oxidative stress-induced skin injury via adaptive regulation of the NRF2 defense system. *Biomaterials* **2020**, *257*, 120264. [CrossRef] [PubMed]
7. Plumas, J.; Chaperot, L.; Richard, M.J.; Molens, J.P.; Bensa, J.C.; Favrot, M.C. Mesenchymal stem cells induce apoptosis of activated T cells. *Leukemia* **2005**, *19*, 1597–1604. [CrossRef]
8. Zhao, M.; Liu, S.; Wang, C.; Wang, Y.; Wan, M.; Liu, F.; Gong, M.; Yuan, Y.; Chen, Y.; Cheng, J.; et al. Mesenchymal Stem Cell-Derived Extracellular Vesicles Attenuate Mitochondrial Damage and Inflammation by Stabilizing Mitochondrial DNA. *ACS Nano* **2021**, *15*, 1519–1538. [CrossRef] [PubMed]
9. Dutra Silva, J.; Su, Y.; Calfee, C.S.; Delucchi, K.L.; Weiss, D.; McAuley, D.F.; O'Kane, C.; Krasnodembskaya, A.D. Mesenchymal stromal cell extracellular vesicles rescue mitochondrial dysfunction and improve barrier integrity in clinically relevant models of ARDS. *Eur. Respir. J.* **2021**, *58*, 2002978. [CrossRef]
10. Rabelink, T.J.; Little, M.H. Stromal cells in tissue homeostasis: Balancing regeneration and fibrosis. *Nat. Rev. Nephrol.* **2013**, *9*, 747–753. [CrossRef]
11. Penn, P.E.; Jiang, D.Z.; Fei, R.G.; Sitnicka, E.; Wolf, N.S. Dissecting the hematopoietic microenvironment. IX. Further characterization of murine bone marrow stromal cells. *Blood* **1993**, *81*, 1205–1213. [CrossRef] [PubMed]
12. Nakagami, H.; Maeda, K.; Morishita, R.; Iguchi, S.; Nishikawa, T.; Takami, Y.; Kikuchi, Y.; Saito, Y.; Tamai, K.; Ogihara, T.; et al. Novel autologous cell therapy in ischemic limb disease through growth factor secretion by cultured adipose tissue-derived stromal cells. *Arterioscler. Thromb. Vasc. Biol.* **2005**, *25*, 2542–2547. [CrossRef]
13. Panepucci, R.A.; Siufi, J.L.C.; Silva, W.A.; Proto-Siquiera, R.; Neder, L.; Orellana, M.; Rocha, V.; Covas, D.T.; Zago, M.A. Comparison of gene expression of umbilical cord vein and bone marrow-derived mesenchymal stem cells. *Stem Cells* **2004**, *22*, 1263–1278. [CrossRef] [PubMed]
14. Kim, D.-W.; Staples, M.; Shinozuka, K.; Pantcheva, P.; Kang, S.-D.; Borlongan, C.V. Wharton's jelly-derived mesenchymal stem cells: Phenotypic characterization and optimizing their therapeutic potential for clinical applications. *Int. J. Mol. Sci.* **2013**, *14*, 11692–11712. [CrossRef] [PubMed]
15. Abumaree, M.H.; Al Jumah, M.A.; Kalionis, B.; Jawdat, D.; Al Khaldi, A.; AlTalabani, A.A.; Knawy, B.A. Phenotypic and functional characterization of mesenchymal stem cells from chorionic villi of human term placenta. *Stem Cell Rev. Rep.* **2013**, *9*, 16–31. [CrossRef] [PubMed]
16. Miura, M.; Gronthos, S.; Zhao, M.; Lu, B.; Fisher, L.W.; Robey, P.G.; Shi, S. SHED: Stem cells from human exfoliated deciduous teeth. *Proc. Natl. Acad. Sci. USA* **2003**, *100*, 5807–5812. [CrossRef] [PubMed]
17. Park, J.H.; Hwang, I.; Hwang, S.H.; Han, H.; Ha, H. Human umbilical cord blood-derived mesenchymal stem cells prevent diabetic renal injury through paracrine action. *Diabetes Res. Clin. Pract.* **2012**, *98*, 465–473. [CrossRef]
18. Nagaishi, K.; Mizue, Y.; Chikenji, T.; Otani, M.; Nakano, M.; Konari, N.; Fujimiya, M. Mesenchymal stem cell therapy ameliorates diabetic nephropathy via the paracrine effect of renal trophic factors including exosomes. *Sci. Rep.* **2016**, *6*, 34842. [CrossRef] [PubMed]
19. Lee, R.H.; Seo, M.J.; Reger, R.L.; Spees, J.L.; Pulin, A.A.; Olson, S.D.; Prockop, D.J. Multipotent stromal cells from human marrow home to and promote repair of pancreatic islets and renal glomeruli in diabetic NOD/scid mice. *Proc. Natl. Acad. Sci. USA* **2006**, *103*, 17438–17443. [CrossRef]
20. Wang, B.; Yao, K.; Huuskes, B.M.; Shen, H.-H.; Zhuang, J.; Godson, C.; Brennan, E.P.; Wilkinson-Berka, J.L.; Wise, A.F.; Ricardo, S.D. Mesenchymal Stem Cells Deliver Exogenous MicroRNA-let7c via Exosomes to Attenuate Renal Fibrosis. *Mol. Ther. J. Am. Soc. Gene Ther.* **2016**, *24*, 1290–1301. [CrossRef] [PubMed]
21. Rani, S.; Ryan, A.E.; Griffin, M.D.; Ritter, T. Mesenchymal Stem Cell-derived Extracellular Vesicles: Toward Cell-free Therapeutic Applications. *Mol. Ther. J. Am. Soc. Gene Ther.* **2015**, *23*, 812–823. [CrossRef] [PubMed]
22. Li, H.; Rong, P.; Ma, X.; Nie, W.; Chen, C.; Yang, C.; Zhang, J.; Dong, Q.; Wang, W. Paracrine effect of mesenchymal stem cell as a novel therapeutic strategy for diabetic nephropathy. *Life Sci.* **2018**, *215*, 113–118. [CrossRef] [PubMed]
23. Lu, L.-L.; Liu, Y.-J.; Yang, S.-G.; Zhao, Q.-J.; Wang, X.; Gong, W.; Han, Z.-B.; Xu, Z.-S.; Lu, Y.-X.; Liu, D.; et al. Isolation and characterization of human umbilical cord mesenchymal stem cells with hematopoiesis-supportive function and other potentials. *Haematologica* **2006**, *91*, 1017–1026. [PubMed]
24. Yuan, Y.; Li, L.; Zhu, L.; Liu, F.; Tang, X.; Liao, G.; Liu, J.; Cheng, J.; Chen, Y.; Lu, Y. Mesenchymal stem cells elicit macrophages into M2 phenotype via improving transcription factor EB-mediated autophagy to alleviate diabetic nephropathy. *Stem Cells* **2020**, *38*, 639–652. [CrossRef] [PubMed]
25. Sávio-Silva, C.; Soinski-Sousa, P.E.; Simplício-Filho, A.; Bastos, R.M.C.; Beyerstedt, S.; Rangel, É.B. Therapeutic Potential of Mesenchymal Stem Cells in a Pre-Clinical Model of Diabetic Kidney Disease and Obesity. *Int. J. Mol. Sci.* **2021**, *22*, 1546. [CrossRef] [PubMed]
26. Konari, N.; Nagaishi, K.; Kikuchi, S.; Fujimiya, M. Mitochondria transfer from mesenchymal stem cells structurally and functionally repairs renal proximal tubular epithelial cells in diabetic nephropathy in vivo. *Sci. Rep.* **2019**, *9*, 5184. [CrossRef]
27. Bai, Y.; Huang, L.; Fan, Y.; Li, Y. Marrow mesenchymal stem cell mediates diabetic nephropathy progression via modulation of Smad2/3/WTAP/m6A/ENO1 axis. *FASEB J.* **2024**, *38*, e23729. [CrossRef] [PubMed]

28. Zhang, F.; Wang, C.; Wen, X.; Chen, Y.; Mao, R.; Cui, D.; Li, L.; Liu, J.; Chen, Y.; Cheng, J.; et al. Mesenchymal stem cells alleviate rat diabetic nephropathy by suppressing CD103+ DCs-mediated CD8+ T cell responses. *J. Cell Mol. Med.* **2020**, *24*, 5817–5831. [CrossRef]
29. Lin, L.; Lin, H.; Wang, D.; Bao, Z.; Cai, H.; Zhang, X. Bone marrow mesenchymal stem cells ameliorated kidney fibrosis by attenuating TLR4/NF-κB in diabetic rats. *Life Sci.* **2020**, *262*, 118385. [CrossRef] [PubMed]
30. Bai, Y.; Wang, J.; He, Z.; Yang, M.; Li, L.; Jiang, H. Mesenchymal Stem Cells Reverse Diabetic Nephropathy Disease via Lipoxin A4 by Targeting Transforming Growth Factor β (TGF-β)/smad Pathway and Pro-Inflammatory Cytokines. *Med. Sci. Monit.* **2019**, *25*, 3069–3076. [CrossRef]
31. Tang, L.-X.; Wei, B.; Jiang, L.-Y.; Ying, Y.-Y.; Li, K.; Chen, T.-X.; Huang, R.-F.; Shi, M.-J.; Xu, H. Intercellular mitochondrial transfer as a means of revitalizing injured glomerular endothelial cells. *World J. Stem Cells* **2022**, *14*, 729–743. [CrossRef]
32. Liu, L.; Zhou, Y.; Zhao, X.; Yang, X.; Wan, X.; An, Z.; Zhang, H.; Tian, J.; Ge, C.; Song, X. Bone Marrow Mesenchymal Stem Cell-Derived Exosomes Alleviate Diabetic Kidney Disease in Rats by Inhibiting Apoptosis and Inflammation. *Front. Biosci. (Landmark Ed.)* **2023**, *28*, 203. [CrossRef] [PubMed]
33. Ebrahim, N.; Ahmed, I.A.; Hussien, N.I.; Dessouky, A.A.; Farid, A.S.; Elshazly, A.M.; Mostafa, O.; Gazzar, W.B.E.; Sorour, S.M.; Seleem, Y.; et al. Mesenchymal Stem Cell-Derived Exosomes Ameliorated Diabetic Nephropathy by Autophagy Induction through the mTOR Signaling Pathway. *Cells* **2018**, *7*, 226. [CrossRef]
34. Grange, C.; Tritta, S.; Tapparo, M.; Cedrino, M.; Tetta, C.; Camussi, G.; Brizzi, M.F. Stem cell-derived extracellular vesicles inhibit and revert fibrosis progression in a mouse model of diabetic nephropathy. *Sci. Rep.* **2019**, *9*, 4468. [CrossRef] [PubMed]
35. An, X.; Liao, G.; Chen, Y.; Luo, A.; Liu, J.; Yuan, Y.; Li, L.; Yang, L.; Wang, H.; Liu, F.; et al. Intervention for early diabetic nephropathy by mesenchymal stem cells in a preclinical nonhuman primate model. *Stem Cell Res. Ther.* **2019**, *10*, 363. [CrossRef]
36. Duan, Y.; Luo, Q.; Wang, Y.; Ma, Y.; Chen, F.; Zhu, X.; Shi, J. Adipose mesenchymal stem cell-derived extracellular vesicles containing microRNA-26a-5p target TLR4 and protect against diabetic nephropathy. *J. Biol. Chem.* **2020**, *295*, 12868–12884. [CrossRef] [PubMed]
37. Jin, J.; Shi, Y.; Gong, J.; Zhao, L.; Li, Y.; He, Q.; Huang, H. Exosome secreted from adipose-derived stem cells attenuates diabetic nephropathy by promoting autophagy flux and inhibiting apoptosis in podocyte. *Stem Cell Res. Ther.* **2019**, *10*, 95. [CrossRef]
38. Hao, Y.; Miao, J.; Liu, W.; Cai, K.; Huang, X.; Peng, L. Mesenchymal Stem Cell-Derived Exosomes Carry MicroRNA-125a to Protect Against Diabetic Nephropathy by Targeting Histone Deacetylase 1 and Downregulating Endothelin-1. *Diabetes Metab. Syndr. Obes. Targets Ther.* **2021**, *14*, 1405–1418. [CrossRef] [PubMed]
39. Zhao, T.; Jin, Q.; Kong, L.; Zhang, D.; Teng, Y.; Lin, L.; Yao, X.; Jin, Y.; Li, M. microRNA-15b-5p shuttled by mesenchymal stem cell-derived extracellular vesicles protects podocytes from diabetic nephropathy via downregulation of VEGF/PDK4 axis. *J. Bioenerg. Biomembr.* **2022**, *54*, 17–30. [CrossRef] [PubMed]
40. Takemura, S.; Shimizu, T.; Oka, M.; Sekiya, S.; Babazono, T. Transplantation of adipose-derived mesenchymal stem cell sheets directly into the kidney suppresses the progression of renal injury in a diabetic nephropathy rat model. *J. Diabetes Investig.* **2020**, *11*, 545–553. [CrossRef]
41. Lee, S.E.; Jang, J.E.; Kim, H.S.; Jung, M.K.; Ko, M.S.; Kim, M.-O.; Park, H.S.; Oh, W.; Choi, S.J.; Jin, H.J.; et al. Mesenchymal stem cells prevent the progression of diabetic nephropathy by improving mitochondrial function in tubular epithelial cells. *Exp. Mol. Med.* **2019**, *51*, 1–14. [CrossRef]
42. Chen, L.; Xiang, E.; Li, C.; Han, B.; Zhang, Q.; Rao, W.; Xiao, C.; Wu, D. Umbilical Cord-Derived Mesenchymal Stem Cells Ameliorate Nephrocyte Injury and Proteinuria in a Diabetic Nephropathy Rat Model. *J. Diabetes Res.* **2020**, *2020*, 8035853. [CrossRef] [PubMed]
43. Xiang, E.; Han, B.; Zhang, Q.; Rao, W.; Wang, Z.; Chang, C.; Zhang, Y.; Tu, C.; Li, C.; Wu, D. Human umbilical cord-derived mesenchymal stem cells prevent the progression of early diabetic nephropathy through inhibiting inflammation and fibrosis. *Stem Cell Res. Ther.* **2020**, *11*, 336. [CrossRef] [PubMed]
44. Zhang, Y.; Le, X.; Zheng, S.; Zhang, K.; He, J.; Liu, M.; Tu, C.; Rao, W.; Du, H.; Ouyang, Y.; et al. MicroRNA-146a-5p-modified human umbilical cord mesenchymal stem cells enhance protection against diabetic nephropathy in rats through facilitating M2 macrophage polarization. *Stem Cell Res. Ther.* **2022**, *13*, 171. [CrossRef] [PubMed]
45. Nie, P.; Bai, X.; Lou, Y.; Zhu, Y.; Jiang, S.; Zhang, L.; Tian, N.; Luo, P.; Li, B. Human umbilical cord mesenchymal stem cells reduce oxidative damage and apoptosis in diabetic nephropathy by activating Nrf2. *Stem Cell Res. Ther.* **2021**, *12*, 450. [CrossRef] [PubMed]
46. Yue, Y.; Yeh, J.-N.; Chiang, J.Y.; Sung, P.-H.; Chen, Y.-L.; Liu, F.; Yip, H.-K. Intrarenal arterial administration of human umbilical cord-derived mesenchymal stem cells effectively preserved the residual renal function of diabetic kidney disease in rat. *Stem Cell Res. Ther.* **2022**, *13*, 186. [CrossRef] [PubMed]
47. Li, X.; Guo, L.; Chen, J.; Liang, H.; Liu, Y.; Chen, W.; Zhou, L.; Shan, L.; Wang, H. Intravenous injection of human umbilical cord-derived mesenchymal stem cells ameliorates not only blood glucose but also nephrotic complication of diabetic rats through autophagy-mediated anti-senescent mechanism. *Stem Cell Res. Ther.* **2023**, *14*, 146. [CrossRef]
48. Zhang, H.; Wang, X.; Hu, B.; Li, P.; Abuduaini, Y.; Zhao, H.; Jieensihan, A.; Chen, X.; Wang, S.; Guo, N.; et al. Human umbilical cord mesenchymal stem cells attenuate diabetic nephropathy through the IGF1R-CHK2-p53 signalling axis in male rats with type 2 diabetes mellitus. *J. Zhejiang Univ. Sci. B* **2024**, *25*, 568–580. [CrossRef]

49. He, J.; Liu, B.; Du, X.; Wei, Y.; Kong, D.; Feng, B.; Guo, R.; Asiamah, E.A.; Griffin, M.D.; Hynes, S.O.; et al. Amelioration of diabetic nephropathy in mice by a single intravenous injection of human mesenchymal stromal cells at early and later disease stages is associated with restoration of autophagy. *Stem Cell Res. Ther.* **2024**, *15*, 66. [CrossRef] [PubMed]
50. Li, H.; Rong, P.; Ma, X.; Nie, W.; Chen, Y.; Zhang, J.; Dong, Q.; Yang, M.; Wang, W. Mouse Umbilical Cord Mesenchymal Stem Cell Paracrine Alleviates Renal Fibrosis in Diabetic Nephropathy by Reducing Myofibroblast Transdifferentiation and Cell Proliferation and Upregulating MMPs in Mesangial Cells. *J. Diabetes Res.* **2020**, *2020*, 3847171. [CrossRef] [PubMed]
51. Cui, C.; Zang, N.; Song, J.; Guo, X.; He, Q.; Hu, H.; Yang, M.; Wang, Y.; Yang, J.; Zou, Y.; et al. Exosomes derived from mesenchymal stem cells attenuate diabetic kidney disease by inhibiting cell apoptosis and epithelial-to-mesenchymal transition via miR-424-5p. *FASEB J.* **2022**, *36*, e22517. [CrossRef]
52. Wang, Y.; Liu, J.; Wang, H.; Lv, S.; Liu, Q.; Li, S.; Yang, X.; Liu, G. Mesenchymal Stem Cell-Derived Exosomes Ameliorate Diabetic Kidney Disease Through the NLRP3 Signaling Pathway. *Stem Cells* **2023**, *41*, 368–383. [CrossRef] [PubMed]
53. Wang, Y.; Lu, D.; Lv, S.; Liu, X.; Liu, G. Mesenchymal stem cell-derived exosomes ameliorate diabetic kidney disease through NOD2 signaling pathway. *Ren. Fail.* **2024**, *46*, 2381597. [CrossRef] [PubMed]
54. Yin, S.; Liu, W.; Ji, C.; Zhu, Y.; Shan, Y.; Zhou, Z.; Chen, W.; Zhang, L.; Sun, Z.; Zhou, W.; et al. hucMSC-sEVs-Derived 14-3-3ζ Serves as a Bridge between YAP and Autophagy in Diabetic Kidney Disease. *Oxidative Med. Cell. Longev.* **2022**, *2022*, 3281896. [CrossRef] [PubMed]
55. Li, Q.; Liu, J.; Su, R.; Zhen, J.; Liu, X.; Liu, G. Small extracellular vesicles-shuttled miR-23a-3p from mesenchymal stem cells alleviate renal fibrosis and inflammation by inhibiting KLF3/STAT3 axis in diabetic kidney disease. *Int. Immunopharmacol.* **2024**, *139*, 112667. [CrossRef] [PubMed]
56. Zhang, K.; Zheng, S.; Wu, J.; He, J.; Ouyang, Y.; Ao, C.; Lang, R.; Jiang, Y.; Yang, Y.; Xiao, H.; et al. Human umbilical cord mesenchymal stem cell-derived exosomes ameliorate renal fibrosis in diabetic nephropathy by targeting Hedgehog/SMO signaling. *FASEB J.* **2024**, *38*, e23599. [CrossRef]
57. Han, X.; Wang, J.; Li, R.; Huang, M.; Yue, G.; Guan, L.; Deng, Y.; Cai, W.; Xu, J. Placental Mesenchymal Stem Cells Alleviate Podocyte Injury in Diabetic Kidney Disease by Modulating Mitophagy via the SIRT1-PGC-1alpha-TFAM Pathway. *Int. J. Mol. Sci.* **2023**, *24*, 4696. [CrossRef]
58. Wang, J.; Liu, H.; Yue, G.; Deng, Y.; Cai, W.; Xu, J. Human placenta-derived mesenchymal stem cells ameliorate diabetic kidney disease by modulating the T helper 17 cell/ regulatory T-cell balance through the programmed death 1/programmed death-ligand 1 pathway. *Diabetes Obes. Metab.* **2024**, *26*, 32–45. [CrossRef] [PubMed]
59. Liu, H.; Wang, J.; Yue, G.; Xu, J. Placenta-derived mesenchymal stem cells protect against diabetic kidney disease by upregulating autophagy-mediated SIRT1/FOXO1 pathway. *Ren. Fail.* **2024**, *46*, 2303396. [CrossRef]
60. Rao, N.; Wang, X.; Xie, J.; Li, J.; Zhai, Y.; Li, X.; Fang, T.; Wang, Y.; Zhao, Y.; Ge, L. Stem Cells from Human Exfoliated Deciduous Teeth Ameliorate Diabetic Nephropathy In Vivo and In Vitro by Inhibiting Advanced Glycation End Product-Activated Epithelial-Mesenchymal Transition. *Stem Cells Int.* **2019**, *2019*, 2751475. [CrossRef]
61. Ni, Y.; Chen, Y.; Jiang, X.; Pu, T.; Zhang, L.; Li, S.; Hu, L.; Bai, B.; Hu, T.; Yu, L.; et al. Transplantation of Human Amniotic Mesenchymal Stem Cells Up-Regulates Angiogenic Factor Expression to Attenuate Diabetic Kidney Disease in Rats. *Diabetes Metab. Syndr. Obes. Targets Ther.* **2023**, *16*, 331–343. [CrossRef]
62. Packham, D.K.; Fraser, I.R.; Kerr, P.G.; Segal, K.R. Allogeneic Mesenchymal Precursor Cells (MPC) in Diabetic Nephropathy: A Randomized, Placebo-controlled, Dose Escalation Study. *EBioMedicine* **2016**, *12*, 263–269. [CrossRef]
63. Perico, N.; Remuzzi, G.; Griffin, M.D.; Cockwell, P.; Maxwell, A.P.; Casiraghi, F.; Rubis, N.; Peracchi, T.; Villa, A.; Todeschini, M.; et al. Safety and Preliminary Efficacy of Mesenchymal Stromal Cell (ORBCEL-M) Therapy in Diabetic Kidney Disease: A Randomized Clinical Trial (NEPHSTROM). *J. Am. Soc. Nephrol.* **2023**, *34*, 1733–1751. [CrossRef] [PubMed]
64. Chen, Y.; Xu, Y.; Chi, Y.; Sun, T.; Gao, Y.; Dou, X.; Han, Z.; Xue, F.; Li, H.; Liu, W.; et al. Efficacy and safety of human umbilical cord-derived mesenchymal stem cells in the treatment of refractory immune thrombocytopenia: A prospective, single arm, phase I trial. *Signal Transduct. Target. Ther.* **2024**, *9*, 102. [CrossRef]
65. Salemkour, Y.; Yildiz, D.; Dionet, L.; t Hart, D.C.; Verheijden, K.A.T.; Saito, R.; Mahtal, N.; Delbet, J.-D.; Letavernier, E.; Rabant, M.; et al. Podocyte Injury in Diabetic Kidney Disease in Mouse Models Involves TRPC6-mediated Calpain Activation Impairing Autophagy. *J. Am. Soc. Nephrol.* **2023**, *34*, 1823–1842. [CrossRef]
66. Lv, S.-S.; Liu, G.; Wang, J.-P.; Wang, W.-W.; Cheng, J.; Sun, A.-L.; Liu, H.-Y.; Nie, H.-B.; Su, M.-R.; Guan, G.-J. Mesenchymal stem cells transplantation ameliorates glomerular injury in streptozotocin-induced diabetic nephropathy in rats via inhibiting macrophage infiltration. *Int. Immunopharmacol.* **2013**, *17*, 275–282. [CrossRef] [PubMed]
67. Wise, A.F.; Williams, T.M.; Rudd, S.; Wells, C.A.; Kerr, P.G.; Ricardo, S.D. Human mesenchymal stem cells alter the gene profile of monocytes from patients with Type 2 diabetes and end-stage renal disease. *Regen. Med.* **2016**, *11*, 145–158. [CrossRef]
68. Su, W.; Yin, Y.; Zhao, J.; Hu, R.; Zhang, H.; Hu, J.; Ren, R.; Zhang, Y.; Wang, A.; Lyu, Z.; et al. Exosomes derived from umbilical cord-derived mesenchymal stem cells exposed to diabetic microenvironment enhance M2 macrophage polarization and protect against diabetic nephropathy. *FASEB J.* **2024**, *38*, e23798. [CrossRef]
69. Lin, R.-Z.; Im, G.-B.; Luo, A.C.; Zhu, Y.; Hong, X.; Neumeyer, J.; Tang, H.-W.; Perrimon, N.; Melero-Martin, J.M. Mitochondrial transfer mediates endothelial cell engraftment through mitophagy. *Nature* **2024**, *629*, 660–668. [CrossRef]

70. Yuan, Y.; Yuan, L.; Li, L.; Liu, F.; Liu, J.; Chen, Y.; Cheng, J.; Lu, Y. Mitochondrial transfer from mesenchymal stem cells to macrophages restricts inflammation and alleviates kidney injury in diabetic nephropathy mice via PGC-1α activation. *Stem Cells* **2021**, *39*, 913–928. [CrossRef]
71. Paliwal, S.; Chaudhuri, R.; Agrawal, A.; Mohanty, S. Regenerative abilities of mesenchymal stem cells through mitochondrial transfer. *J. Biomed. Sci.* **2018**, *25*, 31. [CrossRef] [PubMed]
72. Del Vecchio, V.; Rehman, A.; Panda, S.K.; Torsiello, M.; Marigliano, M.; Nicoletti, M.M.; Ferraro, G.A.; De Falco, V.; Lappano, R.; Lieto, E.; et al. Mitochondrial transfer from Adipose stem cells to breast cancer cells drives multi-drug resistance. *J. Exp. Clin. Cancer Res.* **2024**, *43*, 166. [CrossRef] [PubMed]
73. Simonson, M.S. Phenotypic transitions and fibrosis in diabetic nephropathy. *Kidney Int.* **2007**, *71*, 846–854. [CrossRef] [PubMed]
74. Liles, J.T.; Corkey, B.K.; Notte, G.T.; Budas, G.R.; Lansdon, E.B.; Hinojosa-Kirschenbaum, F.; Badal, S.S.; Lee, M.; Schultz, B.E.; Wise, S.; et al. ASK1 contributes to fibrosis and dysfunction in models of kidney disease. *J. Clin. Investig.* **2018**, *128*, 4485–4500. [CrossRef] [PubMed]
75. Lang, H.; Dai, C. Effects of Bone Marrow Mesenchymal Stem Cells on Plasminogen Activator Inhibitor-1 and Renal Fibrosis in Rats with Diabetic Nephropathy. *Arch. Med. Res.* **2016**, *47*, 71–77. [CrossRef]
76. Wei, L.; Gao, J.; Wang, L.; Tao, Q.; Tu, C. Hippo/YAP signaling pathway: A new therapeutic target for diabetes mellitus and vascular complications. *Ther. Adv. Endocrinol. Metab.* **2023**, *14*, 20420188231220134. [CrossRef]
77. Yang, D.; Livingston, M.J.; Liu, Z.; Dong, G.; Zhang, M.; Chen, J.-K.; Dong, Z. Autophagy in diabetic kidney disease: Regulation, pathological role and therapeutic potential. *Cell Mol. Life Sci.* **2018**, *75*, 669–688. [CrossRef] [PubMed]
78. Ceccariglia, S.; Cargnoni, A.; Silini, A.R.; Parolini, O. Autophagy: A potential key contributor to the therapeutic action of mesenchymal stem cells. *Autophagy* **2020**, *16*, 28–37. [CrossRef]
79. Regmi, S.; Raut, P.K.; Pathak, S.; Shrestha, P.; Park, P.-H.; Jeong, J.-H. Enhanced viability and function of mesenchymal stromal cell spheroids is mediated via autophagy induction. *Autophagy* **2021**, *17*, 2991–3010. [CrossRef]
80. Verma, A.; Patel, A.B. Finerenone: A Non-steroidal Mineralocorticoid Receptor Blocker for Diabetic Kidney Disease. *Trends Endocrinol. Metab.* **2021**, *32*, 261–263. [CrossRef] [PubMed]
81. Habib, H.A.; Heeba, G.H.; Khalifa, M.M.A. Effect of combined therapy of mesenchymal stem cells with GLP-1 receptor agonist, exenatide, on early-onset nephropathy induced in diabetic rats. *Eur. J. Pharmacol.* **2021**, *892*, 173721. [CrossRef] [PubMed]
82. Yang, C.C.; Chen, Y.L.; Sung, P.H.; Chiang, J.Y.; Chen, C.H.; Li, Y.C.; Yip, H.K. Repeated administration of adipose-derived mesenchymal stem cells added on beneficial effects of empagliflozin on protecting renal function in diabetic kidney disease rat. *Biomed. J.* **2024**, *47*, 100613. [CrossRef]
83. Meng, J.; Gao, X.; Liu, X.; Zheng, W.; Wang, Y.; Wang, Y.; Sun, Z.; Yin, X.; Zhou, X. Effects of xenogeneic transplantation of umbilical cord-derived mesenchymal stem cells combined with irbesartan on renal podocyte damage in diabetic rats. *Stem Cell Res. Ther.* **2024**, *15*, 239. [CrossRef] [PubMed]
84. Liu, Q.; Lv, S.; Liu, J.; Liu, S.; Wang, Y.; Liu, G. Mesenchymal stem cells modified with angiotensin-converting enzyme 2 are superior for amelioration of glomerular fibrosis in diabetic nephropathy. *Diabetes Res. Clin. Pract.* **2020**, *162*, 108093. [CrossRef]
85. Jeong, J.-O.; Han, J.W.; Kim, J.-M.; Cho, H.-J.; Park, C.; Lee, N.; Kim, D.-W.; Yoon, Y.-S. Malignant tumor formation after transplantation of short-term cultured bone marrow mesenchymal stem cells in experimental myocardial infarction and diabetic neuropathy. *Circ. Res.* **2011**, *108*, 1340–1347. [CrossRef]
86. Hickson, L.J.; Eirin, A.; Conley, S.M.; Taner, T.; Bian, X.; Saad, A.; Herrmann, S.M.; Mehta, R.A.; McKenzie, T.J.; Kellogg, T.A.; et al. Diabetic Kidney Disease Alters the Transcriptome and Function of Human Adipose-Derived Mesenchymal Stromal Cells but Maintains Immunomodulatory and Paracrine Activities Important for Renal Repair. *Diabetes* **2021**, *70*, 1561–1574. [CrossRef] [PubMed]
87. Su, W.; Yu, S.; Yin, Y.; Li, B.; Xue, J.; Wang, J.; Gu, Y.; Zhang, H.; Lyu, Z.; Mu, Y.; et al. Diabetic microenvironment preconditioning of adipose tissue-derived mesenchymal stem cells enhances their anti-diabetic, anti-long-term complications, and anti-inflammatory effects in type 2 diabetic rats. *Stem Cell Res. Ther.* **2022**, *13*, 422. [CrossRef]
88. Nagaishi, K.; Mizue, Y.; Chikenji, T.; Otani, M.; Nakano, M.; Saijo, Y.; Tsuchida, H.; Ishioka, S.; Nishikawa, A.; Saito, T.; et al. Umbilical cord extracts improve diabetic abnormalities in bone marrow-derived mesenchymal stem cells and increase their therapeutic effects on diabetic nephropathy. *Sci. Rep.* **2017**, *7*, 8484. [CrossRef]
89. Nguyen, L.T.; Hoang, D.M.; Nguyen, K.T.; Bui, D.M.; Nguyen, H.T.; Le, H.T.A.; Hoang, V.T.; Bui, H.T.H.; Dam, P.T.M.; Hoang, X.T.A.; et al. Type 2 diabetes mellitus duration and obesity alter the efficacy of autologously transplanted bone marrow-derived mesenchymal stem/stromal cells. *Stem Cells Transl. Med.* **2021**, *10*, 1266–1278. [CrossRef] [PubMed]
90. Lau, T.T.; Wang, D.-A. Stromal cell-derived factor-1 (SDF-1): Homing factor for engineered regenerative medicine. *Expert Opin. Biol. Ther.* **2011**, *11*, 189–197. [CrossRef]
91. Wu, S.; Li, L.; Wang, G.; Shen, W.; Xu, Y.; Liu, Z.; Zhuo, Z.; Xia, H.; Gao, Y.; Tan, K. Ultrasound-targeted stromal cell-derived factor-1-loaded microbubble destruction promotes mesenchymal stem cell homing to kidneys in diabetic nephropathy rats. *Int. J. Nanomed.* **2014**, *9*, 5639–5651. [CrossRef]
92. Zhang, Y.; Ye, C.; Wang, G.; Gao, Y.; Tan, K.; Zhuo, Z.; Liu, Z.; Xia, H.; Yang, D.; Li, P. Kidney-targeted transplantation of mesenchymal stem cells by ultrasound-targeted microbubble destruction promotes kidney repair in diabetic nephropathy rats. *BioMed Res. Int.* **2013**, *2013*, 526367. [CrossRef]

93. Wang, G.; Zhuo, Z.; Yang, B.; Wu, S.; Xu, Y.; Liu, Z.; Tan, K.; Xia, H.; Wang, X.; Zou, L.; et al. Enhanced Homing Ability and Retention of Bone Marrow Stromal Cells to Diabetic Nephropathy by Microbubble-Mediated Diagnostic Ultrasound Irradiation. *Ultrasound Med. Biol.* **2015**, *41*, 2977–2989. [CrossRef] [PubMed]
94. Wang, K.; Liu, T.; Zhang, Y.; Lv, H.; Yao, H.; Zhao, Y.; Li, J.; Li, X. Combined Placental Mesenchymal Stem Cells with Guided Nanoparticles Effective Against Diabetic Nephropathy in Mouse Model. *Int. J. Nanomed.* **2024**, *19*, 901–915. [CrossRef]
95. Zhang, C.; Shang, Y.; Chen, X.; Midgley, A.C.; Wang, Z.; Zhu, D.; Wu, J.; Chen, P.; Wu, L.; Wang, X.; et al. Supramolecular Nanofibers Containing Arginine-Glycine-Aspartate (RGD) Peptides Boost Therapeutic Efficacy of Extracellular Vesicles in Kidney Repair. *ACS Nano* **2020**, *14*, 12133–12147. [CrossRef]
96. Peng, H.; Li, Y.; Ji, W.; Zhao, R.; Lu, Z.; Shen, J.; Wu, Y.; Wang, J.; Hao, Q.; Wang, J.; et al. Intranasal Administration of Self-Oriented Nanocarriers Based on Therapeutic Exosomes for Synergistic Treatment of Parkinson's Disease. *ACS Nano* **2022**, *16*, 869–884. [CrossRef] [PubMed]
97. Li, X.; Li, Y.; Yu, C.; Bao, H.; Cheng, S.; Huang, J.; Zhang, Z. ROS-Responsive Janus Au/Mesoporous Silica Core/Shell Nanoparticles for Drug Delivery and Long-Term CT Imaging Tracking of MSCs in Pulmonary Fibrosis Treatment. *ACS Nano* **2023**, *17*, 6387–6399. [CrossRef] [PubMed]
98. Yuan, Z.; Kolluri, K.K.; Gowers, K.H.C.; Janes, S.M. TRAIL delivery by MSC-derived extracellular vesicles is an effective anticancer therapy. *J. Extracell. Vesicles* **2017**, *6*, 1265291. [CrossRef]

MDPI AG
Grosspeteranlage 5
4052 Basel
Switzerland
Tel.: +41 61 683 77 34

International Journal of Molecular Sciences Editorial Office
E-mail: ijms@mdpi.com
www.mdpi.com/journal/ijms

www.ingramcontent.com/pod-product-compliance
Lightning Source LLC
Chambersburg PA
CBHW061554210926
51850CB00009B/2102
9783725858798